T0211131

Oxidative Stress and Inflammatory Mechanisms in Obesity, Diabetes, and the Metabolic Syndrome

OXIDATIVE STRESS AND DISEASE

Series Editors

LESTER PACKER, PH.D.
ENRIQUE CADENAS, M.D., PH.D.

University of Southern California School of Pharmacy
Los Angeles, California

Oxidative Stress and Inflammatory Mechanisms in Obesity, Diabetes, and the Metabolic Syndrome

edited by

Lester Packer
Helmut Sies

CRC Press
Taylor & Francis Group
Boca Raton London New York

CRC Press is an imprint of the
Taylor & Francis Group, an **informa** business

CRC Press
Taylor & Francis Group
6000 Broken Sound Parkway NW, Suite 300
Boca Raton, FL 33487-2742

First issued in paperback 2019

ISBN-13: 978-1-4200-4378-5 (hbk)
ISBN-13: 978-0-367-38878-2 (pbk)

Library of Congress Cataloging-in-Publication Data

Oxidative stress and inflammatory mechanisms in obesity, diabetes, and the
    metabolic syndrome / edited by Lester Packer and Helmut Sies.
        p. ; cm. -- (Oxidative stress and disease ; 22)
    "A CRC title."
    Includes bibliographical references and index.
    ISBN-13: 978-1-4200-4378-5 (hardcover : alk. paper)
    ISBN-10: 1-4200-4378-1 (hardcover : alk. paper)
    1. Oxidative stress. 2. Metabolic syndrome. 3. Obesity. 4. Diabetes. I. Packer,
Lester. II. Sies, H. (Helmut), 1942- III. Series.
    [DNLM: 1. Metabolic Syndrome X--physiopathology. 2. Diabetes Mellitus, Type
2--physiopathology. 3. Inflammation--physiopathology. 4. Obesity--physiopathology.
5. Oxidative Stress--physiology. W1 OX626 v.22 2007 / WK 820 O98 2007]

QP177.O935 2007
616.3'9--dc22   2007005664

Visit the Taylor & Francis Web site at
http://www.taylorandfrancis.com

and the CRC Press Web site at
http://www.crcpress.com

Table of Contents

SECTION I Oxidative Stress, Metabolic Syndrome, Obesity, Diabetes, and Uncoupling Proteins

Chapter 1
The Metabolic Syndrome Defined 3
Neil J. Stone and Jennifer Berliner

Chapter 2
The Metabolic Syndrome: The Question of Balance between the
Pro-Inflammatory Effect of Macronutrients and the Anti-Inflammatory
Effect of Insulin 15
Paresh Dandona, Ajay Chaudhuri, Priya Mohanty, and Husam Ghanim

Chapter 3
The Role of Oxidative Stress in Diseases Associated with
Overweight and Obesity 33
Ginger L. Milne, Ling Gao, Joshua D. Brooks, and Jason D. Morrow

Chapter 4
Metabolic Syndrome Due to Early Life Nutritional Modifications 47
Malathi Srinivasan, Paul Mitrani, and Mulchand S. Patel

Chapter 5
Oxidative Stress and Antioxidants in the Perinatal Period 71
Hiromichi Shoji, Yuichiro Yamashiro, and Berthold Koletzko

SECTION II Influence of Dietary Factors, Micronutrients, and Metabolism

Series Preface

OXYGEN BIOLOGY AND MEDICINE

Through evolution, oxygen — itself a free radical — was chosen as the terminal electron acceptor for respiration. The two unpaired electrons of oxygen spin in the same direction; thus, oxygen is a biradical. Other oxygen-derived free radicals such as superoxide anion or hydroxyl radicals formed during metabolism or by ionizing radiation are stronger *oxidants*, i.e., endowed with higher chemical reactivities. Oxygen-derived free radicals are generated during metabolism and energy production in the body and are involved in regulation of signal transduction and gene expression, activation of receptors and nuclear transcription factors, oxidative damage to cell components, antimicrobial and cytotoxic actions of immune system cells, as well as in aging and age-related degenerative diseases. Conversely, cells conserve antioxidant mechanisms to counteract the effects of oxidants; these *antioxidants* may remove oxidants either in a highly specific manner (for example, by superoxide dismutases) or in a less specific manner (for example, through small molecules such as vitamin E, vitamin C, and glutathione). *Oxidative stress* as classically defined is an *imbalance between oxidants and antioxidants*. Overwhelming evidence indicates that oxidative stress can lead to cell and tissue injury. However, the same free radicals that are generated during oxidative stress are produced during normal metabolism and, as a corollary, are involved in both human health and disease.

UNDERSTANDING OXIDATIVE STRESS

In recent years, the research disciplines interested in oxidative stress have grown and enormously increased our knowledge of the importance of the cell redox status and the recognition of oxidative stress as a process with implications for many pathophysiological states. From this multi- and inter-disciplinary interest in oxidative stress emerges a concept that attests to the vast consequences of the complex and dynamic interplay of oxidants and antioxidants in cellular and tissue settings. Consequently, our view of oxidative stress is growing in scope and new future directions. Likewise, the term *reactive oxygen species* — adopted at some stage in order to highlight non-radical oxidants such as H_2O_2 and $^1O_2$ — now fails to reflect the rich variety of other reactive species in free radical biology and medicine encompassing nitrogen- , sulfur- , oxygen- , and carbon-centered radicals. With the discovery of nitric oxide, nitrogen-centered radicals gathered momentum and have matured into an area of enormous importance in biology and medicine. Nitric oxide or nitrogen monoxide (NO), a free radical generated in a variety of cell types by nitric oxide synthases (NOSs), is involved in a wide array of physiological and pathophysiological phenomena such as vasodilation, neuronal signaling, and inflammation. Of great importance is the radical–radical reaction of nitric oxide with superoxide anion. This is among the most rapid non-enzymatic reactions in biology (well over the diffusion-controlled limits) and yields the potent non-radical oxidant, peroxynitrite. The involvement of this species in tissue injury through oxidation and nitration reactions is well documented.

Virtually all diseases thus far examined involve free radicals. In most cases, free radicals are secondary to the disease process, but in some instances causality is established by free radicals. Thus, there is a delicate balance between oxidants and antioxidants in health and disease. Their proper balance is essential for ensuring healthy aging.

Both reactive oxygen and nitrogen species are involved in the redox regulation of cell functions. Oxidative stress is increasingly viewed as a major upstream component in the signaling cascade involved in inflammatory responses, stimulation of cell adhesion molecules, and chemoattractant production and as an early component in age-related neurodegenerative disorders such as Alzheimer's, Parkinson's, and Huntington's diseases, and amyotrophic lateral sclerosis. Hydrogen peroxide is probably the most important redox signaling molecule that, among others, can activate NFκB, Nrf2, and other universal transcription factors. Increasing steady-state levels of hydrogen peroxide have been linked to a cell's redox status with clear involvement in adaptation, proliferation, differentiation, apoptosis, and necrosis.

The identification of oxidants in regulation of redox cell signaling and gene expression was a significant breakthrough in the field of oxidative stress: the classical definition of oxidative stress as an *imbalance between the production of oxidants and the occurrence of cell antioxidant defenses* proposed by Sies in 1985 now seems to provide a limited concept of oxidative stress, but it emphasizes the significance of cell redox status. Because individual signaling and control events occur through discrete redox pathways rather than through global balances, a new definition of oxidative stress was advanced by Dean P. Jones (*Antioxidants & Redox Signaling* [2006]) as a disruption of redox signaling and control that recognizes the occurrence of compartmentalized cellular redox circuits. Recognition of discrete thiol redox circuits led Jones to provide this new definition of oxidative stress. Measurements of GSH/GSSG, cysteine/cystine, or thioredoxin$_{reduced}$/thioredoxin$_{oxidized}$ provide a quantitative definition of oxidative stress. Redox status is thus dependent on the degree to which tissue-specific cell components are in the oxidized state.

In general, the reducing environments inside cells help to prevent oxidative damage. In this reducing environment, disulfide bonds (S–S) do not spontaneously form because sulfhydryl groups are maintained in the reduced state (SH), thus preventing protein misfolding or aggregation. The reducing environment is maintained by metabolism and by the enzymes involved in maintenance of thiol/disulfide balance and substances such as glutathione, thioredoxin, vitamins E and C, and enzymes such as superoxide dismutases, catalase, and the selenium-dependent glutathione reductase and glutathione and thioredoxin-dependent hydroperoxidases (periredoxins) that serve to remove reactive oxygen species (hydroperoxides). Also of importance is the existence of many tissue- and cell compartment-specific isoforms of antioxidant enzymes and proteins.

Compelling support for the involvement of free radicals in disease development originates from epidemiological studies showing that enhanced antioxidant status is associated with reduced risk of several diseases. Of great significance is the role that micronutrients play in modulation of redox cell signaling; this establishes a strong linking of diet and health and disease centered on the abilities of micronutrients to regulate redox cell signaling and modify gene expression.

These concepts are anticipated to serve as platforms for the development of tissue-specific therapeutics tailored to discrete, compartmentalized redox circuits. This, in essence, dictates principles of drug development-guided knowledge of mechanisms of oxidative stress. Hence, successful interventions will take advantage of new knowledge of compartmentalized redox control and free radical scavenging.

OXIDATIVE STRESS IN HEALTH AND DISEASE

Oxidative stress is an underlying factor in health and disease. In this series of books, the importance of oxidative stress and diseases associated with organ systems of the body is highlighted by exploring the scientific evidence and clinical applications of this knowledge. This series is intended for researchers in the basic biomedical sciences and clinicians. The potential of such knowledge for healthy aging and disease prevention warrants further knowledge about how oxidants and antioxidants modulate cell and tissue function.

Lester Packer
Enrique Cadenas

Preface

Metabolic syndrome is a multicomponent disorder characterized by obesity, insulin resistance, dyslipidemia, and hypertension that is associated with the risks of type 2 diabetes mellitus and cardiovascular disease. Obesity plays a central role and is a principal causative factor in the progression of metabolic syndrome leading to type 2 diabetes mellitus and cardiovascular disease. Increased systemic oxidative stress may be an important mechanism by which obesity increases the incidence of atherosclerotic cardiovascular disease, the progression of which entails oxidative (up-regulation of oxidative enzymes) and inflammatory (increased production of tumor necrosis factors and other cytokines) components. The significance of inflammatory processes in accumulated fat appears to be an early initiator of metabolic syndrome, thus strengthening an association with oxidative stress and supporting the development of drugs targeted at ameliorating cellular redox status. Likewise, the more active angiotensin system in obesity may contribute to expanded oxidative stress that serves as a key signaling event in vascular remodeling.

A greater understanding of the molecular mechanisms that underlie inflammation and oxidative stress with implications for metabolic stress must be achieved so that evidence-based nutritional and pharmacological therapies can be developed to attenuate the impacts of obesity-induced insulin resistance and ensuing metabolic syndrome.

The chapters in this book report cutting-edge research exploring the intracellular events mediating or preventing oxidative stress and pro-inflammatory processes in obesity and type 2 diabetes as well as the molecular mechanisms inherent in the progression of metabolic stress, phenotypic perspectives, dietary factors, and micronutrients. They attempt to provide perspectives on the different components of metabolic stress and obesity and their associations with oxidative stress and inflammation. Pharmacological interventions in disease management are not emphasized and are covered elsewhere.

The editors gratefully appreciate the initial stimulus for this book — a workshop on obesity, oxidative stress related to metabolic syndrome, uncoupling proteins, and micronutrient action conducted March 15 through 18, 2006 as part of the Twelfth Annual Meeting of the Oxygen Club of California at Santa Barbara. We would like to acknowledge the sponsorship of that workshop by the Human Nutrition Group of BASF, Ludwigshafen, Germany, and, in particular, the input and help of Dr. Ute Obermüller-Jevic.

About the Editors

Lester Packer received his PhD in microbiology and biochemistry from Yale University. For many years he was professor and senior researcher at the University of California at Berkeley. Currently he is an adjunct professor in the Department of Pharmacology and Pharmaceutical Sciences at the University of Southern California. Recently, he was appointed distinguished professor at the Institute of Nutritional Sciences of the Chinese Academy of Sciences, Shanghai, China. His research interests are related to the molecular, cellular, and physiological role of oxidants, free radicals, antioxidants, and redox regulation in health and disease.

Professor Packer is the recipient of numerous scientific achievement awards including three honorary doctoral degrees. He has served as president of the International Society of Free Radical Research (SRRRI), president of the Oxygen Club of California (OCC), and vice-president of UNESCO—the United Nations Global Network on Molecular and Cell Biology (MCBN).

Helmut Sies earned an MD from the University of Munich, Germany, and an honorary PhD from the University of Buenos Aires, Argentina. He is a professor and chairman at the Institute of Biochemistry and Molecular Biology I at Heinrich Heine University at Dusseldorf, Germany. He served as a visiting professor at the University of California (Berkeley) and is an adjunct professor at the University of Southern California. He is a fellow of the National Foundation for Cancer Research based in Bethesda, Maryland and a fellow of the Royal College of Physicians of London.

Dr. Sies has been president of the Society for Free Radical Research International and is also the president of the Oxygen Club of California. His research interests focus on the field of oxidative stress, oxidants, and antioxidants.

Contributors

Farhad Amiri, PhD
Lady Davis Institute for Medical
  Research
Sir Mortimer B. Davis-Jewish General
  Hospital
McGill University
Montreal, Canada

Anna Aronis
School of Nutritional Sciences
Institute of Biochemistry, Food
  Science, and Nutrition
The Hebrew University of Jerusalem
Rehovot, Israel

Karim Benkirane, PhD
Lady Davis Institute for Medical
  Research
Sir Mortimer B. Davis-Jewish General
  Hospital
McGill University
Montreal, Canada

Jennifer Berliner, MD
Cardiology Section
Feinberg School of Medicine
Northwestern University
Chicago, Illinois

T. Bobbert
Abteilung fur Endokrinologie,
  Diabetes und Ernahrung
Charite Universitatsmedizin
Berlin, Germany

Joshua D. Brooks
Division of Clinical Pharmacology
Vanderbilt University Medical
  Center
Nashville, Tennessee

Antonio Ceriello, MD
Warwick Medical School
University of Warwick
Coventry, United Kingdom

Catherine B. Chan
Department of Biomedical
  Sciences
Atlantic Veterinary College
University of Prince Edward Island
Charlottetown, Canada

Ajay Chaudhuri, MD
School of Medicine and Biomedical
  Sciences
State University of New York
and
Kaleida Health
Buffalo, New York

An-Sik Chung
Department of Biological
  Sciences
Korea Advanced Institute of Science
  and Technology
Daejeon, Republic of Korea

Paresh Dandona, MD, PhD
School of Medicine and Biomedical
  Sciences
State University of New York
and
Kaleida Health
Buffalo, New York

Sridevi Devaraj
Laboratory for Atherosclerosis and
  Metabolic Research
University of California at Davis
  Medical Center
Sacramento, California

Jingyu Diao
Department of Physiology
Faculty of Medicine
University of Toronto
Toronto, Canada

John Alan Farmer
Baylor College of Medicine
Houston, Texas

Ling Gao
Division of Clinical Pharmacology
Vanderbilt University Medical Center
Nashville, Tennessee

Husam Ghanim, PhD
School of Medicine and Biomedical
  Sciences
State University of New York
and
Kaleida Health
Buffalo, New York

Barry J. Goldstein
Dorrance Hamilton Research
  Laboratories
Jefferson Medical College
Thomas Jefferson University
Philadelphia, Pennsylvania

**David Heber, MD, PhD, FACP,
  FACN**
Center for Human Nutrition
David Geffen School of Medicine
University of California
Los Angeles, California

Erik J. Henriksen
Department of Physiology
College of Medicine
University of Arizona
Tucson, Arizona

Sushil K. Jain
Department of Pediatrics
Health Sciences Center
Louisiana State University
Shreveport, Louisiana

Ishwarlal Jialal, MD, PhD
Laboratory for Atherosclerosis and
  Metabolic Research
University of California at Davis
  Medical Center
Sacramento, California

Janet C. King
Children's Hospital
Oakland Research Institute
Oakland, California

Berthold Koletzko, MD
Division of Metabolic Disorders and
  Nutrition
Dr. von Hauner Children's Hospital
Ludwig Maximilians University
Munich, Germany

Ki-Up Lee
Department of Internal Medicine
University of Ulsan College of
  Medicine
Seoul, Republic of Korea

Woo Je Lee
Department of Internal Medicine
Inje University College of Medicine
Seoul, Republic of Korea

Kalyankar Mahadev
Dorrance Hamilton Research
  Laboratories
Jefferson Medical College
Thomas Jefferson University
Philadelphia, Pennsylvania

Ginger L. Milne
Division of Clinical Pharmacology
Vanderbilt University Medical
  Center
Nashville, Tennessee

Paul Mitrani
Department of Biochemistry
School of Medicine and Biomedical
  Sciences
State University of New York
Buffalo, New York

Priya Mohanty, MD
School of Medicine and Biomedical
  Sciences
State University of New York
and
Kaleida Health
Buffalo, New York

Jason D. Morrow, MD
Division of Clinical Pharmacology
Vanderbilt University Medical
  Center
Nashville, Tennessee

Jong-Min Park
Department of Biological Sciences
Korea Advanced Institute of Science
  and Technology
Daejeon, Republic of Korea

Joong-Yeol Park
Department of Internal Medicine
University of Ulsan College of
  Medicine
Seoul, Republic of Korea

Mulchand S. Patel
Department of Biochemistry
School of Medicine and Biomedical
  Sciences
State University of New York
Buffalo, New York

Helen M. Roche
Institute of Molecular Medicine
Trinity Health Sciences Centre
St. James Hospital
Dublin, Republic of Ireland

**Ernesto L. Schiffrin, MD, PhD,
  FRSC, FRCPC, FACP**
Lady Davis Institute for Medical
  Research
Sir Mortimer B. Davis-Jewish General
  Hospital
McGill University
Montreal, Canada

Hiromichi Shoji
Department of Pediatrics
Juntendo University School of
  Medicine
Tokyo, Japan
and
Division of Metabolic Disorders and
  Nutrition
Dr. von Hauner Children's
  Hospital
Ludwig Maximilians University
Munich, Germany

Uma Singh
Laboratory for Atherosclerosis and
  Metabolic Research
University of California at Davis
  Medical Center
Sacramento, California

Joachim Spranger
Department of Clinical Nutrition
German Institute of Human Nutrition
Potsdam-Rehbruecke
Nuthetal, Germany

Malathi Srinivasan
Department of Biochemistry
School of Medicine and Biomedical
  Sciences
State University of New York
Buffalo, New York

Neil J. Stone, MD
Cardiology Section
Feinberg School of Medicine
Northwestern University
Chicago, Illinois

Oren Tirosh
School of Nutritional Sciences
Institute of Biochemistry, Food
  Science, and Nutrition
The Hebrew University of Jerusalem
Rehovot, Israel

Michael B. Wheeler
Department of Physiology
Faculty of Medicine
University of Toronto
Toronto, Canada

Xiangdong Wu
Dorrance Hamilton Research
  Laboratories
Jefferson Medical College
Thomas Jefferson University
Philadelphia, Pennsylvania

Yuichiro Yamashiro
Department of Pediatrics
Juntendo University School of
  Medicine
Tokyo, Japan

Section I

Oxidative Stress, Metabolic Syndrome, Obesity, Diabetes, and Uncoupling Proteins

1 The Metabolic Syndrome Defined

Neil J. Stone and Jennifer Berliner

CONTENTS

RATIONALE FOR DEFINING METABOLIC SYNDROME

In the past decade, sedentary lifestyles, atherogenic high calorie diets, and weight gains have characterized adolescents and adults in the United States and in many countries across the globe.[1,2] Indeed, a recent report[2] estimated the prevalences of overweight and obese people in the U.S. above 60 and 30%, respectively. This is not a unique burden for the U.S., but reflects a worldwide trend demonstrating an increased prevalence in metabolic risk factors[3,4] including visceral obesity, insulin resistance, dyslipidemia with abnormal values for triglycerides and high density lipoprotein cholesterol (HDL-c), hypertension, and (if measured) pro-thrombotic and inflammatory markers.

Although metabolic risk factors track with body mass index (BMI) calculated as weight/height squared, this has not proven the best correlate of coronary heart disease (CHD) events. More than 50 years ago, a French investigator drew attention to fat distribution in a graphic way showing the difference between those with android or abdominal patterns and those with gynecoid or female patterns of fat distribution.[5] In 2004, the global INTERHEART project demonstrated that the measure of obesity most directly related to myocardial infarction (MI) was not BMI, but a measure of body fatness, the waist:hip ratio.[6] Although not as good as the waist:hip ratio, waist measurements were more predictive than BMI.

In addition to the appreciation that weight gain and obesity are increasing global problems and that the location of body fat has both prognostic and therapeutic importance is the recognition of atherogenic risk factor clustering. This concept was noted by Framingham investigators who found it was common in both men and women, worsened with weight gain, and appreciably increased the risk of CHD.[7] The odds ratios for individuals with this risk factor clustering was more than two-fold for men and more than six-fold for women. Subsequent studies identified factors that can intensify these metabolic abnormalities. They include aging (higher prevalence at higher age), ethnic subgroups, endocrine dysfunction (increased prevalence in polycystic ovary syndrome), and physical inactivity.[8] The distinction from those unaffected is not trivial. Characteristic clinical features of risk factor clustering identify individuals who are more likely to become diabetic and/or experience a cardiac event. This spurred interest in viewing this constellation of clustered metabolic risk factors as a syndrome. It is important to recognize that metabolic syndrome is not a disease entity with a single etiologic cause. Nonetheless, the metabolic syndrome designation seems apt if you define a syndrome as "a set of symptoms or conditions that occur together and suggest specific underlying factors, prognosis, or guide to treatment."

Like the clinical syndrome known as heart failure, there may be great utility in recognizing patients who meet diagnostic criteria. Although a consensus has not emerged, current definitions identify patients with metabolic risk factors that can be seen over their lifespans, from an early stage when risk factors are emerging until they are fully established later. Although definitions vary, some believe metabolic syndrome is a useful construct even for those who have developed diabetes. Clinicians are increasingly recognizing metabolic syndrome among those with CHD and stroke. Accordingly, this syndrome may be progressive in nature. Insulin resistance, along with weight gain and obesity, is a major underlying risk factor for this syndrome, but as will be seen, most definitions do not consider metabolic syndrome strictly synonymous with insulin resistance. Moreover, although metabolic syndrome predicts CHD, it is not a replacement for Framingham risk scoring which, according to the ATP III panel, is required for setting LDL-c goals to determine intensity of treatment.[4] In clinical practice, recognition of metabolic syndrome is a useful tool for helping both patients and clinical care teams understand the adverse prognosis for both diabetes and CHD and direct therapeutic interventions. Furthermore, those with metabolic syndrome face increased risks for CHD that exceed the sum of the risk factors.[8] This should not be surprising given the systemic factors such as inflammation and hypercoagulability that accompany metabolic syndrome and its attendant visceral obesity.[3] Most important, several well controlled clinical trials have established that in individuals at high risk for progression to diabetes (and most likely with metabolic syndrome), a therapeutic lifestyle regimen directed at diet, regular exercise, and modest weight loss reduces the progression to type 2 diabetes by almost 60%.[9,10]

Next we will consider various published definitions of metabolic syndrome in the hope that it will be possible to gain perspective on the merits and potential shortcomings of each definition.

METABOLIC SYNDROME DEFINITIONS

WORLD HEALTH ORGANIZATION (WHO) DEFINITION

In 1998, the WHO proposed a set of criteria[11] to define metabolic syndrome. Its definition required the presence of insulin resistance as a component of the diagnosis. Insulin resistance was defined as the diagnosis of type 2 diabetes mellitus, impaired fasting glucose, impaired glucose tolerance or, for those with normal fasting glucose levels (<100 mg/dL), glucose uptake below the lowest quartile for the background population under investigation under hyperinsuline-mic, euglycemic conditions. In addition to the presence of insulin resistance, the WHO criteria require the presence of two additional risk factors that may include hypertension or treatment with antihypertensive medications, hypertriglyceri-demia, low HDL cholesterol, BMI >30 kg/m^2, and urinary albumin excretion rate >20 µg/min (Table 1.1).

ATP III INITIAL REPORT

In 2001, the National Cholesterol Education Program (NCEP) introduced defini-tions of metabolic syndrome, a constellation of major risk factors, life-habit risk factors, and emerging risk factors for CHD in its guidelines. The ATP III Expert Panel[12] recognized metabolic syndrome as a secondary target of risk reduction therapy, after the primary target of LDL cholesterol was attained. Its definition of metabolic syndrome included abdominal obesity, triglyceride levels, HDL cholesterol, blood pressure, and fasting glucose. The diagnosis of metabolic syndrome was made when three or more of the risk determinants were present (Table 1.2).

AMERICAN ASSOCIATION OF CLINICAL ENDOCRINOLOGISTS GUIDELINES

The guidelines of the American Association of Clinical Endocrinologists[13] appear to be a hybrid between the ATP III and WHO definitions of metabolic syndrome, but no defined number of risk factors is specified. Diagnosis is left to clinical judgment. The criteria include obesity, elevated triglycerides, low HDL choles-terol, elevated blood pressure, elevated fasting glucose and other risk factors, including polycystic ovary syndrome, sedentary lifestyle, advancing age, belong-ing to an ethnic group at high risk for type 2 diabetes or cardiovascular disease, family history of type 2 diabetes, hypertension, and cardiovascular disease (Table 1.3).

TABLE 1.1
WHO Diagnostic Criteria for Metabolic Syndrome

Component	Criteria	Comments
Insulin resistance as identified by type 2 diabetes, impaired fasting glucose, or impaired glucose tolerance		For those with normal fasting glucose levels (<110 mg/dL), glucose uptake below the lowest quartile for background population under investigation under hyperinsulinemic, euglycemc conditions
Any two of the following: body mass index; waist:hip ratio	>30 kg/m²; >0.9 in men or >0.85 in women	Waist circumference at umbilicus divided by circumference of hips at their widest point
Plasma triglycerides	≥150 mg/dL	≥1.7 mmol/L
HDL cholesterol	<35 mg/dL in men or <39 mg/dL in women	<0.9 mmol/L for men; <1.0 mmol/L for women
Antihypertensive medication and/or systolic or diastolic blood pressure	≥160 mm Hg or ≥90 mm Hg	
Urinary albumin excretion rate or urinary albumin:creatinine ratio	>20 µg/min or >20 mg/g	Urinary albumin:creatinine ratio can be done on spot urine

Source: From Alberti, K.G. and Zimmet, P.Z. *Diabet Med* 15, 539, 1998.

TABLE 1.2
ATP III Metabolic Syndrome

Risk Factor (Three or More for Diagnosis)	Defining Level
Abdominal obesity (waist circumference)	Men: >102 cm (>40 inches) Women: >88 cm (>35 inches)
Triglycerides	≥150 mg/dL
High-density lipoprotein cholesterol	Men: <40 mg/dL Women: <50 mg/dL
Blood pressure	≥130/≥85 mm Hg
Fasting glucose	≥110 mg/dL

Source: From Adult Treatment Panel (ATP) III. *JAMA* 285, 2486, 2001.

TABLE 1.3
American Association of Clinical Endocrinologists Diagnostic Criteria for Insulin Resistance Syndrome

Risk Factor	Defining Level
Elevated triglycerides	>150 mg/dL
Low HDL	Men: <40 mg/dL
	Women: <50 mg/dL
Elevated blood pressure	>130/85 mm Hg
Two-hour post-glucose challenge	>140 mg/dL
Fasting glucose	110 to 125 mg/dL
Other risk factors	Family history of type 2 diabetes, hypertension, or cardiovascular disease; history of CVD, HTN, NAFLD, aeonthosis nigricans, history of glucose intolerance or gestational dialoetes mellitus, BMI >25 kg/m2, polycystic ovary syndrome; sedentary lifestyle; >40 age, membership in ethnic group having high risk for type 2 diabetes mellitus or cardiovascular disease

Source: From Einhorn, D. et al., *Endocr Pract* 9, 237, 2003.

EUROPEAN GROUP FOR STUDY OF INSULIN RESISTANCE SYNDROME

In 1999, the European Group for Study of Insulin Resistance (EGIR) proposed a modification of the WHO definition.[14] This group used the term *insulin resistance syndrome* rather than *metabolic syndrome*. They likewise assumed that insulin resistance is the major cause, and required evidence of it for diagnosis. EGIR defines insulin resistance syndrome as the presence of insulin resistance or fasting hyperinsulinemia (highest 25%) and two of the following: hyperglycemia, hypertension, dyslipidemia, and central obesity. All these criteria must be measured before it is possible to evaluate the presence of the syndrome. Hyperglycemia should be defined at fasting values to provide reproducible criteria (Table 1.4).

REVISED ATP III GUIDELINES

In an updated version of the original ATP III guidelines,[15] the threshold for impaired fasting glucose (IFG) was reduced from ≥110 to ≥100 mg/dL. This adjustment corresponded to the recently modified American Diabetes Association (ADA) criteria for IFG. The rest of the criteria were essentially the same although several points were clarified. A diagnosis of the metabolic syndrome required three of the following diagnoses: abdominal obesity (as waist circumference), elevated triglycerides, reduced HDL cholesterol, hypertension, and elevated fasting glucose. As noted in ATP III, some people will manifest features of insulin resistance and metabolic syndrome with only moderate increases in waist

TABLE 1.4
European Group for Study of Insulin Resistance Syndrome Criteria

Risk Factor	Defining Level
Defined by presence of insulin resistance or fasting hyperinsulinemia (highest 25%) AND two of the following:*	
Hyperglycemia	Fasting plasma glucose ≥6.1 mmol/L, but nondiabetic
Hypertension	Systolic/diastolic blood pressure ≥140/90 mm Hg or treated for hypertension
Dyslipidemia	Triglycerides >2.0 mmol/L or HDL-cholesterol <1.0 mmol/L or treated for dyslipidemia
Central obesity	Waist circumference ≥94 cm in men and ≥80 cm in women

* All these criteria must be measured before it is possible to evaluate presence of the syndrome. Hyperglycemia should be defined at fasting (fasting plasma glucose ≥6.1 mmol/L or impaired fasting glucose) in non-diabetic individuals.

Source: From Balkau, B. and Charles, M.A., *Diabet Med* 16, 442, 1999.

circumference (94 to 101 cm in men and 80 to 87 cm in women). Among the characteristics that may predispose to insulin resistance and metabolic syndrome in such individuals are type 2 diabetes mellitus in first degree relatives before age 60, polycystic ovary disease, fatty liver, C-reactive protein >3 mg/L, microalbuminuria, impaired glucose tolerance, and elevated total apolipoprotein B.

Some populations are predisposed to insulin resistance, metabolic syndrome, and type 2 diabetes mellitus, with only moderate increases in waist circumference (i.e., populations from South Asia, China, Japan, and other Asian countries). None of these phenotypic features or ethnic differences was included in the ATP III diagnostic criteria, but if individuals with such characteristics have only moderate elevations of waist circumference plus at least two ATP III metabolic syndrome features, the writing group suggests that consideration should be given to managing them in the same way people with three ATP III risk factors are managed (Table 1.5).

INTERNATIONAL DIABETES FEDERATION (IDF) DEFINITION

The IDF clinical definition[16] requires the presence of abdominal obesity for diagnosis. When abdominal obesity is present, two additional factors originally listed in the ATP III definition are sufficient for diagnosis: raised concentration of triglycerides, reduced concentration of HDL cholesterol, hypertension, and raised fasting plasma glucose concentration or previously diagnosed type 2 diabetes. IDF recognized and emphasized ethnic differences in the correlation of abdominal obesity and other metabolic syndrome risk factors. For this reason, criteria of abdominal obesity were specified by nationality or ethnicity based on

TABLE 1.5
Revised ATP III Criteria

Risk Factor (Any Three for Diagnosis)	Defining Level
Elevated waist circumference*	Men: ≥102 cm (≥40 inches)
	Women: ≥88 cm (≥35 inches)
Elevated triglycerides	≥150 mg/dL (1.7 mmol/L) or on drug treatment for elevated TG
Reduced HDL cholesterol	Men: <40 mg/dL (1.03 mmol/L)
	Women: <50 mg/dL (1.3 mmol/L)
	Drug treatment for reduced HDL cholesterol
Elevated blood pressure	≥130 mm Hg systolic blood pressure OR ≥85 mm Hg diastolic blood pressure OR on antihypertensive drug treatment in patient with a history of HTN
Elevated fasting glucose	≥100 mg/dL or on drug treatment for elevated glucose

* Some U.S. adults of non-Asian origin (white, black, Hispanic) with marginally increased waist circumferences (94 to 101 cm [37 to 39 inches] in men and 80 to 87 cm [31 to 34 inches] in women) may have strong genetic contributions to insulin resistance and should benefit from changes in lifestyle habits, similar to men with categorical increases in waist circumference. Lower waist circumference cut point (≥90 cm [35 inches] in men and ≥80 cm in women) appears appropriate for Asian Americans.

Source: From Grundy, S.M. et al., *Circulation* 112, 2735, 2005.

best available population estimates. Further data defining cut points for abdominal circumference from various population groups should be especially useful, although as more multiracial individuals are seen, it may be difficult to know which cut point to use (Table 1.6 and Table 1.7).

IS METABOLIC SYNDROME USEFUL AND WHICH DEFINITION SHOULD BE USED?

Some have argued that metabolic syndrome as a concept is not needed and have decried the fact that there is no single, accepted definition.[17] Others point to the usefulness of the construct in clinical practice because at a minimum, it focuses both patients and healthcare teams on both the root causes and individual metabolic risk factors that are present. This seems pertinent today with the twin worldwide epidemics of obesity and diabetes.

In addition, there is a growing awareness of the importance of lifestyle variables as determinants of progressive atherosclerosis in individuals and applying this to the primary and secondary prevention of MI. The global case-control INTERHEART study found that nine factors explained more than 90% of the risk of MI in both sexes, at all ages, and in all regions.[18] Its list of predictive factors included familiar factors found in the Framingham risk score: abnormal

TABLE 1.6
International Diabetes Federation Criteria for Metabolic Syndrome

Risk Factor	Defining Level
Abdominal obesity* AND two additional factors:	
Triglycerides	≥150 mg/dL or specific treatment for this abnormality
Reduced concentration of HDL cholesterol	<40 mg/dL in men and <50 mg/dL in women or specific treatment for this abnormality
Elevated blood pressure	Systolic blood pressure ≥130 mm Hg or diastolic blood pressure ≥85 mm Hg or treatment of previously diagnosed hypertension
Raised fasting plasma glucose concentration	≥100 mg/dL or previously diagnosed type 2 diabetes

* Defined as waist circumference ≥94 cm for Europid men and ≥80 cm for Europid women, with ethnicity-specific values for other groups.

Source: From Alberti, K.G. et al. *Lancet* 366, 1059, 2005.

TABLE 1.7
Thresholds for Abdominal Obesity (IDF Criteria)

National Origin	Threshold for Men	Threshold for Women
Europid*	≥94 cm	≥80 cm
Chinese	≥90 cm	≥80 cm
South Asian	≥90 cm	≥80 cm
Japanese	≥85 cm	≥90 cm
Ethnic South and Central American	Use South Asian recommendations until more specific data are available	
Sub-Saharan African	Use European data until more specific data are available	
Eastern Mediterranean and Middle-Eastern	Use European data until more specific data are available	

* In the U.S., ATP III values (102 cm for males and 88 cm for females) are likely to continue to be used for clinical purposes.

Source: From Alberti, K.G. et al. *Lancet* 366, 1059, 2005.

lipids, diabetes, smoking, hypertension, metabolic and lifestyle factors such as abdominal obesity, psychosocial factors, levels of consumption of fruits and vegetables, alcohol intake, and regular physical activity.

Interestingly, these investigators created a study definition of metabolic syndrome by using self-reported diabetes and hypertension as surrogates for measured fasting blood sugar and blood pressure, measured apo B and apo A1 as surrogates for fasting triglycerides and HDL-c, and measured waist:hip ratio instead of waist circumference. Not surprisingly they found that their dichotomous metabolic syndrome variables were not as good as predictors of acute MI as the nine variables listed above that predicted over 90% of cases of acute MI. This has been a criticism of the metabolic syndrome, but in fairness, ATP III always emphasized that patients (especially those having two or more risk factors) need Framingham risk scoring with its *continuous* variables to calculate global risk and suggested attention to metabolic variables *after* LDL-c goals had been realized. For example, the ATP III algorithm notes that if triglycerides are 200 mg/dL or more despite attainment of the LDL-c goal, then treatment to achieve non-HDL-c goals (some may prefer apo B) is recommended.

It is hoped that with further investigation, consensus will eventually be reached on a single definition for metabolic syndrome.

Table 1.8 summarizes unique features, drawbacks, and comments regarding the definitions discussed in this chapter. Endocrinologists and investigators may be drawn to definitions that focus on insulin resistance. Clinicians may prefer the easily identified clinical features of ATP III that improve with lifestyle changes and the resultant modest degrees of weight loss. Further studies and discussion hopefully will clarify the final form of the metabolic syndrome definition. Until then, regardless of the criteria utilized, a systematic identification of metabolic variables leads invariably to an appropriate strong focus on the need for improved diet, regular physical activity, and weight reduction that most would agree is an important first step in the comprehensive approach to reducing the burden of diabetes and CHD worldwide.

TABLE 1.8
Features, Drawbacks, and Comments on Various Definitions of Metabolic Syndrome

Defining Organization	Unique Features	Drawbacks	Comments
WHO	Requires insulin resistance; BP requirements higher than ATP III; HDL cholesterol requirements lower; uses BMI instead of waist circumference; microalbuminuria is criterion	Requires glucose testing that may require a separate office or hospital visit	Demonstrating insulin resistance in a patient without type 2 diabetes usually involves oral glucose tolerance testing or hyperinsulinemic/euglycemic clamp testing; considered costly or inconvenient in many clinical practices
2001 ATP III and update	Uses abdominal obesity rather than BMI as criterion; simple criteria for diagnosis; update clarified several diagnosis issues	No measure of insulin resistance; this would detract from ease of use of these criteria	Weight loss achieved by better diet and regular exercise has potential to improve all components of the metabolic syndrome definition
AACE	Uses BMI rather than abdominal obesity as criterion	Reproducibility; no defined number of risk factors specified	Diagnosis left to clinical judgment; this makes estimates of prognosis from clinical studies difficult
EGIR	Requires insulin resistance for diagnosis; excludes patients with type 2 diabetes mellitus; BP requirements higher than ATP III; uses abdominal obesity rather than BMI as criterion; simple criteria for diagnosis	Insulin levels not reliable in clinical practice setting	Unlike some definitions, it excludes type 2 diabetes mellitus; some would argue that therapeutic implications of metabolic syndrome may be valuable in diabetics with increased waist circumferences
IDF	Requires abdominal obesity for diagnosis; simple criteria (same as ATP III) for diagnosis	Physician must have data on waist circumference for ethnic subgroups	Uses a recognizable and easily measurable clinical feature (increased waist circumference) to initiate consideration of diagnosis

REFERENCES

1. Janssen, I. et al. Health Behaviour in School-Aged Children Obesity Working Group: comparison of overweight and obesity prevalence in school-aged youth from 34 countries and their relationships with physical activity and dietary patterns. *Obes Rev* 6, 123, 2005.
2. Flegal, K.M. et al. Prevalence and trends in obesity among U.S. adults, 1999–2000. *JAMA* 288, 1723, 2002.
3. Eckel, R., Grundy, S., and Zimmet, P. The metabolic syndrome. *Lancet* 365, 1415, 2005.
4. Grundy, S.M. Metabolic syndrome scientific statement by the American Heart Association and the National Heart, Lung, and Blood Institute. *Arterioscler Thromb Vasc Biol* 25, 2243, 2005.
5. Vague, J. La differenciation sexuelle, facteur determinant des formes de l'obesite. *Presse Med* 30, 339, 1947.
6. Yusuf, S. et al. INTERHEART Study Investigators: obesity and the risk of myocardial infarction in 27,000 participants from 52 countries: a case-control study. *Lancet* 366, 1640, 2005.
7. Wilson, P.F.F., Kannel, W.B., Silbershatz, H., and D'Agostino, R.B. Clustering of metabolic risk factors and coronary heart disease. *Arch Int Med* 159, 1104, 1999.
8. Grundy, S. Metabolic syndrome: connecting and reconciling cardiovascular and diabetes worlds. *JACC* 47, 1093, 2006.
9. Diabetes Prevention Program Research Group. Reduction in the incidence of type 2 diabetes with lifestyle intervention or metformin. *New Engl J Med* 346, 393, 2002.
10. Tuomilehto, J. et al. Finnish Diabetes Prevention Study Group: prevention of type 2 diabetes mellitus by changes in lifestyle among subjects with impaired glucose tolerance. *New Engl J Med* 344, 1343, 2001.
11. Alberti, K.G. and Zimmet, P.Z. Definition, diagnosis and classification of diabetes mellitus and its complications. Part 1. Diagnosis and classification of diabetes mellitus provisional report of a WHO consultation. *Diabet Med* 15, 539, 1998.
12. Adult Treatment Panel (ATP) III. Executive summary of third report of the National Cholesterol Education Program (NCEP) Expert Panel on Detection, Evaluation, and Treatment of High Blood Cholesterol in Adults. *JAMA* 285, 2486, 2001.
13. Einhorn, D. et al. American College of Endocrinology position statement on the insulin resistance syndrome. *Endocr Pract* 9, 237, 2003.
14. Balkau, B. and Charles, M.A. Comment on the provisional report from the WHO consultations, European Group for the Study of Insulin Resistance (EGIR). *Diabet Med* 16, 442, 1999.
15. Grundy, S.M. et al. Diagnosis and management of the metabolic syndrome: an American Heart Association/National Heart, Lung, and Blood Institute scientific statement. *Circulation* 112, 2735, 2005.
16. Alberti, K.B., Zimmet, P., and Shaw, J. The metabolic syndrome: a new worldwide definition. *Lancet* 366, 1059, 2005.
17. Kahn, R., Buse, J., Ferrannini, E., and Stern, M. The metabolic syndrome: time for a critical appraisal. A joint statement from the American Diabetes Association and the European Association for the Study of Diabetes. *Diabet Care* 28, 2289, 2005.

18. Yusuf, S. et al. Effect of potentially modifiable risk factors associated with myocardial infarction in 52 countries (the INTERHEART study): case-control study. *Lancet* 364, 937, 2004.

2 The Metabolic Syndrome: The Question of Balance between the Pro-Inflammatory Effect of Macronutrients and the Anti-Inflammatory Effect of Insulin

Paresh Dandona, Ajay Chaudhuri,
Priya Mohanty, and Husam Ghanim

CONTENTS

INTRODUCTION

The common occurrence of the combination of obesity, insulin resistance, hypertension, hypertriglyceridemia, low HDL cholesterol, and hyperinsulinemia was first described by Reaven as insulin resistance syndrome or metabolic syndrome. The syndrome was recognized as a pro-atherogenic risk causing coronary heart disease (CHD). More recently, other features like elevated plasma PAI-1 and CRP have been added to the combination. In view of recent data demonstrating that insulin exerts an anti-inflammatory effect while macronutrients exert pro-inflammatory effects, we are in a better position to explain why an insulin-resistant state such as metabolic syndrome is pro-inflammatory and also explain how it develops. This focused review discusses the relevance of these recent observations, puts into perspective the pathogenesis of various features of metabolic syndrome, and also predicts some features that may be incorporated into it in the future.

INSULIN RESISTANCE SYNDROME: METABOLIC PERSPECTIVE

Reaven's original description of the metabolic syndrome consisted of obesity, insulin resistance, hypertension, impaired glucose tolerance or diabetes, hyperinsulinemia, and dyslipidemia characterized by elevated triglyceride and low HDL concentrations.[1] All these features serve as risk factors for atherosclerosis and thus metabolic syndrome constitutes a significant risk for coronary heart disease[2-5] (Table 2.1). The features of obesity or overweight and insulin resistance also provide significant risks for developing type 2 diabetes.[5,6] The risks for CHD and diabetes with metabolic syndrome are greater than those for simple obesity alone without insulin resistance and therefore the understanding of the pathogenesis and, through it, a rational approach to its therapy are of prime importance.

TABLE 2.1
Classic Biological Effects of Insulin and Classic Metabolic Syndrome Based on Resistance to Metabolic Effects of Insulin

Nutrients	Normal Insulin Action	Insulin-Resistant State
Carbohydrates	Hepatic glucose production	Hyperglycemia
	Glucose utilization	Hyperinsulinemia
	Glycogenesis	
Lipids	Lipolysis	Lipolysis
	FFA and glycerol	FFA and glycerol
	Lipogenesis	Hepatic triglyceride and apo B synthesis
	HDL	Hypertriglyceridemia
	Triglycerides	HDL
		Small dense LDL
Proteins	Gluconeogenesis	Glyconeogenesis
	Amino acids	Protein catabolism
	Protein synthesis	Protein synthesis
Purines	Uric acid clearance	Hyperuricemia
	Uric acid formation	

Source: From Dandona, P. et al., *Circulation* 111, 1448, 2005.

The original conceptualization of this syndrome was based on resistance to the metabolic actions of insulin. Thus, hyperinsulinemia, glucose intolerance, type 2 diabetes, hypertriglyceridemia, and low HDL concentrations could be accounted for by resistance to the actions of insulin on carbohydrate and lipid metabolism. While the features described above explain the atherogenesis to some extent, Reaven maintained that hyperinsulinemia contributed to atherogenicity and thus insulin is atherogenic, leading to CHD and cerebrovascular disease associated with this syndrome. However, as our understanding of the action of insulin evolves to comprehensively include recent discoveries,[7] we can better see that insulin resistance is the basis of most if not all the features of this syndrome.

Obesity probably leads to hypertension through (1) increased vascular tone created by a reduced bioavailability of nitric oxide (NO) due to increased oxidative stress,[8] (2) increased asymmetric dimethyl arginine (ADMA) concentrations,[9] (3) increased sympathetic tone,[10] and (4) increased expression of angiotensinogen by adipose tissue leading to an activation of the renin–angiotensin system.[11] The last factor requires further critical investigation.

Metabolic syndrome is characterized by low HDL in association with an elevated triglyceride concentration. This is believed to be due to an increased triglyceride load in HDL particles that are acted upon by hepatic lipase that hydrolyzes the triglyceride. The loss of the triglyceride results in a small HDL particle filtered by the kidney, resulting in decreases in apolipoprotein A (apo A) and HDL concentrations. Apart from an increase in the loss of apo A, data demonstrate that insulin may promote apo A gene transcription.[12] Therefore, insulin-resistant states may be associated with diminished apo A biosynthesis.[13]

An increase in plasma free fatty acid (FFA) concentrations plays a key role in the pathogenesis of insulin resistance through specific actions that block insulin signal transduction. Increases of plasma FFA concentrations in normal subjects to levels comparable to those in the obese also resulted in the induction of oxidative stress, inflammation, and subnormal vascular reactivity along with insulin resistance.[14] Because resistance to insulin also results in the relative non-suppression of adipocyte hormone-sensitive lipase, there is further enhancement of lipolysis and an increase in FFA concentration, leading to a vicious cycle of lipolysis, increased FFA, insulin resistance, and inflammation.

Several new features have been added to the syndrome over time. These include elevated plasminogen activator inhibitor-1 (PAI-1) concentrations and elevated C-reactive protein (CRP) concentrations. These features were added on the basis that they were found frequently in association with metabolic syndrome and no rational explanation indicated why they actually occurred. These features are probably related to both insulin resistance and obesity. The relationship of inflammation to obesity and insulin resistance still needs to be explained.[15]

NOVEL NON-METABOLIC ACTIONS OF INSULIN

The nonmetabolic actions of insulin are readily explained by the recent observations that insulin is an anti-inflammatory hormone and that macronutrient intake is pro-inflammatory. Insulin has been shown to suppress several pro-inflammatory transcription factors such as nuclear factor kappa-B (NFκB), early growth response-1 (Egr-1), activator protein-1 (AP-1), and the corresponding genes they regulate that mediate inflammation.[16,17] An impairment of the action of insulin due to insulin resistance would thus result in the activation of these pro-inflammatory transcription factors and an increase in the expression of the corresponding genes.

Insulin has been shown to suppress NFκB binding activity, reactive oxygen species (ROS) generation, p47[phox] expression, increase inhibitor kappa-B (IκB) expression in mononuclear cells (MNCs), and suppress plasma concentrations of intracellular adhesion molecule-1 (ICAM-1) and monocyte chemoattractant protein-1 (MCP-1).[16] In addition, insulin suppresses AP-1 and Egr-1, two pro-inflammatory transcription factors and their respective genes, matrix metalloproteinase-9 (MMP-9), tissue factor (TF), and PAI-1.[17-19] Thus, insulin exerts comprehensive anti-inflammatory effects and also has anti-oxidant effects as reflected in the suppression of ROS generation and p47[phox] expression (Figure 2.1).[16,20]

Two further pieces of evidence demonstrating the anti-inflammatory action of insulin have emerged recently. First, the treatment of type 2 diabetes with insulin for 2 weeks caused reductions in CRP and MCP-1.[21] Second, the treatment of severe hyperglycemia associated with marked increases in inflammatory mediators with insulin resulted in rapid marked decreases in the concentrations of inflammatory mediators.[22] Most recently, in a rat model in which inflammation was induced with endotoxin, insulin suppressed the concentration of inflammatory mediators including

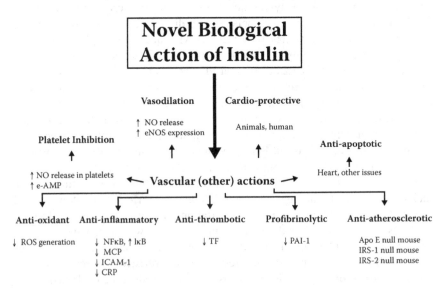

FIGURE 2.1 Novel biological effects of insulin targeted at endothelial cells, platelets, and leucocytes resulting in vasodilation, anti-aggregatory effects on platelets, anti-inflammatory effects, and other related effects. (*Source:* From Dandona, P. et al., *Circulation* 111, 1448, 2005.)

interleukin-1 beta (IL-1β), IL-6, macrophage inhibitory factor (MIF), and tumor necrosis factor alpha (TNFα).[23] Insulin also suppressed the expression of pro-inflammatory transcription factor CEBP and cytokines in the livers of the experimental animals. Similar reductions in inflammatory mediators were observed in rats with thermal injuries treated with insulin.[24] Finally, insulin has been shown to suppress the increases in cytokine concentrations in pigs challenged with endotoxin.[23]

The anti-inflammatory, anti-oxidant (ROS-suppressive), anti-thrombotic, and pro-fibrinolytic effects of insulin have recently been shown to occur in patients with acute myocardial infarctions when they were treated with low dose infusions of insulin independently of decreases in glucose concentrations. These patients demonstrated impressive 40% reductions in plasma CRP and serum amyloid A (SAA) concentrations at 24 hours. This reduction was maintained at 48 hours of insulin infusion.[25]

These anti-inflammatory effects of insulin have also been shown in patients undergoing coronary artery bypass grafts in association with extracorporeal circulation.[26] The increase in plasma CRP concentration occurring within 16 hours of surgery is 30 times greater than that in patients with ST-Elevation Myocardial Infarction (STEMI).[26] The reduction in the magnitude of increase in CRP and SAA is 40% — very similar to that observed following insulin infusion in patients with STEMI.[26]

Another novel anti-apoptotic effect of insulin has recently been described. In experimental acute myocardial infarction in the rat heart, the addition of insulin to the reperfusion fluid led to a 50% reduction in infarct size.[27] More recently, a similar cardio-protective effect of insulin was shown in human acute myocardial infarction when insulin at a low dose was infused with a thrombolytic agent and heparin.[20] Conversely, insulin-resistant states of obesity and type 2 diabetes have been shown to be associated with larger infarcts than those observed in non-diabetic subjects. Further work is required to establish this feature as an integral component of metabolic syndrome.

It should also be mentioned that insulin administration suppresses atherogenesis in apolipoprotein E null mice.[28] Conversely, interference with insulin signal transduction, as in IRS-2 null mice, resulted in atherosclerosis.[29] The IRS-1 null mouse also has a tendency toward atherosclerosis. It is relevant that a mutation of IRS-1 (arginine at 792) leads to abnormal vascular reactivity, a decrease in eNOS expression in endothelial cells, and an increased incidence of coronary heart disease.[30] It is interesting that the pro-atherogenic effects of insulin proposed primarily on the basis of *in vitro* studies are being challenged by evidence generated in the past 6 years.[7] This debate has been further fueled by two recent articles showing that knocking out the insulin receptor in myelogenic cells that are precursors of MNCs (cells that play a crucial role in the pathogenesis of atherosclerosis) is anti-atherogenic in the background of LDL receptor deficiency and pro-atherogenic in apo E-deficient animals.[31,32]

Consistent with the anti-inflammatory effects of insulin, insulin sensitizers of the thiazolidinediones class (troglitazone[33,34] and rosiglitazone[35]) have been shown to exert anti-inflammatory effects in addition to their glucose lowering effects in patients with diabetes. Troglitazone has been shown to suppress the development of diabetes in patients at high risk of developing this condition.[36] Trials are under way to determine whether rosiglitazone and pioglitazone prevent both type 2 diabetes and atherosclerotic complications. Positive results from those trials would support the concept that inflammatory mechanisms underlie the pathogenesis both of insulin resistance and atherosclerosis. It is of interest in this regard that metformin causes reductions in plasma concentrations of MIF in obese subjects.[37] Obese individuals have elevated plasma concentrations of this cytokine and increases in the expression of this cytokine in MNCs.[37] While evidence indicates that TZDs exert direct anti-inflammatory effects on macrophages *in vitro*, it is possible that their effects *in vivo* may arise through insulin sensitization.

OBESITY AND INFLAMMATION

The above data explain why an insulin-resistant state may be pro-inflammatory. They do not, however, explain the origin of insulin resistance. Mutations of the genes involved in insulin signal transduction provide one approach to the study of this issue in humans and in mice with specific gene knockouts. Such lesions are of interest but are too infrequent to provide a basis for the understanding of the pathogenesis of insulin resistance at large in humans. Thus, some recent observations on the interference of insulin signal transduction by inflammatory mechanisms are of great interest because obesity is a pro-inflammatory state (Figure 2.2).

Even if we accept that inflammatory mechanisms are involved in the pathogenesis of interference with insulin signal transduction and of insulin resistance itself, how does inflammation arise? Over the past decade, obesity has been associated with inflammation. This association was first proposed in a landmark paper by Hotamisligil et al. in which TNFα was shown to be constitutively expressed by adipose tissue, to be hyper-expressed in obesity, and to mediate insulin resistance in the major animal models of obesity.[38] This seminal paper also demonstrated that the neutralization of TNFα with soluble TNFα receptors resulted in the restoration of insulin sensitivity. Thus, the TNFα pro-inflammatory cytokine was the mediator of insulin resistance. Although the infusion of soluble TNFα receptors in humans has not reproduced the results observed in mice,[39] Hotamisligil's paper laid the foundation of the concept that inflammatory mechanisms may have a role to play in the pathogenesis of insulin resistance.

More data have now accumulated to reinforce the concept that obesity is an inflammatory state in humans. Increased plasma concentrations of TNFα, IL-6, CRP, MIF, and other inflammatory mediators were demonstrated in obese subjects.[37,40–44] Adipose tissue has been shown to express most of these pro-inflammatory mediators. It has also been shown that macrophages residing in adipose tissue may also be sources of pro-inflammatory factors and may also modulate the secretory activities of adipocytes.[45]

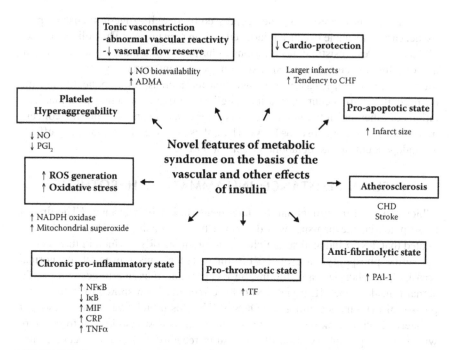

FIGURE 2.2 Extension of metabolic syndrome based on novel actions of insulin. (*Source: From Dandona, P. et al., Circulation* 111, 1448, 2005.)

Tissue macrophages are derived from monocytes in blood. Recently, the mononuclear cells of obese patients, of which monocytes constitute a fraction, have also been shown to be in an inflammatory state, expressing increased amounts of pro-inflammatory cytokines and related factors.[46] In addition, these cells have been shown to have significantly increased binding of NFκB, the key pro-inflammatory transcription factor, and an increase in the intranuclear expression of p65 (Rel A), the major protein component of NFκB. These cells also express diminished amounts of IκBβ, the inhibitor of NFκB. Evidence of inflammation clearly exists in various cells and in plasma in obesity.

In addition to TNFα and IL-6, the major adipocyte cytokines, three other important proteins, leptin, adiponectin, and resistin, need mention. While leptin is known for its function as a satiety signal that inhibits feeding, it has additional roles as a regulator of sexual function and as an immune modulator. It is also pro-inflammatory and induces platelet aggregation.[47–49] Thus, its elevated concentrations may contribute to the pro-inflammatory state of obesity and to atherogenesis in the long term. On the other hand, adiponectin, secreted in abundance by adipocytes in normal subjects, is anti-inflammatory and thus potentially anti-atherogenic. In contrast to leptin, its concentration falls with weight gain and in obesity.[50,51] It has been suggested that a low adiponectin concentration may be a marker for atherosclerosis and coronary heart disease.[52] Furthermore, in several experimental models, it has been shown to be protective to the arterial endothelium.

Resistin, discovered as a gene suppressed by rosiglitazone in mouse adipose tissue, earned its name because it induced insulin resistance.[53] Thiazolidenediones (TZDs) suppress resistin concentrations in humans while inducing an increase in adiponectin concentration consistent with the anti-inflammatory effects of these drugs. Furthermore, it has been shown that the increase in adiponectin and the decrease in resistin occur early after rosiglitazone treatment along with manifestations of other anti-inflammatory effects prior to any changes in plasma concentrations of insulin, glucose, or FFAs. Thus, these early anti-inflammatory effects are independent of the metabolic effects of TZDs.[54]

INSULIN RESISTANCE: INFLAMMATORY HYPOTHESIS

Which aspect of the inflammatory state results in insulin resistance? The first of these potential mechanisms was described by Hotamisligil et al., who demonstrated that TNFα induced serine phosphorylation of IRS-1 which in turn caused the serine phosphorylation of the insulin receptor. This prevented the normal tyrosine phosphorylation of the insulin receptor and thus interfered with insulin signal transduction.[55] IL-6 and TNFα have recently been shown to induce suppressor of cytokine signaling-3 (SOCS-3).[56,57] This protein was hitherto thought to interfere with cytokine signal transduction but is now also known to interfere with tyrosine phosphorylation of the insulin receptor and insulin receptor substrate-1 (IRS-1) and cause ubiquitination and proteosomal degradation of IRS-1.[58] This in turn reduces the activation of Akt (protein kinase B) which normally causes the translocation of the Glut-4 insulin-responsive glucose transporter to the plasma membrane. It also induces the phosphorylation of the nitric oxide synthase (NOS) enzyme and its activation to generate NO.[59]

A newly described protein known as TRB3 has also been shown to interfere with the activation of Akt and thus to interfere with insulin action.[60] However, the association of TRB3 with inflammatory mechanisms has not been demonstrated.

Recent data indicate that Akt2, a key protein involved in insulin signal transduction that mediates the phosphorylation and activation of e-NOS and NO secretion, also prevents the mobilization of Rac-1 to cell membranes, thus preventing superoxide generation. Superoxide generation is dependent upon the translocation of essential elements of NADPH oxidase (e.g., p47^{phox}) from the cytosol to the membrane. This is mediated by Rac.[61] In the absence of Akt2, there will be an increase in the translocation of Rac-1 to the membrane, greater formation of NADPH oxidase complex, increased superoxide generation, and oxidative stress. It has been shown that Akt2 null mice develop insulin resistance and mild hyperglycemia in association with hyperinsulinemia.[62]

MACRONUTRIENTS AND ORIGIN OF INFLAMMATION

If indeed obesity is a pro-inflammatory state and inflammatory mechanisms interfere with insulin signal transduction, what is the origin of this pro-inflammatory state? The answer comes mainly from recent observations demonstrating that macronutrient intake may induce oxidative stress and inflammatory responses. Thus a 75-g glucose challenge has been shown to induce an increase in superoxide generation by leucocytes by 140% over the basal in addition to increasing p47[phox] expression, a subunit of NADPH oxidase, the enzyme that converts molecular O_2 to superoxide radical.[63]

Equicaloric amounts of cream (fat) intake result in similar amounts of oxidative stress.[64] Glucose intake also results in comprehensive inflammation as reflected in an increase in intranuclear NFκB binding, a decrease in IκB expression, and an increase in IKKα and IKKβ, the two kinases that phosphorylate IκBα and IκBβ and result in their ubiquitination and proteosomal degradation.[65] Glucose intake also causes increases in two other pro-inflammatory transcription factors: AP-1 and Egr-1.[66] AP-1 regulates the transcription of matrix metalloproteinases while Egr-1 modulates the transcription of tissue factor (TF) and PAI-1. Thus, glucose intake increases the expression of matrix metalloproteinases 2 and 9 as well as expression of TF and PAI-1.

A mixed meal from a fast food chain was also shown to induce the activation of NFκB, a reduction in IκBα, an increase in IKKα and IKKβ, and an increase in superoxide radical generation by MNCs.[67] It is also of interest that an intravenous infusion of triglyceride with heparin in normal subjects with elevations of FFA concentrations to levels comparable to those found in obese subjects resulted in inflammatory responses.[14]

All genes that are stimulated by acute nutritional intake have also been shown to be activated in the basal states of obese subjects such that the concentrations of these gene products were elevated. Consistent with this, reductions in macronutrient intake in obese subjects (1000 kcal daily for 4 weeks) were shown to reduce both oxidative stress and inflammatory mediators.[8] Similarly, a 48-hour fast has been shown to reduce ROS generation by over 50% in normal subjects; the expression of p47[phox] was also reduced.[68]

Clearly, macronutrient intake is a major regulator of oxidative stress. It is relevant that the superoxide radical generated during oxidative stress is an activator of at least two major pro-inflammatory transcription factors, NFκB and AP-1. NFκB regulates the transcriptional activities of at least 200 genes, most of which are pro-inflammatory.[69–72] Thus, it is not surprising that obesity is a pro-inflammatory condition. Indeed, the MNCs of obese individuals are in a pro-inflammatory state, expressing an excess of a series of pro-inflammatory genes in addition to demonstrating increased NFκB binding, p65 expression, and decreased IκBβ protein.[46]

In addition to obesity and increased macronutrient intake, genetic and other environmental factors may induce the activation of inflammatory mechanisms and the induction of oxidative stress. These factors may be relevant in those ethnic groups in whom metabolic syndrome has been shown to occur in the absence of obesity. In these groups, migration to western countries like the U.S. and U.K. results in increased adiposity with a sedentary lifestyle that results in a phenotype of the metabolic syndrome against an appropriate genetic background (Figure 2.3).

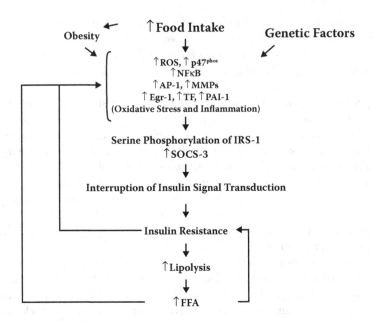

FIGURE 2.3 Pathogenesis of metabolic syndrome. (*Source:* From Dandona, P. et al., *Circulation* 111, 1448, 2005.)

When considering macronutrient-induced inflammation, it can be argued that the foods consumed now were always consumed, so why are their pro-inflammatory effects suddenly becoming relevant? The reason is that the amounts of food consumed now are far greater; furthermore, larger portions of the average diet consist of fast foods and lack sufficient fiber, fruits, and vegetables. Furthermore, the relative contents of refined carbohydrates and saturated fats have increased. This combination results in the inability of endogenously secreted insulin in response to meal intake to suppress the inflammation generated by the meal.

It is of interest in this regard that a 900-kcal AHA step 2 diet-based meal rich in fruit and fiber does not cause significant oxidative stress or inflammation in contrast to the effect of an isocaloric fast food meal.[73] Furthermore, orange juice taken in amounts equivalent to 75 g glucose (= 300 cal) does not induce any increase in $O_2^\bullet$ generation or an increase in NFB binding by MNCs. This may be due partly to the anti-oxidant and anti-inflammatory effects of ascorbic acid and flavanoids contained in orange juice. However, it has also been shown that fructose, which accounts for 50% of carbohydrates in orange juice, does not cause $O_2^\bullet$ generation or increased NFB binding. It is also of interest that orange juice induced a greater increase in insulin for a certain increase in glucose concentration when compared to glucose challenge.[74] This raises a novel concept when considering post-prandial hyperglycemia, oxidative stress, and inflammation following ingestion of various foods. The recent demonstration that M3 muscarinic receptors in the β cells of the pancreatic islets enhanced the insulinogenic response to a given glucose load or concentration suggests such an avenue of investigation. It is possible that certain foods like orange juice activate and/or increase the expression of M3 muscarinic receptors on β cells to enhance insulin secretion in response to glucose challenge.[75]

Both glucose and cream intake induce the expression of a series of pro-inflammatory genes in peripheral blood MNCs as reflected in the mRNA. These genes include TNFα, IL-6, and other related genes. Among the genes induced are SOCS-3, the molecule considered a potential candidate for the mediation of insulin resistance through serine phosphorylation, ubiquitination, and proteasomal degradation of IRS-1.

The increase in superoxide radical generation also results in diminished bioavailability of NO since NO binds to superoxide radical to form peroxynitrate.[76] In addition to the fact that Akt is inhibited due to insulin resistance, and thus NOS is also inhibited, the reduction in NO bioavailability can result in a marked reduction in NO action. Furthermore, TNFα suppresses the expression of NOS. The factors result in abnormalities in endothelium-mediated vasodilatation and vascular reactivity.[77] Interestingly, the abnormalities in vascular reactivity in the obese insulin-resistant population can be reproduced acutely by a 900-kcal fast food meal, just as the pro-inflammatory changes in obesity can be reproduced by a similar meal.[67] It is noteworthy that the plasma concentrations of asymmetric dimethylarginine (ADMA) are elevated in obesity and inhibit NOS activity, thus reducing the synthesis and secretion of NO.[78] Rosiglitazone suppresses plasma ADMA concentrations while improving the impaired vascular reactivity in the obese and those with type 2 diabetes.[79]

While the initial work on macronutrient intake with glucose, cream, and fast food meals showed a pro-inflammatory effect associated with oxidative stress, data are now emerging to demonstrate that some macronutrients may be safe and non-inflammatory. Thus, a 900-kcal breakfast rich in fruit and fiber does not cause oxidative stress or inflammation. The intake of vitamin E prior to glucose challenge also suppresses oxidative stress and inflammation. Similarly, alcohol[65] and orange juice in equicaloric amounts do not cause oxidative stress or inflammation. Because orange juice is rich in flavanoids and vitamin C, it is possible that the presence of macronutrients in food may alter or suppress oxidative stress or inflammation. Furthermore, data show that vitamin E administration to patients with insulin resistance reduced cytokine production by MNCs.[80]

Other strategies to prevent post-prandial oxidative stress and inflammation may involve drug therapies. One initial attempt in this area indicates that rosiglitazone treatment for 6 weeks (8 mg/day) led to the total suppression of oxidative and inflammatory stress caused by a 75-g glucose challenge in type 2 diabetes.[81]

CONCLUSION: INFLAMMATION HYPOTHESIS OF METABOLIC SYNDROME

In conclusion, the pro-inflammatory state of obesity and metabolic syndrome originates with excessive caloric intake and is probably due to over-nutrition in a majority of patients in the U.S. The pro-inflammatory state induces insulin resistance leading to clinical and biochemical manifestations of the metabolic syndrome. This resistance to insulin action further promotes inflammation through increases in lipolysis and plasma FFA concentrations on one hand and interference with the anti-inflammatory effect of insulin on the other.

While these factors may be the most important ones in a majority of patients with metabolic syndrome, it is possible that genetic factors may also contribute to the inflammatory stress in metabolic syndrome. These factors may be important in ethnic groups such as Asian Indians who may have increased amounts of upper abdominal fat in spite of normal BMI values.[82] Since excessive nutritional intake probably accounts for the inflammation at least in obesity-associated metabolic syndrome, the most rational way to suppress such inflammation is through caloric restriction.

The other lifestyle change that affects inflammation is exercise. Exercise produces a decrease in the indices of inflammation such as plasma CRP concentration.[83] The mechanism underlying this effect is not known. However, it is noteworthy that a lifestyle change is a very effective way to reduce the rate of development of diabetes in a pre-diabetic population as shown by the diabetes prevention study.[84,85] Exercise and reductions in macronutrient intake both cause a reduction in inflammation. Several drugs have also been shown to reduce oxidative stress and inflammation and may therefore be used in the treatment of metabolic syndrome. Among them are thiazolidinediones, angiotensin II receptor blockers, and carvedilol, a β-blocker with some α-blocking activity.[33,35,54,86–88]

REFERENCES

1. Reaven GM. Banting Lecture 1988: role of insulin resistance in human disease. *Diabetes* 37, 1595, 1988.
2. Pyorala K. Relationship of glucose tolerance and plasma insulin to the incidence of coronary heart disease: results from two population studies in Finland. *Diabetes Care* 2, 131, 1979.
3. Ninomiya JK, L'Italien G, Criqui MH et al. Association of the metabolic syndrome with history of myocardial infarction and stroke in the third national health and nutrition examination survey. *Circulation* 109, 42, 2004.
4. Lakka HM, Laaksonen DE, Lakka TA et al. The metabolic syndrome and total and cardiovascular disease mortality in middle-aged men. *JAMA* 288, 2709, 2002.
5. Festa A, D'Agostino R, Jr, Howard G et al. Chronic subclinical inflammation as part of the insulin resistance syndrome: the Insulin Resistance Atherosclerosis Study (IRAS). *Circulation* 102, 42, 2000.
6. Klein BE, Klein R, and Lee KE. Components of the metabolic syndrome and risk of cardiovascular disease and diabetes in beaver dam. *Diabetes Care* 25, 1790, 2002.
7. Dandona P, Aljada A, and Mohanty P. The anti-inflammatory and potential anti-atherogenic effect of insulin: a new paradigm. *Diabetologia* 45, 924, 2002.

8. Dandona P, Mohanty P, Ghanim H et al. The suppressive effect of dietary restriction and weight loss in the obese on the generation of reactive oxygen species by leukocytes, lipid peroxidation, and protein carbonylation. *J Clin Endocrinol Metab* 86, 355, 2001.

9. Lin KY, Ito A, Asagami T et al. Impaired nitric oxide synthase pathway in diabetes mellitus: role of asymmetric dimethylarginine and dimethylarginine dimethylaminohydrolase. *Circulation* 106, 987, 2002.

10. Esler M, Rumantir M, Kaye D et al. The sympathetic neurobiology of essential hypertension: disparate influences of obesity, stress, and noradrenaline transporter dysfunction? *Am J Hypertens* 14, 139S, 2001.

11. Giacchetti G, Faloia E, Sardu C et al. Gene expression of angiotensinogen in adipose tissue of obese patients. *Int J Obes Relat Metab Disord* 24, Suppl 2, S142, 2000.

12. Groenendijk M, Cantor RM, Blom NH et al. Association of plasma lipids and apolipoproteins with the insulin response element in the apoC-III promoter region in familial combined hyperlipidemia. *J Lipid Res* 40, 1036, 1999.

13. Mooradian AD, Haas MJ, and Wong NC. Transcriptional control of apolipoprotein A-I gene expression in diabetes. *Diabetes* 53, 513, 2004.

14. Tripathy D, Mohanty P, Dhindsa S et al. Elevation of free fatty acids induces inflammation and impairs vascular reactivity in healthy subjects. *Diabetes* 52, 2882, 2003.

15. Dandona P, Aljada A, and Bandyopadhyay A. Inflammation: the link between insulin resistance, obesity and diabetes. *Trends Immunol* 25, 4, 2004.

16. Dandona P, Aljada A, Mohanty P et al. Insulin inhibits intranuclear nuclear factor kappa-B and stimulates I-kappa-B in mononuclear cells in obese subjects: evidence for an anti-inflammatory effect? *J Clin Endocrinol Metab* 86, 3257, 2001.

17. Aljada A, Ghanim H, Mohanty P et al. Insulin inhibits the pro-inflammatory transcription factor early growth response gene-1 (Egr)-1 expression in mononuclear cells (MNC) and reduces plasma tissue factor (TF) and plasminogen activator inhibitor-1 (PAI-1) concentrations. *J Clin Endocrinol Metab* 87, 1419, 2002.

18. Ghanim H, Mohanty P, Aljada A et al. Insulin reduces the pro-inflammatory transcription factor, activation protein-1 (AP-1), in mononuclear cells (MNC) and plasma matrix metalloproteinase-9 (MMP-9) concentration. *Diabetes* 50, Suppl 2, A408, 2001.

19. Dandona P, Aljada A, Mohanty P et al. Insulin suppresses plasma concentration of vascular endothelial growth factor and matrix metalloproteinase-9. *Diabetes Care* 26, 3310, 2003.

20. Chaudhuri A, Janicke D, Wilson MF et al. Anti-inflammatory and profibrinolytic effect of insulin in acute ST-segment-elevation myocardial infarction. *Circulation* 109, 849, 2004.

21. Takebayashi K, Aso Y, and Inukai T. Initiation of insulin therapy reduces serum concentrations of high-sensitivity C-reactive protein in patients with type 2 diabetes. *Metabolism* 53, 693, 2004.

22. Stentz FB, Umpierrez GE, Cuervo R et al. Proinflammatory cytokines, markers of cardiovascular risks, oxidative stress, and lipid peroxidation in patients with hyperglycemic crises. *Diabetes* 53, 2079, 2004.

23. Jeschke MG, Klein D, Bolder U et al. Insulin attenuates the systemic inflammatory response in endotoxemic rats. *Endocrinology* 145, 4084, 2004.

24. Jeschke MG, Einspanier R, Klein D et al. Insulin attenuates the systemic inflammatory response to thermal trauma. *Mol Med.* 8, 443, 2002.

25. Chaudhuri A, Janicke D, Wilson MF et al. Anti-inflammatory and pro-fibrinolytic effect of insulin in acute ST-elevation myocardial infarction. *Circulation* 109, 849, 2004.

26. Visser L, Zuurbier CJ, Hoek FJ et al. Glucose, insulin and potassium applied as perioperative hyperinsulinaemic normoglycaemic clamp: effects on inflammatory response during coronary artery surgery. *Br J Anaesth* 95, 448, 2005.

27. Jonassen AK, Sack MN, Mjos OD et al. Myocardial protection by insulin at reperfusion requires early administration and is mediated via Akt and p70s6 kinase cell-survival signaling. *Circ Res* 89, 1191, 2001.

28. Shamir R, Shehadeh N, Rosenblat M et al. Oral insulin supplementation attenuates atherosclerosis progression in apolipoprotein E-deficient mice. *Arterioscler Thromb Vasc Biol* 23, 104, 2003.

29. Kubota T, Kubota N, Moroi M et al. Lack of insulin receptor substrate-2 causes progressive neointima formation in response to vessel injury. *Circulation* 107, 3073, 2003.

30. Federici M, Pandolfi A, De Filippis EA et al. G972R IRS-1 variant impairs insulin regulation of endothelial nitric oxide synthase in cultured human endothelial cells. *Circulation* 109, 399, 2004.

31. Han S, Liang CP, DeVries-Seimon T et al. Macrophage insulin receptor deficiency increases ER stress-induced apoptosis and necrotic core formation in advanced atherosclerotic lesions. *Cell Metab* 3, 257, 2006.

32. Baumgartl J, Baudler S, Scherner M et al. Myeloid lineage cell-restricted insulin resistance protects apolipoprotein E-deficient mice against atherosclerosis. *Cell Metab* 3, 247, 2006.

33. Ghanim H, Garg R, Aljada A et al. Suppression of nuclear factor kappa-B and stimulation of inhibitor kappa-B by troglitazone: evidence for an anti-inflammatory effect and a potential anti-atherosclerotic effect in the obese. *J Clin Endocrinol Metab* 86, 1306, 2001.

34. Aljada A, Garg R, Ghanim H et al. Nuclear factor-kappa-B suppressive and inhibitor-kappa-B stimulatory effects of troglitazone in obese patients with type 2 diabetes: evidence of an anti-inflammatory action? *J Clin Endocrinol Metab* 86, 3250, 2001.

35. Mohanty P, Aljada A, Ghanim H et al. Evidence for a potent antiinflammatory effect of rosiglitazone. *J Clin Endocrinol Metab* 89, 2728, 2004.

36. Buchanan TA, Xiang AH, Peters RK et al. Preservation of pancreatic beta-cell function and prevention of type 2 diabetes by pharmacological treatment of insulin resistance in high-risk Hispanic women. *Diabetes* 51, 2796, 2002.

37. Dandona P, Aljada A, Ghanim H et al. Increased plasma concentration of macrophage migration inhibitory factor (MIF) and MIF mRNA in mononuclear cells in the obese and the suppressive action of metformin. *J Clin Endocrinol Metab* 89, 5043, 2004.

38. Hotamisligil GS, Shargill NS, and Spiegelman BM. Adipose expression of tumor necrosis factor-alpha: direct role in obesity-linked insulin resistance. *Science* 259, 87, 1993.

39. Ofei F, Hurel S, Newkirk J et al. Effects of an engineered human anti-TNF-alpha antibody (CDP571) on insulin sensitivity and glycemic control in patients with NIDDM. *Diabetes* 45, 881, 1996.

40. Kern PA, Ranganathan S, Li C et al. Adipose tissue tumor necrosis factor and interleukin-6 expression in human obesity and insulin resistance. *Am J Physiol Endocrinol Metab* 280, E745, 2001.

41. Dandona P, Weinstock R, Thusu K et al. Tumor necrosis factor-alpha in sera of obese patients: fall with weight loss. *J Clin Endocrinol Metab* 83, 2907, 1998.

42. Vozarova B, Weyer C, Hanson K et al. Circulating interleukin-6 in relation to adiposity, insulin action, and insulin secretion. *Obes Res* 9, 414, 2001.

43. Wakabayashi I. Age-related change in relationship between body-mass index, serum sialic acid, and atherogenic risk factors. *J Atheroscler Thromb* 5, 60, 1998.

44. Pradhan AD, Manson JE, Rifai N et al. C-reactive protein, interleukin-6, and risk of developing type 2 diabetes mellitus. *JAMA* 286, 327, 2001.

45. Xu H, Barnes GT, Yang Q et al. Chronic inflammation in fat plays a crucial role in the development of obesity-related insulin resistance. *J Clin Invest* 112, 1821, 2003.

46. Ghanim H, Aljada A, Hofmeyer D et al. The circulating mononuclear cells in the obese are in a pro-inflammatory state. *Circulation* 110, 1564, 2004.

47. La Cava A, Alviggi C, and Matarese G. Unraveling the multiple roles of leptin in inflammation and autoimmunity. *J Mol Med* 82, 4, 2004.

48. Huang L and Li C. Leptin: a multifunctional hormone. *Cell Res.* 10, 81, 2000.

49. Nakata M, Yada T, Soejima N et al. Leptin promotes aggregation of human platelets via the long form of its receptor. *Diabetes* 48, 426, 1999.

50. Kubota N, Terauchi Y, Yamauchi T et al. Disruption of adiponectin causes insulin resistance and neointimal formation. *J Biol Chem* 277, 25863, 2002.

51. Ukkola O and Santaniemi M. Adiponectin: a link between excess adiposity and associated comorbidities? *J Mol Med* 80, 696, 2002.

52. Pischon T, Girman CJ, Hotamisligil GS et al. Plasma adiponectin levels and risk of myocardial infarction in men. *JAMA* 291, 1730, 2004.

53. Steppan CM, Bailey ST, Bhat S et al. The hormone resistin links obesity to diabetes. *Nature* 409, 307, 2001.

54. Ghanim H, Dhindsa S, Aljada A et al. Low dose rosiglitazone exerts an anti-inflammatory effect with an increase in adiponectin independently of free fatty acid (FFA) fall and insulin sensitization in obese type 2 diabetics. *J Clin Endocrinol Metab* June 27, 2006.

55. Hotamisligil GS, Peraldi P, Budavari A et al. IRS-1-mediated inhibition of insulin receptor tyrosine kinase activity in TNF-alpha- and obesity-induced insulin resistance. *Science* 271, 665, 1996.

56. Senn JJ, Klover PJ, Nowak IA et al. Suppressor of cytokine signaling-3 (SOCS-3), a potential mediator of interleukin-6-dependent insulin resistance in hepatocytes. *J Biol Chem* 278, 13740, 2003.

57. Emanuelli B, Peraldi P, Filloux C et al. SOCS-3 inhibits insulin signaling and is up-regulated in response to tumor necrosis factor-alpha in the adipose tissue of obese mice. *J Biol Chem* 276, 47944, 2001.

58. Rui L, Yuan M, Frantz D et al. SOCS-1 and SOCS-3 block insulin signaling by ubiquitin-mediated degradation of IRS1 and IRS2. *J Biol Chem* 277, 42394, 2002.

59. Dimmeler S, Fleming I, Fisslthaler B et al. Activation of nitric oxide synthase in endothelial cells by Akt-dependent phosphorylation. *Nature* 399, 601, 1999.

60. Du K, Herzig S, Kulkarni RN et al. TRB3: a tribbles homolog that inhibits Akt/PKB activation by insulin in liver. *Science* 300, 1574, 2003.

61. Ozaki M, Haga S, Zhang HQ et al. Inhibition of hypoxia/reoxygenation-induced oxidative stress in HGF-stimulated antiapoptotic signaling: role of PI3-K and Akt kinase upon rac1. *Cell Death Differ* 10, 508, 2003.

62. Cho H, Mu J, Kim JK et al. Insulin resistance and a diabetes mellitus-like syndrome in mice lacking the protein kinase Akt2 (PKB beta). *Science* 292, 1728, 2001.

63. Mohanty P, Hamouda W, Garg R et al. Glucose challenge stimulates reactive oxygen species (ROS) generation by leucocytes. *J Clin Endocrinol Metab* 85, 2970, 2000.

64. Mohanty P, Ghanim H, Hamouda W et al. Both lipid and protein intakes stimulate increased generation of reactive oxygen species by polymorphonuclear leukocytes and mononuclear cells. *Am J Clin Nutr* 75, 767, 2002.

65. Dhindsa S, Tripathy D, Mohanty P et al. Differential effects of glucose and alcohol on reactive oxygen species generation and intranuclear nuclear factor-kappa-B in mononuclear cells. *Metabolism* 53, 330, 2004.

66. Aljada A, Ghanim H, Mohanty P et al. Glucose intake induces an increase in AP-1 and Egr-1 binding activities and tissue factor and matrix metalloproteinase expressions in mononuclear cells and plasma tissue factor and matrix metalloproteinase concentrations. *Am J Clin Nutr* 80, 51, 2004.

67. Aljada A, Mohanty P, Ghanim H et al. Increase in intranuclear nuclear factor kappaB and decrease in inhibitor kappaB in mononuclear cells after a mixed meal: evidence for a proinflammatory effect. *Am J Clin Nutr* 79, 682, 2004.

68. Dandona P, Mohanty P, Hamouda W et al. Inhibitory effect of a two-day fast on reactive oxygen species (ROS) generation by leucocytes and plasma ortho-tyrosine and meta-tyrosine concentrations. *J Clin Endocrinol Metab* 86, 2899, 2001.

69. Woronicz JD, Gao X, Cao Z et al. I-kappa-B kinase-beta: NF-kappa-B activation and complex formation with I-kappa-B kinase-alpha and NIK. *Science* 278, 866, 1997.

70. Wang S, Leonard SS, Castranova V et al. The role of superoxide radical in TNF-alpha induced NF-kappa-B activation. *Ann Clin Lab Sci* 29, 192, 1999.

71. Verma IM, Stevenson JK, Schwarz EM et al. Rel/NF-kappa B/I-kappa B family: intimate tales of association and dissociation. *Genes Dev* 9, 2723, 1995.

72. Velasco M, Diaz-Guerra MJ, Martin-Sanz P et al. Rapid up-regulation of I-kappa-B and abrogation of NF-kappa-B activity in peritoneal macrophages stimulated with lipopolysaccharide. *J Biol Chem* 272, 23025, 1997.

73. Mohanty P, Daoud N, Ghanim H et al. Absence of oxidative stress and inflammation following the intake of a 900 kcal meal rich in fruit and fiber. *Diabetes* 53, Suppl 2, A405, 2004.

74. Ghanim H, Mohanty P, Pathak R et al. Caloric intake from orange juice does not induce oxidative and inflammatory response. Boston, Endocrine Society, 88th Annual Meeting, 2006.

75. Gautam D, Han SJ, Hamdan FF et al. A critical role for beta cell M3 muscarinic acetylcholine receptors in regulating insulin release and blood glucose homeostasis *in vivo*. *Cell Metab* 3, 449, 2006.

76. Koppenol WH, Moreno JJ, Pryor WA et al. Peroxynitrite, a cloaked oxidant formed by nitric oxide and superoxide. *Chem Res Toxicol* 5, 834, 1992.

77. Fard A, Tuck CH, Donis JA et al. Acute elevations of plasma asymmetric dimethylarginine and impaired endothelial function in response to a high-fat meal in patients with type 2 diabetes. *Arterioscler Thromb Vasc Biol* 20, 2039, 2000.

78. Chan NN and Chan JC. Asymmetric dimethylarginine (ADMA): a potential link between endothelial dysfunction and cardiovascular diseases in insulin resistance syndrome? *Diabetologia* 45, 1609, 2002.

79. Stuhlinger MC, Abbasi F, Chu JW et al. Relationship between insulin resistance and an endogenous nitric oxide synthase inhibitor. *JAMA* 287, 1420, 2002.

80. Devaraj S and Jialal I. Low-density lipoprotein postsecretory modification, monocyte function, and circulating adhesion molecules in type 2 diabetic patients with and without macrovascular complications: the effect of alpha-tocopherol supplementation. *Circulation* 102, 191, 2000.

81. Ahmad-Aljada SD, Ghanim H, Garg R et al. Rosiglitazone treatment reduces glucose-induced oxidative stress and inflammation. Presented at Endocrine Society 87th Annual Meeting, San Diego, 2005.

82. Banerji MA, Faridi N, Atluri R et al. Body composition, visceral fat, leptin, and insulin resistance in Asian Indian men. *J Clin Endocrinol Metab* 84, 137, 1999.

83. Church TS, Barlow CE, Earnest CP et al. Associations between cardiorespiratory fitness and C-reactive protein in men. *Arterioscler Thromb Vasc Biol* 22, 1869, 2002.

84. Tuomilehto J, Lindstrom J, Eriksson JG et al. Prevention of type 2 diabetes mellitus by changes in lifestyle among subjects with impaired glucose tolerance. *New Engl J Med* 344, 1343, 2001.

85. Knowler WC, Barrett-Connor E, Fowler SE et al. Reduction in the incidence of type 2 diabetes with lifestyle intervention or metformin. *New Engl J Med* 346, 393, 2002.

86. Dandona P, Karne R, Ghanim H et al. Carvedilol inhibits reactive oxygen species generation by leukocytes and oxidative damage to amino acids. *Circulation* 101, 122, 2000.

87. Garg R, Kumbkarni Y, Aljada A et al. Troglitazone reduces reactive oxygen species generation by leukocytes and lipid peroxidation and improves flow-mediated vasodilatation in obese subjects. *Hypertension* 36, 430, 2000.

88. Dandona P, Kumar V, Aljada A et al. Angiotensin II receptor blocker valsartan suppresses reactive oxygen species generation in leukocytes, nuclear factor-kappa B, in mononuclear cells of normal subjects: evidence of an antiinflammatory action. *J Clin Endocrinol Metab* 88, 4496, 2003.

3 The Role of Oxidative Stress in Diseases Associated with Overweight and Obesity

Ginger L. Milne, Ling Gao, Joshua D. Brooks, and Jason D. Morrow

CONTENTS

INTRODUCTION

The marked increase in the incidence of overweight and obese persons is recognized as perhaps the most serious public health issue in the United States. It is estimated that two-thirds of American adults are overweight and nearly 30% are obese.[1,2] Additionally, the incidence of overweight and obesity in children and adolescents is rising; in 2004, it was estimated that 16% of youth are either overweight or obese.[1,3] Both morbidity and mortality increase with excessive body weight.[4–6]

Multiple studies have shown that the risks of developing cardiovascular disease, type 2 diabetes mellitus, hypertension, dyslipidemia, gallbladder disease, osteoarthritis, stroke, and certain types of cancers increase with degree of overweight in both men and women.[2,7] Furthermore, early onset of symptoms of these diseases that were once only associated with adulthood are now observed in overweight adolescents and include high blood pressure, atherosclerosis, and type 2 diabetes mellitus.[8] Despite the well characterized association of overweight and disease incidence, the mechanisms by which overweight contributes to disease pathology are poorly understood. Nonetheless, several reports provide evidence that elevated systemic oxidant stress may be an important mechanism by which obesity promotes the development of chronic human disease.[9–11]

This chapter will examine the measurement of F_2-isoprostanes (IsoPs), a class of oxidized lipids generated from the peroxidation of arachidonic acid, as a marker of oxidant stress in overweight, obesity, and associated diseases. It will also assess the impacts of select therapeutic interventions on F_2-IsoP formation in this population.

F_2-ISOPROSTANES AS MARKERS OF OXIDANT STRESS *IN VIVO*

F_2-IsoPs represent one class of oxidized lipids formed *in vivo* in humans.[12–14] These compounds that were first identified by our laboratory in 1990 are prostaglandin (PG)-like molecules generated non-enzymatically from the free radical-initiated peroxidation of arachidonic acid, a ubiquitous polyunsaturated fatty acid.[15] Figure 3.1 shows the mechanism of formation of the F_2-IsoPs. Briefly, after abstraction of a bis-allylic hydrogen atom, one molecule of oxygen adds to arachidonic acid to form a peroxyl radical. This peroxyl radical then undergoes 5-*exo* cyclization and subsequent addition of another molecule of oxygen to form PGG_2-like compounds.

The unstable bicyclic endoperoxide intermediates are then reduced to the F_2-IsoPs, isomers of the cyclooxygenase-derived product $PGF_{2\alpha}$. Based on this mechanism of formation, four series of F_2-IsoP regioisomers are generated. In total, 64 different F_2-IsoP stereoisomers are formed from the oxidation of arachidonic acid.

In more recent years since their discovery, several methods have been developed to quantify the formation of F_2-IsoPs *in vivo*.[16] The method our laboratory and others found to be most sensitive, specific, and reliable employs the use of mass spectrometry and stable isotope dilution techniques.[17] Employing this technique, F_2-IsoPs have been detected in all human tissues and fluids examined, which is of particular importance in that it allows for an assessment of the effects of diseases on endogenous oxidant tone.[14] Additionally, defining levels of F_2-IsoPs *in vivo* is important because it allows for the determination of the extent to which various therapeutic interventions affect levels of oxidant stress.

For human studies, the quantification of F_2-IsoPs in body fluids such as urine and plasma is significantly more convenient and less invasive than measuring these compounds in organ tissue. Based on available data, quantification of these compounds in either plasma or urine is representative of their endogenous

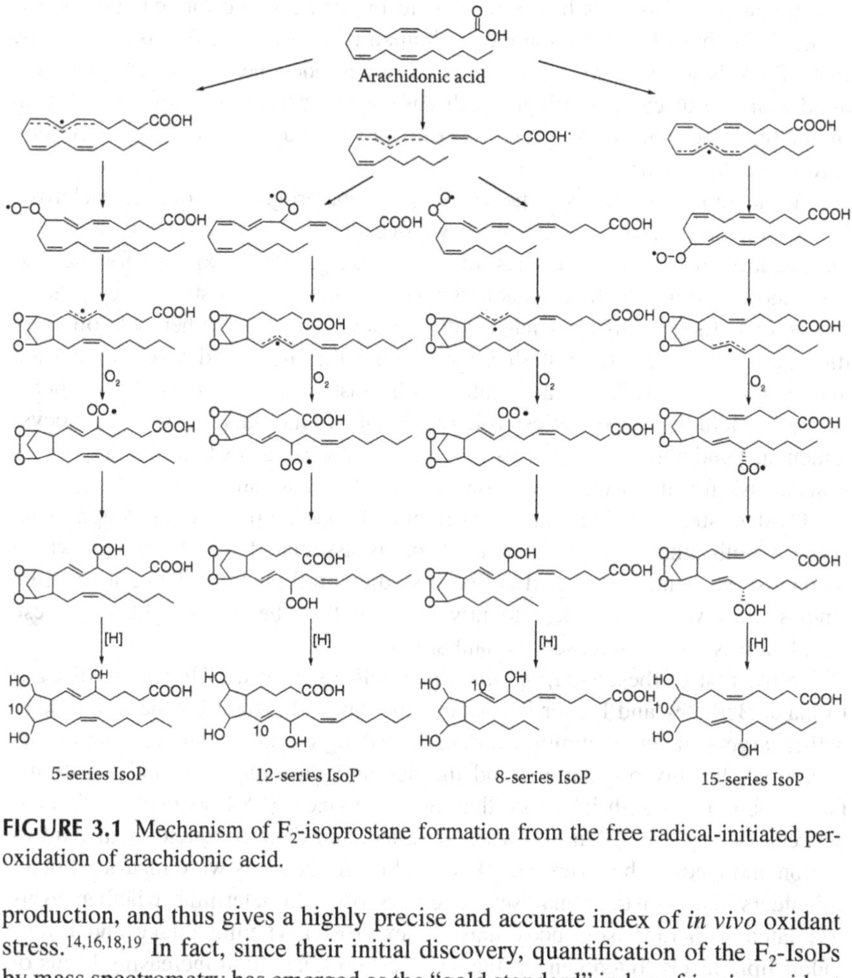

5-series IsoP 12-series IsoP 8-series IsoP 15-series IsoP

FIGURE 3.1 Mechanism of F_2-isoprostane formation from the free radical-initiated peroxidation of arachidonic acid.

production, and thus gives a highly precise and accurate index of *in vivo* oxidant stress.[14,16,18,19] In fact, since their initial discovery, quantification of the F_2-IsoPs by mass spectrometry has emerged as the "gold standard" index of *in vivo* oxidant stress status.

Recently, this assertion was independently confirmed by a National Institutes of Health-sponsored program termed the Biomarkers of Oxidative Stress Study (BOSS) that directly compared measurements of plasma and urinary F_2-IsoPs with other well known but less robust biomarkers of oxidant stress, including malondialdehyde (MDA) and other measures of lipid peroxidation, plasma glutathione, plasma antioxidant levels, protein carbonyls, and 8-hydroxy-deoxyguanosine.[20,21]

F_2-ISOPROSTANES AND OVERWEIGHT AND OBESITY

As discussed above, measurement of plasma or urinary F_2-IsoPs allows for an assessment of the effects of diseases on oxidant tone *in vivo*. In the past decade, quantification of F_2-IsoPs has been used to implicate a role for oxidative stress in the pathophysiology of a number of human conditions and diseases. Notably, F_2-IsoP levels were shown to be increased in neurodegenerative conditions such as Alzheimer's disease, Huntington's disease, aging, certain types of cancers, and, of notable importance to consequences of overweight and obesity, atherosclerotic cardiovascular disease.[14,22–26]

A role for oxidant stress in the development and progression of atherosclerosis has been hypothesized for more than two decades.[27–29] In recent years, however, the quantification of F_2-IsoPs has allowed investigators to explore, for the first time, the extent to which humans undergo enhanced oxidant stress under pathophysiological situations associated with the development of atherosclerotic cardiovascular disease. These studies have found that increased levels of plasma and/or urinary F_2-IsoPs are associated with most of the risk factors for atherosclerosis, including hypercholesterolemia,[30] diabetes mellitus,[31–33] hyperhomocysteinemia,[34] and chronic cigarette smoking.[35–37] This suggests that certain populations at risk for the disease are under increased oxidant stress.

Oxidant stress levels in the overweight and obese, a population at significant risk for both atherosclerosis and also for its associated risk factors in which oxidant stress is increased, had not been studied until recently. Three interesting studies, however, have independently reported that the overweight and obese population is under increased oxidant stress.

In the first of these studies, Block and colleagues at the University of California at Berkeley and Kaiser Permanente in Oakland sought for the first time to gather large-scale epidemiological data describing oxidative damage that occurs in normal, healthy populations and the demographic, physical, and nutritional factors associated with it.[9] More than 300 volunteers (55% women, 45% men) between the ages of 19 and 80 were recruited, and their complete dietary information and medical histories were known. Plasma F_2-IsoPs were measured in all volunteers and statistical analyses were performed to determine whether levels correlated with race, sex, body mass index (BMI), vitamin intake and levels, and/or lipid levels. Interestingly, the strongest correlate with increasing levels of F_2-IsoPs was increasing BMI (Figure 3.2).

In a second study reported in that same year, Davi and colleagues at the University of Rome reported a smaller study (93 volunteers) in which they specifically investigated lipid peroxidation in obese women by measuring levels of urinary F_2-IsoPs.[10] As in the first study, F_2-IsoP levels increased significantly with increasing BMI. Furthermore, these investigators went on to characterize the cause-and-effect relationship of this association by examining the effect of short-term, diet-induced weight loss on urinary F_2-IsoPs. Interestingly, with weight loss occurred a parallel decrease in BMI and a significant reduction in F_2-IsoP levels (Figure 3.3). Despite the small size of this weight loss study (12 volunteers), these findings have interesting clinical implications in the possible role of antioxidants in the treatment of overweight and obese individuals.

In the third and largest study by Keaney et al., urinary F_2-IsoPs were quantified in nearly 3000 participants in the Framingham Heart Study, and the authors again

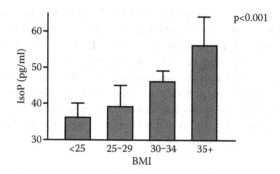

FIGURE 3.2 Endogenous levels of F_2-isoprostanes increase with increasing body mass index.

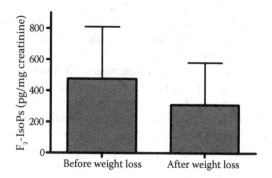

FIGURE 3.3 Urinary levels of F_2-isoprostanes decrease after diet-induced weight loss. (Adapted from Davi, G. et al., *JAMA* 288, 2008, 2002.)

found that enhanced IsoP formation was strongly correlated with increased BMI.[11] Interestingly, they showed that obesity, smoking, and diabetes were independently associated with increased oxidant stress. These findings add credence to the two smaller studies because this study was carried out with a large community-based population of otherwise healthy individuals. A particularly pertinent aspect of this study with respect to determining the role that obesity-associated oxidant stress plays in atherosclerosis and other diseases is the fact that participants in the trial will be followed over time so that clinical outcomes can be correlated with oxidant stress.

An important question raised by the studies discussed herein relates to mechanisms by which overweight and obesity induce oxidant stress-associated disease progression. It is likely that multiple pathways contribute as overweight and obesity are associated with elevated systemic inflammation and activated coagulation cascades in addition to oxidant stress.[2,38] A number of pathways capable of generating injury-inducing free radicals derived largely from molecular oxygen are known to be perturbed in obesity. For example, the renin–angiotensin system is up-regulated in obesity.[39] Angiotensin II has been shown to induce NADPH oxidase in various tissues with a resulting increase in superoxide production.[40] Angiotensin II has also been shown to increase LDL uptake by macrophages, resulting in enhanced lipoprotein oxidation.[40] Further, obesity has been associated with reduced antioxidant defense mechanisms, including decreased erythrocyte glutathione and glutathione peroxidase.[41] The extent to which these and other mechanisms contribute to obesity-associated oxidant stress needs to be explored and will likely provide key information about the importance of obesity in the development and progression of disease.

THERAPEUTIC TARGETS FOR REDUCING OXIDANT STRESS IN OVERWEIGHT AND OBESE PATIENTS

The findings of Block,[9] Davi,[10] and Keaney[11] are not only important with respect to the study of basic mechanisms underlying oxidant stress associated with obesity, but they also have important public health implications in regard to the treatment of obesity-associated disease. The incidence of overweight and obesity is becoming more prevalent in the United States and weight loss programs are often ineffective.[2] Thus, the number of persons with diseases associated with obesity is going to be a continuing burden to the medical community[42] and novel strategies to prevent and treat these disorders based on the physiological perturbations associated with obesity need to be developed and tested. The findings of the studies discussed herein implicate decreasing *in vivo* levels of oxidant stress as one potential therapeutic target for obesity-associated disesase.

ANTIOXIDANTS

One potential therapy to reduce oxidant stress *in vivo* is antioxidant supplementation. Animal and human epidemiologic studies carried out in the 1980s and 1990s suggested that antioxidants decreased atherosclerosis, presumably by reducing oxidant stress. However, prospective clinical trials of antioxidant supplementation using vitamin E and other agents have been disappointing and yielded conflicting results.[43–47] Three large trials, ATBC, GISSI, and HOPE, involving tens of thousands of subjects, failed to show reductions of cardiovascular events when vitamin E was used at doses ranging from 50 to 400 IU/day.[47–50] On the other hand, two trials, CHAOS and SPACE, involving fewer subjects, reported near 50% reductions in the incidence of cardiovascular events with supplementation with 800 IU/day.[51,52]

All these studies suffer from the limitation of using only cardiovascular events as trial endpoints and not assessing oxidant stress in study participants. Thus, it is impossible to determine whether vitamin E or other antioxidants inhibited oxidant injuries in the populations studied. Small studies performed by our laboratory, however, have shown that in order to reduce levels of F_2-IsoPs *in vivo*, an individual must take at least 800 IU/day of vitamin E for 16 weeks and probably more, implying that vitamin E is a very low potency antioxidant.[53] Further, supplementation with that very large amount of vitamin E is not practical due to safety issues related to consuming that amount of vitamin E. These data coupled with the clinical trial data suggest that supplementation with vitamin E is unlikely to prevent atherosclerotic events in humans. However, it is not possible to conclude from these studies that oxidant stress is not involved in the development and/or progression of atherosclerosis. Further study is needed to determine whether decreasing oxidative stress *in vivo* through antioxidant supplementation can prevent associated diseases.

ω-3 POLYUNSATURATED FATTY ACIDS

Emerging evidence has implicated increased dietary intake of fish oil containing large amounts of polyunsaturated fatty eicosapentaenoic acid (20:5, ω-3, EPA) and docosahexaenoic acid (22:6, ω-3, DHA), as beneficial in the prevention and treatment of a number of diseases in which environmental and lifestyle factors play roles. These include atherosclerotic cardiovascular disease and sudden death, metabolic syndrome, neurodegeneration, and various inflammatory disorders among others, but the mechanisms by which fish oil is protective are unknown.[54–57] Recent data, however, suggest that the anti-inflammatory effects and other biologically relevant properties of ω-3 fatty acids are due, in part, to the generation of various bioactive oxidation products.[58–63]

One potentially important anti-atherogenic and anti-inflammatory mechanism of ω-3 polyunsaturated fatty acids is their interference with the arachidonic acid cascade that generates pro-inflammatory eicosanoids.[54,64,65] EPA can replace arachidonic acid in phospholipid bilayers and is also a competitive inhibitor of COX, reducing the production of 2-series PGs and thromboxane in addition to 4-series leukotrienes. The 3-series-PGs and the 5-series leukotrienes derived from EPA are either less biologically active or inactive compared to the former products and are thus considered to exert effects that are less inflammatory.[66,67] Serhan and colleagues described a group of polyoxygenated DHA and EPA derivatives termed resolvins that are produced in various tissues. These compounds inhibit cytokine expression and other inflammatory responses in microglia, skin cells, and other cell types.[58–61]

In addition to studying the enzymatic oxidation of EPA, significant interest has focused on the biological activities of non-enzymatic free radical-initiated peroxidation products. Sethi and colleagues reported that EPA oxidized in the presence of Cu^{++}, but not native EPA, significantly inhibits human neutrophil and monocyte adhesion to endothelial cells — a process linked to the development of atherosclerosis and other inflammatory disorders.[62,63] This effect was induced via inhibition of endothelial adhesion receptor expression and was modulated by the activation of the peroxisome proliferator-activated receptor-α (PPAR-α) by EPA oxidation products.

Oxidized EPA markedly reduced leukocyte rolling and adhesion to venular endothelium of lipopolysaccharide-treated mice *in vivo* and the effect was not observed in PPAR-α-deficient mice. These studies suggest that the beneficial effects of fish oil may be mediated, in part, by the anti-inflammatory effects of oxidized EPA. Similarly, Vallve and colleagues have shown that various non-enzymatically generated aldehyde oxidation products of EPA (and DHA) decreased the expression of the CD36 receptor in human macrophages.[68] Up-regulation of this receptor has been linked to atherosclerosis.

Additional recent reports have suggested that other related biological effects of fish oil, such as modulation of endothelial inflammatory molecules, are related to the peroxidation products of EPA and DHA.[62,69] Arita and colleagues have also shown that non-enzymatically oxidized EPA enhances apoptosis in HL-60 leukemia cells, supporting the contention that oxidized ω-3 polyunsaturated fatty acids are both anti-proliferative and anti-inflammatory.[70] Similar findings have been reported in HepG2 (human hepatoma) cells and AH109A (rat liver cancer) cells.[71,72] Virtually none of these reports, however, have structurally identified the specific peroxidation products responsible for these effects.

The studies above that attributed some of the beneficial biological activities of fish oil to the oxidation of its ω-3 fatty acids provided our laboratory with a rationale to begin studies to systematically define the oxidation of EPA. Specifically, we were interested in the generation of F-ring IsoP-like compounds (F_3-IsoPs), structural analogs of the F_2-IsoPs containing an additional double bond between carbons 17 and 18, based on the hypothesis that these compounds may contribute to the beneficial biological effects of EPA and fish oil supplementation in that they may exert anti-inflammatory biological activities compared to F_2-IsoPs.

One report states that the EPA-derived IsoP, 15-F_{3t}-IsoP, possesses activity that is different from 15-F_{2t}-IsoP in that it does not affect human platelet shape change or aggregation.[73] Note that 15-F_{2t}-IsoP is a ligand for the Tx receptor and induces platelet shape change and also causes vasoconstriction.[15,74] The lack of activity of 15-F_{3t}-IsoP is consistent with observations regarding EPA-derived PGs in that these latter compounds exert either weaker agonist or no effects in comparison to arachidonate-derived PGs.[55,75,76]

Anggard and colleagues provided limited evidence that F_3-IsoPs could be formed from the oxidation of EPA *in vitro*.[77] We therefore considered the possibility that IsoP-like compounds could be formed by the free radical-induced peroxidation of EPA *in vivo*. Our studies showed that F_3-IsoPs were generated from the oxidation of EPA both *in vitro* and *in vivo*.[78] Unlike results from the previous studies by Anggard, the structural characteristics of these compounds were confirmed by a number of chemical derivatization and mass spectrometric techniques including LC/ESI/MS/MS.

As expected, six series of F_3-IsoP regioisomers were identified from both *in vitro* and *in vivo* sources. The mass spectrometric fragmentation patterns of these regioisomers are similar to F_2-IsoP regioisomers and, indeed, information that we previously acquired with F_2-IsoPs was extremely useful in the characterization of these molecules.[79] Of note, the relative abundance of 5- and 18-series F_3-IsoPs predominated over the other series. Such regioisomeric predominance was also reported for F_2-IsoP regioisomers in which 5- and 15-series compounds were

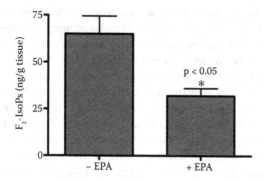

FIGURE 3.4 Supplementation of mice with eicosapentaenoic acid (EPA) in the diet reduces endogenous levels of F_2-isoprostanes in the heart.

formed in greater abundance than 8- and 12-series molecules.[79] At least part of the reason is likely due to the ability of precursors of 8- and 12-series F_2-IsoPs to undergo further oxidation and cyclization to yield a novel class of compounds termed dioxolane-endoperoxides.[80] Although undetermined at present, it is likely that a similar mechanism may account for the predominance of 5- and 18-series F_3-IsoPs.

Interestingly, the levels of these compounds generated from the oxidation of EPA significantly exceeded those of F_2-IsoPs generated from arachidonic acid, perhaps because EPA contains more double bonds and is therefore more easily oxidizable. Furthermore, *in vivo* in mice, levels of F_3-IsoPs in tissues such as heart were virtually undetectable at baseline but supplementation of animals with EPA markedly increased quantities up to 27.4 ± 5.6 ng/g heart.

Of particular note, we found that EPA supplementation markedly reduced levels of arachidonate-derived F_2-IsoPs by up to 64% ($p < 0.05$; Figure 3.4). This observation was significant because, as discussed throughout this chapter, F_2-IsoPs are generally considered to be pro-inflammatory molecules associated with the pathophysiological sequelae of oxidant stress. It is thus intriguing to propose that part of the mechanism by which EPA prevents certain diseases is its ability to decrease F_2-IsoP generation. In addition, it suggests that supplementation with fish oil may be of benefit to populations associated with increased levels of F_2-IsoPs, including the overweight and obese.

CONCLUSIONS

Today, overweight and obesity are two of the most serious public health concerns facing the United States. The incidence of overweight and obesity continues to rise and will likely increase for the next decade. Thus, the health care system will continue to see increasing numbers of persons with diseases associated with obesity including diabetes mellitus, hypertension, and atherosclerosis. Therefore, it is of importance to continue to study the underlying basic mechanisms by which overweight and obesity contribute to disease progression. At the same time, novel strategies to prevent and treat these disorders based upon our understanding of the physiological perturbations associated with obesity need to be developed.

The clinical studies reviewed herein by Block,[9] Davi,[10] and Keaney[11] provide insights into one mechanism, increased oxidant stress, that likely contributes to the pathophysiology of obesity-associated disease progression. Decreasing endogenous oxidant stress may therefore be a target for interventions for these diseases. Interestingly, preliminary animal studies have shown that supplementation with EPA, one of the major polyunsaturated fatty acids in fish oil, reduces *in vivo* levels of F_2-IsoPs, potent mediators of and important biomarkers for endogenous oxidant stress. Based upon these data and well-known findings that consumption of fish and fish oil supplementation reduce the incidence of atherosclerosis and other diseases in humans, fish oil consumption represents a novel potential treatment to decrease obesity-associated disease.

ACKNOWLEDGMENTS

The work reported herein was supported by National Institutes of Health grants GM15431, CA77839, DK48831, and ES13125.

REFERENCES

1. http://www.cdc.gov/nchs/products/pubs/pubd/hestats/obese/obse99.htm
2. Eckel, R.H., Barouch, W.W., and Ershow, A.G., Report of the National Heart, Lung, and Blood Institute–National Institute of Diabetes and Digestive and Kidney Diseases Working Group on the pathophysiology of obesity-associated cardiovascular disease, *Circulation* 105, 2923, 2002.
3. Hedley, A.A. et al., Prevalence of overweight and obesity among U.S. children, adolescents, and adults, 1999–2002, *JAMA* 291, 2847, 2004.
4. Bray, G.A., Overweight, mortality and morbidity, in *Physical Activity and Obesity*, Bouchard, C. et al., Eds., Human Kinetics Publishers, Champaign, IL, 2000, p. 31.
5. National Task Force on the Prevention and Treatment of Obesity, Overweight, obesity, and health risk, *Arch Int Med* 160, 898, 2000.
6. Fontaine, K.R. et al., Years of life lost to obesity, *JAMA* 289, 187, 2003.
7. Grundy, S.M., Obesity, metabolic syndrome and coronary atherosclerosis, *Circulation* 105, 2696, 2002.
8. Daniels, S.R., The consequences of childhood overweight and obesity, *Future Child* 16, 47, 2006.
9. Block, G. et al., Factors associated with oxidative stress in human populations, *Am J Epidemiol* 156, 274, 2002.
10. Davi, G. et al., Platelet activation in obese women: role of inflammation and oxidant stress, *JAMA* 288, 2008, 2002.
11. Keaney, J.F., Jr. et al., Obesity and systemic oxidative stress: clinical correlates of oxidative stress in the Framingham Study, *Arterioscler Thromb Vasc Biol* 23, 434, 2003.

12. Morrow, J.D. and Roberts, L.J., The isoprostanes: unique bioactive products of lipid peroxidation, *Progr Lipid Res* 36, 1, 1997.

13. Morrow, J.D. et al., The isoprostanes: unique prostaglandin-like products of free radical-catalyzed lipid peroxidation, *Drug Metab Rev* 31, 117, 1999.

14. Famm, S.S. and Morrow, J.D., The isoprostanes: unique products of arachidonic acid oxidation: a review, *Curr Med Chem* 10, 1723, 2003.

15. Morrow, J.D. et al., A series of prostaglandin F2-like compounds are produced *in vivo* in humans by a non-cyclooxygenase, free radical-catalyzed mechanism, *Proc Natl Acad Sci USA* 87, 9383, 1990.

16. Roberts, L.J. and Morrow, J.D., Measurement of F_2-isoprostanes: a reliable index of oxidant stress *in vivo*, *Free Radical Biol Med* 28, 505, 2000.

17. Morrow, J.D. and Roberts, L.J., Mass spectrometric quantification of F_2-isoprostanes in biological fluids and tissues as a measure of oxidant stress, *Meth Enzymol* 300, 3, 1999.

18. Morrow, J.D., The isoprostanes: their quantification as an index of oxidant stress status *in vivo*, *Drug Metab Rev* 32, 377, 2000.

19. Griffiths, H.R. et al., Biomarkers, *Mol Aspects Med* 23, 101, 2002.

20. Kadiiska, M.B. et al., Biomarkers of oxidative stress study II: are oxidation products of lipids, proteins, and DNA markers of CCl_4 poisoning? *Free Radical Biol Med* 38, 698, 2005.

21. Kadiiska, M.B. et al., Biomarkers of oxidative stress study III: effects of the nonsteroidal anti-inflammatory agents indomethacin and meclofenamic acid on measurements of oxidative products of lipids in CCl_4 poisoning, *Free Radical Biol Med* 38, 711, 2005.

22. Gniwotta, C. et al., Prostaglandin F2-like compounds, F2-isoprostanes, are present in increased amounts in human atherosclerotic lesions, *Arterioscler Thromb Vasc Biol* 17, 3236, 1997.

23. Gopaul, N.K. et al., Formation of F2-isoprostanes during aortic endothelial cell-mediated oxidation of low density lipoprotein, *FEBS Lett* 348, 297, 1994.

24. Montine, K.S. et al., Isoprostanes and related products of lipid peroxidation in neurodegenerative diseases, *Chem Phys Lipids* 128, 117, 2004.

25. Montine, T.J. et al., Cerebrospinal fluid F2-isoprostanes are elevated in Huntington's disease, *Neurology* 52, 1104, 1999.

26. Montine, T.J. et al., Cerebrospinal fluid F2-isoprostane levels are increased in Alzheimer's disease, *Ann Neurol* 44, 410, 1998.

27. Heinecke, J.W., Oxidants and antioxidants in the pathogenesis of atherosclerosis: implications for the oxidized low density lipoprotein hypothesis, *Atherosclerosis* 141, 1, 1998.

28. Berliner, J. et al., Oxidized lipids in atherogenesis: formation, destruction and action, *Thromb Haemost* 78, 195, 1997.

29. Chisholm, G.M. and Steinberg, D., The oxidative modification hypothesis of atherogenesis: an overview, *Free Radical Biol Med* 18, 1815, 2000.

30. Davi, G. et al., *In vivo* formation of 8-epi-prostaglandin F2 alpha is increased in hypercholesterolemia, *Arterioscler Thromb Vasc Biol* 17, 3230, 1997.

31. Davi, G. et al., *In vivo* formation of 8-iso-prostaglandin F2 alpha and platelet activation in diabetes mellitus: effects of improved metabolic control and vitamin E supplementation, *Circulation* 99, 224, 1999.

32. Davi, G. et al., Enhanced lipid peroxidation and platelet activation in the early phase of type 1 diabetes mellitus: role of interleukin-6 and disease duration, *Circulation* 107, 3199, 2003.
33. Gopaul, N.K. et al., Plasma 8-epi-PGF2 alpha levels are elevated in individuals with non-insulin dependent diabetes mellitus, *FEBS Lett* 368, 225, 1995.
34. Voutilainen, S. et al., Enhanced *in vivo* lipid peroxidation at elevated plasma total homocysteine levels, *Arterioscler Thromb Vasc Biol* 19, 1263, 1999.
35. Frei, B. et al., Gas phase oxidants of cigarette smoke induce lipid peroxidation and changes in lipoprotein properties in human blood plasma: protective effects of ascorbic acid, *Biochem J* 277, 133, 1991.
36. Morrow, J.D. et al., Increase in circulating products of lipid peroxidation (F2-isoprostanes) in smokers: smoking as a cause of oxidative damage, *New Engl J Med* 332, 1198, 1995.
37. Obwegeser, R. et al., Maternal cigarette smoking increases F2-isoprostanes and reduces prostacyclin and nitric oxide in umbilical vessels, *Prostaglandins Other Lipid Mediat* 57, 269, 1999.
38. Chan, J.C. et al., The central roles of obesity-associated dyslipidaemia, endothelial activation and cytokines in the metabolic syndrome: analysis by structural equation modelling, *Int J Obes Relat Metab Disord* 26, 994, 2002.
39. Barton, M. et al., Obesity is associated with tissue-specific activation of renal angiotensin-converting enzyme *in vivo*: evidence for a regulatory role of endothelin, *Hypertension* 35, 329, 2000.
40. Brasier, A.R., Recinos, A., 3rd, and Eledrisi, M.S., Vascular inflammation and the renin-angiotensin system, *Arterioscler Thromb Vasc Biol* 22, 1257, 2002.
41. Trevisan, M. et al., Correlates of markers of oxidative status in the general population, *Am J Epidemiol* 154, 348, 2001.
42. Manson, J.E. and Bassuk, S.S., Obesity in the United States: a fresh look at its high toll, *JAMA* 289, 229, 2003.
43. Rimm, E.B. et al., Vitamin E consumption and the risk of coronary heart disease in men, *New Engl J Med* 328, 1450, 1993.
44. Stampfer, M.J. et al., Vitamin E consumption and the risk of coronary disease in women, *New Engl J Med* 328, 1444, 1993.
45. Diaz, M.N. et al., Antioxidants and atherosclerotic heart disease, *New Engl J Med* 337, 408, 1997.
46. Pratico, D. et al., Vitamin E suppresses isoprostane generation *in vivo* and reduces atherosclerosis in ApoE-deficient mice, *Nat Med* 4, 1189, 1998.
47. Heinecke, J.W., Is the emperor wearing clothes? Clinical trials of vitamin E and the LDL oxidation hypothesis, *Arterioscler Thromb Vasc Biol* 21, 1261, 2001.
48. Alpha-Tocopherol–Beta Carotene Cancer Prevention Study Group, The effect of vitamin E and beta carotene on the incidence of lung cancer and other cancers in male smokers, *New Engl J Med* 330, 1029, 1994.
49. Gruppo Italiano per lo Studio della Sopravvivenza nell'Infarto Miocardico, Dietary supplementation with n-3 polyunsaturated fatty acids and vitamin E after myocardial infarction: results of the GISSI Prevenzione Trial, *Lancet* 354, 447, 1999.
50. Yusuf, S. et al., Vitamin E supplementation and cardiovascular events in high-risk patients, *New Engl J Med* 342, 154, 2000.
51. Stephens, N.G. et al., Randomised controlled trial of vitamin E in patients with coronary disease: Cambridge Heart Antioxidant Study (CHAOS), *Lancet* 347, 781, 1996.

52. Boaz, M. et al., Secondary prevention with antioxidants of cardiovascular disease in endstage renal disease (SPACE): randomised placebo-controlled trial, *Lancet* 356, 1213, 2000.

53. Morrow, J.D. et al., Alpha-tocopherol supplementation reduces plasma isoprostane levels in hypercholesterolemic humans only at doses of 800 IU or greater, *Circulation* 106, 165, 2002.

54. Kris-Etherton, P.M., Harris, W.S., and Appel, L.J., Fish consumption, fish oil, omega-3 fatty acids, and cardiovascular disease, *Arterioscler Thromb Vasc Biol* 23, 20, 2003.

55. Kris-Etherton, P.M., Harris, W.S., and Appel, L.J., Omega-3 fatty acids and cardiovascular disease: new recommendations from the American Heart Association, *Arterioscler Thromb Vasc Biol* 23, 151, 2003.

56. Ruxton, C., Health benefits of omega-3 fatty acids, *Nurs Stand*, 18, 38, 2004.

57. Ruxton, C.H. et al., The health benefits of omega-3 polyunsaturated fatty acids: a review of the evidence, *J Hum Nutr Diet* 17, 449, 2004.

58. Hong, S. et al., Novel docosatrienes and 17S-resolvins generated from docosahexaenoic acid in murine brain, human blood, and glial cells: autacoids in antiinflammation, *J Biol Chem* 278, 14677, 2003.

59. Serhan, C.N. and Levy, B., Novel pathways and endogenous mediators in antiinflammation and resolution, *Chem Immunol Allergy* 83, 115, 2003.

60. Serhan, C.N. et al., Novel functional sets of lipid-derived mediators with antiinflammatory actions generated from omega-3 fatty acids via cyclooxygenase 2-nonsteroidal antiinflammatory drugs and transcellular processing, *J Exp Med* 192, 1197, 2000.

61. Serhan, C.N. et al., Resolvins: a family of bioactive products of omega-3 fatty acid transformation circuits initiated by aspirin treatment that counter proinflammation signals, *J Exp Med* 196, 1025, 2002.

62. Sethi, S., Eastman, A.Y., and Eaton, J.W., Inhibition of phagocyte–endothelium interactions by oxidized fatty acids: a natural anti-inflammatory mechanism? *J Lab Clin Med* 128, 27, 1996.

63. Sethi, S., Inhibition of leukocyte-endothelial interactions by oxidized omega-3 fatty acids: a novel mechanism for the anti-inflammatory effects of omega-3 fatty acids in fish oil, *Redox Rep* 7, 369, 2002.

64. Uauy, R., Mena, P., and Valenzuela, A., Essential fatty acids as determinants of lipid requirements in infants, children and adults, *Eur J Clin Nutr* 53, Suppl 1, S66, 1999.

65. Smith, W.L., Cyclooxygenases, peroxide tone, and the allure of fish oil, *Curr Opin Cell Biol* 17, 174, 2005.

66. Calder, P.C., Polyunsaturated fatty acids, inflammation, and immunity, *Lipids* 36, 1007, 2001.

67. Yang, P. et al., Formation and antiproliferative effect of prostaglandin E(3) from eicosapentaenoic acid in human lung cancer cells, *J Lipid Res* 45, 1030, 2004.

68. Vallve, J.C. et al., Unsaturated fatty acids and their oxidation products stimulate CD36 gene expression in human macrophages, *Atherosclerosis* 164, 45, 2002.

69. Das, U.N., Essential fatty acids, lipid peroxidation, and apoptosis, *Prostaglandins Leukotr Essent Fatty Acids* 61, 157, 1999.

70. Arita, K. et al., Mechanisms of enhanced apoptosis in HL-60 cells by UV-irradiated n-3 and n-6 polyunsaturated fatty acids, *Free Radic Biol Med* 35, 189, 2003.

71. Kokura, S. et al., Enhancement of lipid peroxidation and of the antitumor effect of hyperthermia upon combination with oral eicosapentaenoic acid, *Cancer Lett* 185, 139, 2002.

72. Kiserud, C.E., Kierulf, P., and Hostmark, A.T., Effects of various fatty acids alone or combined with vitamin E on cell growth and fibrinogen concentration in the medium of HepG2 cells, *Thromb Res* 80, 75, 1995.

73. Pratico, D. et al., Local amplification of platelet function by 8-epi prostaglandin F2-alpha is not mediated by thromboxane receptor isoforms, *J Biol Chem* 271, 14916, 1996.

74. Morrow, J.D., Minton, T.A., and Roberts, L.J., 2nd, The F2-isoprostane, 8-epi-prostaglandin F2 alpha, a potent agonist of the vascular thromboxane/endoperoxide receptor, is a platelet thromboxane/endoperoxide receptor antagonist, *Prostaglandins* 44, 155, 1992.

75. Kulkarni, P.S. and Srinivasan, B.D., Prostaglandins E3 and D3 lower intraocular pressure, *Invest Ophthalmol Vis Sci* 26, 1178, 1985.

76. Balapure, A.K. et al., Structural requirements for prostaglandin analog interaction with the ovine corpus luteum prostaglandin F2 alpha receptor: implications for development of a photoaffinity probe, *Biochem Pharmacol* 38, 2375, 1989.

77. Nourooz-Zadeh, J., Halliwell, B., and Anggard, E.E., Evidence for the formation of F3-isoprostanes during peroxidation of eicosapentaenoic acid, *Biochem Biophys Res Commun* 236, 467, 1997.

78. Gao, L. et al., Formation of F-ring isoprostane-like compounds (F_3-isoprostanes) *in vivo* from eicosapentaenoic acid, *J Biol Chem* 281, 14092, 2006.

79. Waugh, R.J. et al., Identification and relative quantitation of F2-isoprostane regio-isomers formed *in vivo* in the rat, *Free Radic Biol Med* 23, 943, 1997.

80. Yin, H., Morrow, J.D., and Porter, N.A., Identification of a novel class of endoperoxides from arachidonate autoxidation, *J Biol Chem* 279, 3766, 2004.

4 Metabolic Syndrome Due to Early Life Nutritional Modifications

Malathi Srinivasan, Paul Mitrani, and Mulchand S. Patel

CONTENTS

INTRODUCTION

Obesity is a rapidly growing worldwide epidemic that is especially prevalent in the United States. Recent data estimate that 66% of American adults are overweight, with about 32% being obese.[1] Interestingly, dramatic increases in obesity rates are relatively recent phenomena, with a rapid rise beginning in the 1980s.[2] Between 1960 and 1980, the prevalence of obesity in the United States increased less than 2% (from 13 to 15%) while obesity rates in the last 25 years have more than doubled.[3,4] Similar trends are seen internationally, indicating that the obesity epidemic is a global health problem.[5]

Because obesity is associated with increased risks for the development of diabetes, cardiovascular disease, dyslipidemia, and hypertension, the epidemic of obesity will be followed by an epidemic of these diseases.[5,6] Therefore, it is no surprise that in parallel with the obesity epidemic an explosion in the incidence of type 2 diabetes mellitus (T2DM) has ensued. T2DM is one of the biggest health concerns related to obesity, affecting more than 18 million people in the United States and ranks as the sixth leading cause of death.[7] Between 1990 and 1998, the prevalence of diabetes increased by 33% in the United States.[8]

Metabolic syndrome, previously referred to as syndrome X,[9] is defined by the clustering of cardiovascular risk factors including obesity, diabetes, hypertension, and dyslipidemia,[5,9] and is associated with insulin resistance.[5] Metabolic syndrome is now estimated to affect ~34% of American adults and up to 36% of adult Europeans.[10] The American Medical Association estimates that the medical costs associated with the treatment of obesity-associated diseases such as T2DM and cardiovascular disorders now exceed the costs of smoking-related diseases.[11]

The adult obesity epidemic has been accompanied by an analogous epidemic of childhood obesity,[12] especially in the United States.[13] While the prevalence of obesity in children has been increasing since the 1960s, the last 25 years have shown a rapid acceleration in childhood obesity rates.[13] During this period, the proportion of overweight children has almost doubled, while the proportion of overweight adolescents has nearly tripled,[7] leading to the increased incidence of T2DM in this sector of the population in the United States.[14] Additionally, the increase in the incidence of overweight children places them at a greater risk of developing obesity and T2DM as adults.[15] If the epidemics of obesity, diabetes, and metabolic syndrome continue to rise, the prognosis of worldwide health will continue to decline. Hence there is a sense of urgency in the need to further our knowledge about the etiology of obesity, specifically related to the steep increase in its incidence within the past 25 years, in an effort to curb the surge in the occurrence of obesity-enhanced diseases.

The rapid rise in obesity rates since the 1980s suggests that the cause is a combination of various behavioral and environmental influences rather than specific genetic and biological changes.[1] It is commonly believed that the recent trends in obesity are the results of increased consumption of calorie-dense, high fat, and high carbohydrate foods along with decreased physical activity.[7] Although these factors do contribute to the onset of obesity, it is now recognized that obesity also results from programmed impairments of energy balance that increase vulnerability to negative environments that include poor diet and limited exercise.[1]

Recent evidence suggests that both increases in food consumption and changes in the quality of nutrition may play decisive roles. For most of the 20th century, caloric consumption remained virtually constant, but then began to rise in the 1980s due to increased consumption of fats and sugars.[2,16] Of particular interest is the finding that the consumption of carbohydrates as a percentage of total calorie intake (due to increases in the consumption of soft drinks and fruit juices) increased by approximately 20% over the past few decades.[2] In addition, evidence indicates that the quality of consumed carbohydrates is changing as high glycemic simple sugars are replacing low glycemic complex carbohydrates.[13] Compared to high glycemic diets, low glycemic diets have been found to increase satiety and delay return of hunger, lower post-prandial insulin levels, and increase insulin sensitivity.[13]

These studies indicate that changes in the quantity and quality of dietary carbohydrate are important components in the pathogenesis of obesity and metabolic syndrome. However, in addition to changing dietary habits in adulthood, recent evidence indicates that exposure to altered nutrition during very early phases of life (pre-natal and immediate post-natal) can result in metabolic programming that predisposes individuals to the development of metabolic syndrome (Figure 4.1).

ROLE OF METABOLIC PROGRAMMING IN ETIOLOGY OF OBESITY EPIDEMIC

Population-based evidence and studies of early nutritional experiences in animals suggest that different nutritional insults during fetal or neonatal life may result in increased risks of developing metabolic diseases such as obesity and metabolic syndrome later in life.[10] Metabolic programming is a phenomenon in which a stimulus or insult that occurs during a critical period of organogenesis in early life results in permanent alterations in the structures and functions of affected organs and increased susceptibility to adult disease (Figure 4.1).[17,18]

The fetal origins hypothesis originally proposed by Barker posits that metabolic programming occurs in any situation in which a stimulus or insult during development establishes a permanent physiological response.[19] A related theory called the thrifty phenotype hypothesis proposes that a poor nutritional environment *in utero* results in fetal metabolic programming that maximizes uptake and conservation of available nutrients.[20,21]

It is well established that critical periods of development that coincide with periods of rapid cell division are important for the abilities of specific tissues and organs to differentiate and mature in preparation for survival after birth.[22,23] Therefore, poor fetal nutrition results in selective protection of the brain at the expense of other organs such as the pancreas and liver, and results in altered growth and permanent changes in organ structure and function (Figure 4.1).[22,24] Metabolic programming is meant to confer an adaptive advantage to an infant exposed to a post-natal deficiency similar to the one encountered *in utero*. This is also related to the idea of thrifty genes that are involved in the adaptive response to environments with scarce availability of food and function to increase survival in such environments. However, when these genes are introduced to situations where resources are abundant, they begin to lose their adaptive advantages and become detrimental due to increased sensitivity to and utilization of those resources.[20,25]

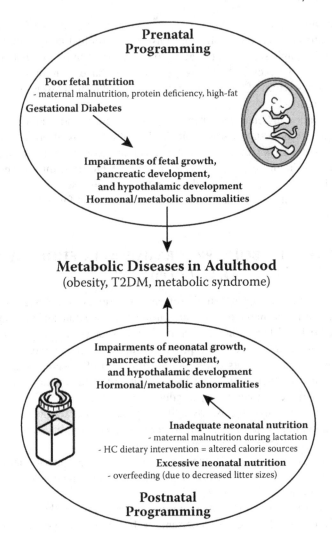

FIGURE 4.1 Metabolic programming due to early life nutritional challenges. Pre-natal programming can be caused by maternal malnutrition, gestational diabetes, or by an overweight/insulin resistant pregnancy. Post-natal programming can occur due to altered neonatal nutrition (under- or over-nutrition, caloric redistribution). The maladaptations in target organs during the period of the nutritional modification persist in later life, resulting in increased risk for metabolic diseases in adulthood.

PRE-NATAL METABOLIC PROGRAMMING

Original studies by Barker et al. showed that adverse intrauterine conditions such as under-nutrition during pregnancy caused disproportionate fetal growth, resulting in low birth weight and adult T2DM[18,26] as well as hypertension and dyslipidemia.[26,28] Data from several other epidemiological studies carried out in different parts of the world support this association between fetal development in an undernourished maternal environment and the later onset of metabolic diseases.[29]

Although data from human epidemiological studies have indicated the connection between impaired fetal development and adult onset diseases such as T2DM, hypertension, cardiovascular problems, and other conditions, intrinsic difficulties surround the conduct of long-term studies in human populations. Therefore, several animal models have been developed to mimic the human situation to further our knowledge on the role of maternal nutritional status during pregnancy and the long-term outcome for progeny. Significant among these models are maternal protein or total caloric restriction and gestational diabetes (Figure 4.1). Several elegant reviews summarize these models in detail.[30-33]

TABLE 4.1

Summary of Immediate and Long-Term Consequences for Progeny Due to Altered Maternal (Pregnancy and Lactation) Nutritional Experience

I.   **Dietary protein restriction during gestation and lactation**
Immediate effects
  Fetal growth retardation
  Impairment in pancreatic islet structure and function
  Malformation of hypothalamic nuclei
Persistent effects
  Increased insulin sensitivity in young adults
  Altered insulin signaling in peripheral tissues
  Renal defects and hypertension
  Development of glucose intolerance with advancing age

II. **Total caloric restriction during gestation and lactation**
30% caloric restriction
  Increased fasting insulin, hyperphagia, adult onset obesity and hypertension
50% caloric restriction
  Impaired islet development, glucose intolerance

III. **Diabetes during pregnancy**
Mild diabetes
  Impaired development of fetal islets
  Impaired glucose tolerance and reduced insulin secretory capacity in adulthood
Severe diabetes
  Fetal growth retardation resulting in smaller-sized adults
  Over-stimulation of fetal β cells due to fetal hyperglycemia
  Onset of insulin resistance in liver, muscle, and adipose tissue in adulthood

IV. **Altered intrauterine environment (maternal hyperinsulinemia/obesity) due to high fat diet consumption**
  Fetal hyperinsulinemia and altered insulin secretory capacity
  Alterations in β and α cell volume and number on post-natal day 1
  Abnormal glucose homeostasis, impaired insulin secretion, increased body adiposity, hyperlipidemia, altered endothelial function and pro-atherogenic lesions in adulthood

---

Maternal protein malnutrition due to a low protein diet results in low protein (LP) progeny that are underweight, hypophagic, hypoglycemic, and hypoinsulinemic throughout life (Table 4.1)[34–36] and show significant alterations in pancreatic islets, hypothalamus, liver, and muscle (Table 4.1).[37–40] Interestingly, offspring of protein-malnourished rats showed reductions in the relative weights of the pancreas, muscle, and liver, while the weights of the brain and lungs were protected[22,41] — consistent with the idea that nutrients are diverted to essential organs when scarce.

While hypoinsulinemia in LP progeny is attributable to lower blood glucose levels, evidence indicates that insufficient β cell stimulation by amino acids during fetal development impairs insulin secretory capacity after birth.[42] Protein malnutrition during pregnancy reduces maternal plasma amino acid concentrations[43] and thus placental transfer of amino acids to the fetus.[42] Reduced fetal levels of taurine have been associated with impaired development of normal fetal insulin secretion,[22,40,44,45] while supplemental taurine given to mothers on LP diets was able to normalize insulin secretion in fetal islets.[22,46]

Reduced insulin secretion by LP progeny is associated with reduced islet proliferation, size, insulin content, and vascularization,[22,34,35,44] as well as a permanent reduction in the number of pancreatic β cells.[35] Neonatal rats showed similar defects in pancreatic development when mothers were kept on low protein diets during lactation.[22,34] While reduced insulin secretion may confer an adaptive response for survival in a protein-restricted environment, the resulting impairment of insulin secretory capacity increases adulthood risk of glucose intolerance and diabetes.[22] Therefore, abnormal stimulation of β cells in early life causes mal-programming of β cell function that results in impaired insulin secretion throughout life and may increase the risk of T2DM.

Maternal protein restriction during gestation and lactation results in offspring (LP) with altered programming not only of pancreatic islets but also of brain areas involved in the regulation of energy homeostasis and increases the risk of adult metabolic diseases, including decreased insulin secretion and increased sympathetic nervous system activity in association with hypertension.[36,47]

Among the affected brain regions are the ventromedial nucleus of the hypothalamus and paraventricular nucleus, which show greater relative volumes, along with increases in neuronal density in LP rats compared to control animals.[36] This increase in neuronal density is associated with increases in the activities of the ventromedial nucleus and paraventricular nucleus, which is consistent with data showing that ventromedial nucleus activity inhibits insulin secretion and stimulates sympathetic nervous system activity,[48] while activity of the paraventricular nucleus stimulates increases in blood pressure.[36,49] In addition, LP rats showed reduced numbers of arcuate neuropeptide Y neurons that may contribute to alterations in neuropeptide Y-mediated regulation of energy homeostasis including sympathetic nervous system activity.[36]

In the case of total caloric restriction, 50% reduction of food availability during the second and third weeks of pregnancy in rats resulted in impaired pancreatic development and glucose intolerance in the progeny at 1 year of age (Table 4.1).[50] Severe caloric restriction (access to only 30% of the food consumed by controls) resulted in increased fasting plasma insulin and leptin, hyperphagia, hypertension, and obesity in the adult progeny.[51]

Diabetic pregnancies are associated with significant structural and functional changes in the fetal endocrine pancreas. For example, a mild diabetic pregnancy in rats resulted in hypertrophy and hyperplasia of fetal islets,[52] as well as enhanced proliferative capacity (Table 4.1).[53,54] In addition, fetal islets have increased insulin content[55] and increased responsiveness to glucose-stimulated insulin secretion.[55,56] A severely diabetic pregnancy, on the other hand, resulted in fetal islets that had lower insulin content[54] and reduced glucose-stimulated insulin secretion (Table 4.1).[56] Further, offspring of rat mothers with gestational diabetes showed significant increases in the density of arcuate neuropeptide Y neurons associated with decreased sympathetic nervous system activity and increased weight gain.[36,57]

Although several studies have focused on the consequences of a malnourished pregnancy that were important during periods of famine and war, the current dietary habits of people (consumption of high energy foods) especially in Westernized societies require investigations into the consequences for the progeny arising from such dietary practices in women. It has been suggested that the consumption of energy-dense foods is a contributing factor in the etiology of the current obesity epidemic and may well be responsible for a significant and increasing proportion of women being overweight during pregnancy.[58] Chronic feeding of a high fat diet to female rats bears a close resemblance to the current dietary habits of humans in Western societies and hence has physiological relevance to the human situation. Long-term consumption of a high fat diet resulted in hyperinsulinemia and increased insulin secretory responses to various secretogogues in term fetuses (Table 4.1).[59] Additionally, the adult progeny of such mothers were obese, glucose-intolerant, hyperinsulinemic, and exhibited abnormal insulin secretory responses to glucose.[59]

Cerf et al.[60] demonstrated that feeding a high fat diet to female rats throughout gestation alone resulted in significant decreases in β cell volume and number and converse changes in α cells, resulting in hyperglycemia in 1-day-old newborn rat pups without changes in serum insulin concentrations. Several reports indicate the long-term consequences of a high fat diet during gestation only or during both gestation and lactation. These include abnormal glucose homeostasis, reduced whole body insulin sensitivity, impaired β cell insulin secretion, and changes in the structure of the pancreas,[61,62] defective mesenteric artery endothelial function,[63] hypertension,[64] alterations in conduit artery and renal functions,[65] increased body adiposity,[61,63] deranged blood lipid profile,[61,64] hyperleptinemia,[63] and pro-artherogenic lesions[66] in adult progeny (Table 4.1). Kozak et al.[67] demonstrated that a high fat diet during gestation and lactation affected body weight regulation in adult progeny via alterations in the functioning of neuropeptide Y.

METABOLIC PROGRAMMING DUE TO ALTERED DIETARY EXPERIENCE IN IMMEDIATE POST-NATAL PERIOD

Critical windows of development of organs and metabolic processes also extend into the immediate post-natal period, suggesting vulnerability of this period to metabolic programming effects via altered nutritional experiences. In the case of mammals, natural rearing by mothers via breast feeding is the nature-made program for rearing of newborns. In humans it is well established that breast feeding is the optimal form of infant nutrition because it confers immunologic, psychological, and developmental benefits to the infant.[68,69] Several studies have indicated that breast feeding is protective against the development of obesity, diabetes, cardiovascular disease, and metabolic syndrome,[70,71] and this protection directly correlated with the duration of breast feeding.[72]

The association of breast feeding with decreased prevalence of childhood obesity[73] is thought to involve protection against metabolic disease through careful regulation of an infant's calorie intake,[74] insulin secretion,[75] and adiposity.[72] It also may help program the complex circuitry involved in the regulation of energy homeostasis that persists throughout life.[72]

McCance was the first researcher to demonstrate from studies in rats that changing the availability of milk during the suckling period had permanent effects on growth trajectories.[76] These studies provided the first evidence indicating a connection between nutritional experiences during the suckling period and long-term metabolic effects. Further studies stemming from adjustment of litter size in rats indicated that rats raised in small litters demonstrated increased body weight gain during the suckling period that persisted into the post-weaning period accompanied by hyperphagia, hyperleptinemia, hyperinsulinemia, and permanent alterations in gene expression and insulin secretory capacity in islets.[57,77–79]

Adult onset obesity in rats raised in small litters is supported by altered programming of the energy circuitry in the hypothalamus (Table 4.2).[80] Post-natal over-nutrition has been shown to alter the functions of the arcuate nucleus and ventromedial hypothalamus. For example, overfed rats showed increases in the density of arcuate neuropeptide Y neurons in association with increases in body weight and food intake.[81,82] In addition, the direction of responses to corticotropin releasing factor is apparently reversed in overfed rats.[83] Also in overfed rats (small litter size), increased inhibition of the ventromedial hypothalamic neurons by agouti-related peptide has been reported.[84] Rats raised in large litters during the suckling period maintained diminished growth patterns throughout life and showed reduced plasma insulin levels compared to rats from small litters of three or four pups per dam (Table 4.2).[76,85]

HIGH CARBOHYDRATE (HC) RAT MODEL

Most studies on metabolic programming using animal models are involved with an altered intrauterine environment induced by alterations in the nutritional status of the mother during pregnancy. Due to difficulties in rearing newborn pups away from their dams, very few studies have examined the effects of nutritional modifications during the suckling period. Adjustment of litter size is an approach in this direction but results only in a decrease or increase in the availability of total calories during the suckling period and does not permit investigation of alterations in the quality of nutrition without affecting total caloric intake. In the rat, the

TABLE 4.2
Summary of Immediate and Long-Term Consequences of Altered Nutritional Experience during Suckling Period

I. **Under- or over-nourishment due to adjustment of litter size**
    Large-sized litters
        Reduced growth, alterations in islet functions
    Small-sized litters
        Increased growth, hyperleptinemia, and hyperinsulinemia from the start and persisting into adulthood
        Molecular and functional alterations in adult islets
        Alterations in hypothalamic neuropeptide expression and function

II. **Caloric redistribution (HC dietary intervention) in suckling period**
    Immediate effects
        Immediate onset of hyperinsulinemia with euglycemia
        Alterations in islet insulin secretory capacity such as increased response to lower glucose concentrations, modified responses to incretins and neuroendocrine effectors, and ability to secrete moderate amounts of insulin in absence of glucose and calcium
        Molecular and structural adaptations in islets such as increased mRNA of preproinsulin gene, reduction in islet size and increase in numbers, increase in islet cell proliferation
    Effects observed in post-weaning period
        Persistence of hyperinsulinemia, abnormal glucose tolerance test (GTT), leftward shift in the insulin secretory response to a glucose stimulus, increased gene expression of preproinsulin and related factors, hyperphagia, significant increases in body weight gain and adult-onset obesity
    Generational effects
        Female rats that underwent HC dietary modification spontaneously transmitted their phenotypes to their progeny
        Programming effects observed in HC progeny included:
        Fetal hyperinsulinemia
        Increased response by fetal islets to glucose and amino acids
        Increased preproinsulin gene expression in fetal islets
        Absence of hyperinsulinemia during suckling period and onset of hyperinsulinemia immediately after weaning
        Increased response to lower glucose concentrations by islets from 28-day-old progeny
        Increases in body weight gain in post-weaning period and full-blown obesity by post-natal day 100

suckling period is an important phase of continued organogenesis, including pancreatic and brain development.[86]

The adaptation of the artificial rearing technique originally described by Hall[87] has enabled our laboratory to investigate the immediate and long-term consequences of a high carbohydrate (HC) dietary modification during the suckling period in rats.[88] Newborn rat pups were raised away from their dams and given a high carbohydrate (HC) milk formula instead of the high fat rat milk received by naturally reared (mother-fed; MF) rat pups.[89,90] In order to establish that the artificial rearing technique *per se* did not cause any metabolic programming effects, newborn rat pups were also artificially reared on a high fat (HF) milk formula, the macronutrient composition of which was identical to that of rat milk.[90,91] Although MF, HF, and HC rat pups received a similar number of calories per day, the primary source of calories was switched from fats to carbohydrates for the HC rats. The mere change in the quality of nutrition during the suckling period (from fat-enriched rat milk to carbohydrate-enriched HC milk formula) resulted in metabolic programming due to the overlap of the HC dietary intervention and the critical post-natal period of pancreatic development in rats.[88,92,93] The high carbohydrate nature of the HC formula (56% carbohydrate compared to 8% in rat milk and in the HF milk formula) increases insulin demand in neonatal rats. This altered nutritional environment induces compensatory adaptations in islet function to ensure that the immediate demand for insulin is met. However, permanent programming of these adaptations results in altered responses to nutritional experiences throughout life and adverse consequences in adulthood (Table 4.2).[88,92,93]

METABOLIC PROGRAMMING IN HC RATS: IMMEDIATE EFFECTS

Artificial rearing of rat pups on the HC milk formula resulted in the immediate onset (within the first 24 hours) of hyperinsulinemia when compared to MF and HF (experimental control) rats and persisted during the entire suckling period in the HC rats (Figure 4.2).[88,94] Interestingly, HC pups showed normal blood glucose levels during this period despite hyperinsulinemia, suggesting that the alterations in islet function are able to compensate for increased insulin demand to maintain

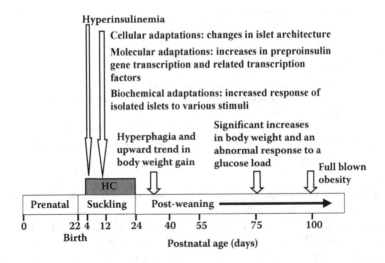

FIGURE 4.2 Overview of immediate and lasting consequences of HC dietary modification in the immediate post-natal period in first generation HC rats.

euglycemia.[94] In contrast, rats artificially reared on an HF formula (caloric distribution: carbohydrate 8%, protein 24%, and fat 68%) as an internal control for the artificial rearing technique were not hyperinsulinemic, suggesting that the metabolic programming of HC rats was the result of a change in the quality of nutrition and not a change in the mode of delivery of nutrition during the suckling period.[90]

In vitro studies of glucose-stimulated insulin secretion (GSIS) in isolated pancreatic islets showed that neonatal HC islets secreted significantly more insulin after both 10 and 60 min of glucose stimulation compared to MF islets at all glucose concentrations studied.[94] In association with this increased insulin secretory response to glucose, HC islets showed increases in (1) the enzyme activities of the low K_m hexokinase, glyceraldehyde-3-phosphate dehydrogenase and pyruvate dehydrogenase complex and (2) increases in levels of glucose transporter protein-2 (Figure 4.2).[94] Up-regulation of glucose metabolism in HC islets suggests that a lowered glucose threshold for insulin secretion may, in part, be responsible for the development of hyperinsulinemia due to programmed hypersensitivity of pancreatic islets to glucose.

In vivo insulin secretion is stimulated by numerous secretogogues including glucose, amino acids, fatty acids, and neuroendocrine and incretin factors.[95] While insulin secretion is primarily induced by glucose stimulation of the K_{ATP} channel-dependent pathway, acetylcholine, norepinephrine, glucagon, glucagon-like peptide-1, and other signals act on K_{ATP} channel-dependent and -independent pathways as well as Ca^{2+} channel-independent pathways to augment insulin secretion.[95] Plasma glucagon-like peptide-1, an incretin hormone that promotes insulin secretion via increases in cAMP levels,[96] was increased in 12-day-old HC rats as were glucagon-like peptide-1 receptor mRNA levels in isolated HC islets.[97] Acetylcholine and glucagon-like peptide-1 potentiation of glucose-stimulated insulin secretion was greater in HC islets compared to MF islets, indicating increased responsiveness to these agonists.[97] Also, the activities of protein kinase C, protein kinase A, and calcium calmodulin kinase II were significantly higher in HC islets compared to MF islets,[86] as were the mRNA levels of adenylyl cyclase type VI.[97] Furthermore, HC islets have reduced sensitivity to norepinephrine stimulation, which inhibits insulin secretion through activation of α-adrenergic receptors on β cells[98] compared to control islets.[97]

In addition to adaptations at the level of the insulin secretory process, molecular adaptations were observed in islets isolated from 12-day-old HC rats. The adaptations included significant increases in insulin biosynthesis and in the mRNA levels of the preproinsulin gene and other components of the cascade implicated in the regulation of the transcription of the preproinsulin gene in neonatal HC islets (Figure 4.2).[99] Gene array analysis of the mRNAs from neonatal HC islets indicated that several clusters of genes involved in a wide array of cellular functions (e.g., cell cycle regulation, protein synthesis, ion channels, and metabolic pathways) were up-regulated in these islets and may contribute to the onset of hyperinsulinemia in these rats.[100]

The overlap of the HC dietary modification with the period of post-natal development of the pancreas also resulted in several changes in the cellular architecture of neonatal HC islets (Table 4.2; Figure 4.2). These include: (1) a reduction in mean islet size and increase in the number of small-sized islets, (2) an increase in the number of islets per unit area, (3) greater apoptotic rates in islets compared to ductal epithelium, and (4) increases in islet cell replication as indicated by the presence of proliferating cell nuclear antigen in a large proportion of β cells in 12-day-old HC islets compared to islets from age-matched MF rats.[101]

Metabolic programming effects were also observed in livers of neonatal HC rats and may be important for the establishment of the HC phenotype observed in adulthood. Precocious induction of the activities of glucokinase and malic enzyme were indicated by substantial increases in their activity levels in 10-day-old HC animals.[89] The lipogenic capacity and glycogen content were also increased in the livers of these rats.[89,102]

The hypothalamus is an important site for energy homeostasis and aberrations in the expression and function of a network of neuropeptides lead to either hyperphagia or hypophagia.[103,104] HC dietary modification resulted in significant increases in the protein content and gene expression of orexigenic signals and converse changes in the anorexigenic signals, suggesting that alterations leading to the eventual onset of obesity in the adulthood of these rats are evident very early in life (M.S. Patel, unpublished observations).

Persistent Effects in Adulthood

HC rats were hyperinsulinemic throughout the period of nutritional modification and continued to be hyperinsulinemic into adulthood even after weaning onto laboratory rodent diets on post-natal day 24 (Figure 4.2).[90,91] Studies of 100-day-old adult rats showed that HC animals had significantly higher plasma insulin levels compared with age-matched MF rats.[105] In addition to persistent hyperinsulinemia, adult HC rats showed impaired glucose tolerance compared with controls, even though they were euglycemic, suggesting a state of insulin resistance (Figure 4.3).[91]

As in the case of 12-day-old HF rats, adult HF rats were not hyperinsulinemic and did not show altered growth or food intake compared to MF rats.[90] While HC and MF rats showed similar body weights during the suckling period,[94] HC rats experienced increased growth rates after weaning in association with hyperphagia that resulted in full-blown obesity by day 100 (Figure 4.2).[90,106]

In addition, islets from adult HC rats showed increases in GSIS, low K_m hexokinase activity, and Ca^{2+}-independent insulin release similar to results from 12-day-old HC islets.[105] Additionally, increases in insulin producing cell masses in adult HC pancreas were noted.[91] At the molecular level, increased preproinsulin gene transcription with concomitant increases in mRNA levels of transcription factors regulating its gene expression and increases in the expression of several clusters of genes as indicated by gene array analysis[105] were observed in

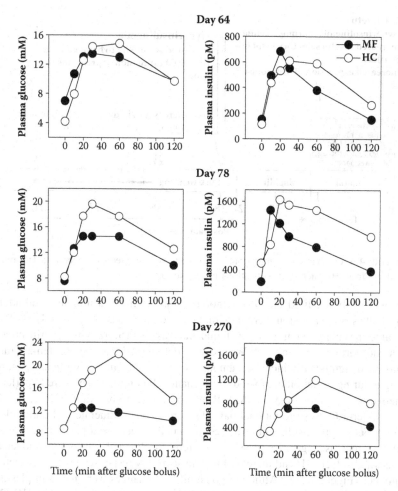

FIGURE 4.3 Plasma glucose and insulin levels in HC male rats after an oral glucose tolerance load (2 g/kg body weight) on post-natal days 64, 78, and 270.[91] Completely overlapping points for plasma glucose of 270-day old MF and HC rats are shown with open circles only.

100-day-old HC islets, suggesting that the molecular effects observed in neonatal islets persist into adulthood. Adult onset obesity was accompanied by increases in the lipogenic capacities of the liver and epididymal adipose tissue, increases in the epididymal adipose tissue weight, and marked increases in the cell sizes of epididymal and omental adipose tissues of 100-day-old male HC rats.[90]

GENERATIONAL EFFECTS

A notable observation from the HC model is that the progeny of HC female rats that underwent the HC dietary modification in their immediate post-natal life

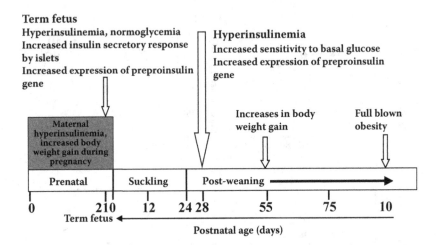

FIGURE 4.4 Overview of generational effects in progeny (second generation HC rats) resulting from HC dietary modifications in female rats.

spontaneously acquired the HC phenotype (chronic hyperinsulinemia and adult onset obesity) without undergoing dietary modifications (Figure 4.4).[106] The intrauterine environment in the HC female rat was characterized by hyperinsulinemia and normoglycemia.[106–108] We recently determined that the HC female rats consumed significantly increased amounts of food and gained significantly more weight during gestation compared to pregnant MF controls; they were hyperleptinemic on gestational day 21 (M.S Patel, unpublished observations).

HC term fetuses (gestational day 21) were hyperinsulinemic and normoglycemic. Fetal hyperinsulinemia was accompanied by increased pancreatic insulin content, increased gene expression of preproinsulin and pancreatic duodenal transcription factor-1, and increased insulin secretory responses to various secretogogues (Figure 4.4).[108] Although no significant changes were noted in plasma insulin levels during the suckling period between the second generation HC (2-HC) and MF rats immediately after weaning, the plasma insulin levels of 2-HC rats were significantly increased (first observed on post-natal day 26; Figure 4.4).[107] As observed in term fetuses, there was a leftward shift in the insulin secretory response to a glucose stimulus and increased preproinsulin gene transcription in islets from 28-day old 2-HC rats.[107]

Similar to the observations in first generation HC rats, no significant differences were noted in the body weights of 2-HC rats compared to age-matched MF rats up to post-natal day 55. After post-natal day 55, significant increases in body weight were observed with full blown obesity evident on approximately post-natal day 100.[106] This correlates with significant increases in the weight of the adipose tissues, increases in the size of the adipocytes, and increases in the activities of the lipogenic enzymes (fatty acid synthase and glucose-6-phosphate dehydrogenase) in both liver and adipose tissue of 100-day old 2-HC rats.[106]

A state of insulin resistance was indicated by decreased content of glycogen in livers and muscles of 100-day-old 2-HC male rats.[109] The decrease was associated with a decrease in the enzyme activity of glycogen synthase and its putative upstream activators.[109] In contrast, the situation was reversed in the epididymal adipose tissues of these rats.[110]

Are the metabolic programming effects induced by early life HC dietary modifications in rat pups reversible phenomena? In order to address this question, we pair-fed HC female rats from the time of weaning to the amount of feed consumed by age-matched MF female rats on a daily basis. Interestingly, such a dietary control in HC female rats resulted in reductions of their plasma insulin levels and body weight gains to the levels observed in age-matched female rats.[108] During pregnancy, the plasma insulin levels and body weights were similar in the pair-fed HC and age-matched MF female rats.[108] Such improvements in pregnancy conditions in the pair-fed HC female rats prevented the transmission of the HC phenotype to their progeny.[108] These results indicate that dietary regulation in HC rats results in the reversal of the metabolic programming effects with good prognosis both for rats of the same generation and the subsequent generation.

The mechanisms responsible for metabolic programming due to altered nutritional experiences in early life are not well understood. Malorganization (functional and/or structural) of target organs has been suggested as a plausible mechanism.[111] In animal models for pre-natal metabolic programming (protein malnourishment, gestational diabetes, and caloric restriction), such adaptations in target organs were reported.[30] In the HC rat model, we observed functional and/or structural changes in neonatal HC rats in target organs such as the islets, gut, hypothalamus, and liver. All of these organs are potentially important for maintaining glucose homeostasis. As indicated in Figure 4.5, it is possible that adaptations in the these organs and the resultant cross-talks among them are necessary for survival of these rat pups in the face of the HC dietary modification but since they occur during the period of immediate post-natal development, such adaptations become permanent and in the post-weaning period, predispose these animals to adult onset obesity and concomitant health issues.

Various interactions between target organs are suggested in Figure 4.5. The HC milk formula may induce alterations in stomach- and gut-derived factors which, through specific receptor-mediated signaling, may alter the functional capabilities of organs such as the islets and the hypothalamus. Similarly, insulin via receptors present in the islets and the hypothalamus may exert altered autocrine and endocrine functions to support the HC phenotypes in these rats. Several neuropeptides enumerated in the hypothalamus have principal roles in the energy homeostasis of the body, insulin secretion by islets, insulin sensitivity of target organs, etc. Therefore, a multifactorial mechanism may be responsible for the onset and persistence of hyperinsulinemia in HC rats which predisposes these rats for insulin resistance leading to altered glucose tolerance and obesity in adulthood.

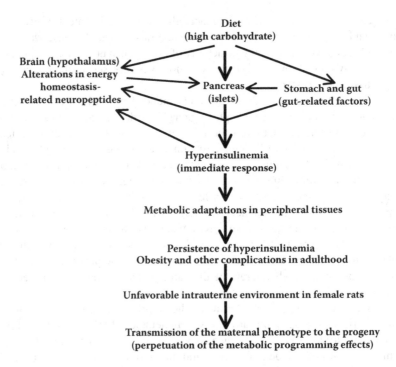

FIGURE 4.5 Postulated scheme leading to development of the HC metabolic phenotype in adulthood due to high-carbohydrate dietary modifications in neonatal rats. The direct effects of the dietary treatment on target organs and possible cross-talks among these organs leading to the adult phenotype are indicated.

CORRELATION OF ALTERED NUTRITIONAL EXPERIENCE IN EARLY LIFE TO SUBSEQUENT HIGH INCIDENCE OF OBESITY AND METABOLIC SYNDROME

As indicated, nutritional experiences *in utero* and/or in the immediate post-natal period (infancy) can to a significant degree influence adult metabolic phenotype as these are periods of rapid development of the organism. Because the main goal is survival, an organism adapts to the altered environment by necessary alterations in target tissues that help it "tide over" the situation but such adaptations lead to unfavorable situations later in life. Both epidemiological data and results from animal studies indicate the importance of adequate nutrition during pregnancy.

Studies of the long-term consequences of an altered nutritional experience in early post-natal life indicate the importance of this phase of life for metabolic programming effects. Dietary habits for all ages have undergone tremendous changes over the past several decades. The present obesity epidemic, to a large measure, may be the result of such changes. Extrapolation of data obtained from HC rat models suggests that post-natal increased consumption of carbohydrates by infants (formula feeding with early introduction of carbohydrate-rich supplements such as cereals, fruits juices, etc.) in Western societies may be partly responsible for the increase in the incidence of obesity. This effect is exacerbated by the mode of feeding (bottle, spoon, etc., resulting in overfeeding). Supplementation of milk (breast or formula) with early introduction of carbohydrate-enriched baby foods and overfeeding may result in malprogramming effects in these babies, leading to adult onset obesity and attendant diseases as observed in the HC rat model.

The spontaneous transfer of the HC phenotype to the progeny and results from our studies of maternal high-fat diet-induced effects in the next generation indicate that maternal obesity primes obesity in progeny. Our results suggest that females that are overnourished or exposed to increased carbohydrate intake in infancy may be obese and insulin resistant during pregnancy and, due to an unfavorable intrauterine environment, are at increased risk for the establishment of a vicious cycle of transmission of their metabolic phenotype to their progeny.

The alarming increase in the numbers of overweight and obese individuals in the United States suggests that a significant number of pregnancies will not be under conditions of optimal health due to increased body weight, moderate hyperinsulinemia, and mild insulin resistance in these females. Fetal development under such conditions may predispose the progeny for the development of obesity in adulthood. This generational effect may be amplified by dietary habits in infancy and lifestyle later in adulthood such that the maternal intrauterine environment becomes more and more favorable for metabolic malprogramming of the progeny from one generation to the next.

ACKNOWLEDGMENT

This work was supported in part by National Institute of Diabetes and Digestive and Kidney Diseases grant DK-61518.

REFERENCES

1. Daniels, J. Obesity: America's epidemic, *Am. J. Nurs.* 106, 40, 2006.
2. Finkelstein, E.A., Ruhm, C.J., and Kosa, K.M. Economic causes and consequences of obesity, *Annu. Rev. Public Health* 26, 239, 2005.
3. Flegal, K.M. et al. Prevalence and trends in obesity among U.S. adults, 1999–2000, *JAMA* 288, 1723, 2002.
4. Flegal, K.M. Epidemiologic aspects of overweight and obesity in the United States, *Physiol. Behav.* 86, 599, 2005.
5. Haffner, S. and Taegtmeyer, H. Epidemic obesity and the metabolic syndrome, *Circulation* 108, 1541, 2003.
6. Zimmet, P., Alberti, K.G., and Shaw, J. Global and societal implications of the diabetes epidemic, *Nature* 414, 782, 2001.
7. Stein, C.J. and Colditz, G.A. The epidemic of obesity, *J. Clin. Endocrinol. Metab.* 89, 2522, 2004.

8. Mokdad, A.H. et al. Prevalence of obesity, diabetes, and obesity-related health risk factors, 2001, *JAMA* 289, 76, 2003.
9. Reaven, G.M. Banting Lecture, 1988: role of insulin resistance in human disease, *Diabetes* 37, 1595, 1988.
10. Armitage, J.A. et al. Developmental programming of the metabolic syndrome by maternal nutritional imbalance: how strong is the evidence from experimental models in mammals? *J. Physiol.* 561, 355, 2004.
11. Schwimmer, J.B. et al. Obesity, insulin resistance, and other clinicopathological correlates of pediatric nonalcoholic fatty liver disease, *J. Pediatr.* 143, 500, 2003.
12. Csabi, G. et al. Presence of metabolic cardiovascular syndrome in obese children, *Eur. J. Pediatr.* 159, 91, 2000.
13. Slyper, A.H. The pediatric obesity epidemic: causes and controversies, *J. Clin. Endocrinol. Metab.* 89, 2540, 2004.
14. Molnar, D. The prevalence of the metabolic syndrome and type 2 diabetes mellitus in children and adolescents, *Int. J. Obes. Relat. Metab. Disord.* 28, Suppl. 3, S70, 2004.
15. Heindel, J.J. Endocrine disruptors and the obesity epidemic, *Toxicol. Sci.* 76, 247, 2003.
16. Trends in intake of energy and macronutrients, United States, 1971–2000, *Morb. Mortal. Wkly. Rep.* 53, 80, 2004.
17. Lucas, A. Programming by early nutrition in man, *Ciba Found. Symp.* 156, 38, 1991.
18. Barker, D.J. et al. Fetal nutrition and cardiovascular disease in adult life, *Lancet* 341, 938, 1993.
19. Barker, D.J. Fetal origins of coronary heart disease, *Brit. Med. J.* 311, 171, 1995.
20. Hales, C.N. and Barker, D.J. Type 2 (non-insulin-dependent) diabetes mellitus: the thrifty phenotype hypothesis, *Diabetologia* 35, 595, 1992.
21. Prentice, A.M. Early influences on human energy regulation: thrifty genotypes and thrifty phenotypes, *Physiol. Behav.* 86, 640, 2005.
22. Holness, M.J., Langdown, M.L., and Sugden, M.C. Early-life programming of susceptibility to dysregulation of glucose metabolism and the development of type 2 diabetes mellitus, *Biochem. J.* 349, Pt 3, 657, 2000.
23. Fowden, A.L., Giussani, D.A., and Forhead, A.J. Intrauterine programming of physiological systems: causes and consequences, *Physiology* 21, 29, 2006.
24. Hales, C.N. and Barker, D.J. Thrifty phenotype hypothesis, *Br. Med. Bull.* 60, 5, 2001.
25. Neel, J.V. Diabetes mellitus: a "thrifty" genotype rendered detrimental by "progress"? *Am. J. Hum. Genet.* 14, 353, 1962.
26. Barker, D.J. et al. Weight in infancy and death from ischaemic heart disease, *Lancet* 2, 577, 1989.
27. Godfrey, K.M. et al. Maternal birthweight and diet in pregnancy in relation to the infant's thinness at birth, *Br. J. Obstet. Gynaecol.* 104, 663, 1997,
28. Barker, D.J. Fetal nutrition and cardiovascular disease in later life, *Br. Med. Bull.* 53, 96, 1997.
29. Petry, C.J. and Hales, C.N. *Intrauterine Development and Its Relationship with Type 2 Diabetes Mellitus*, John Wiley & Sons, Chichester, 1999.
30. Petry, C.J., Ozanne, S.E., and Hales, C.N. Programming of intermediary metabolism, *Mol. Cell. Endocrinol.* 185, 81, 2001.

31. Betram, C.E. and Hanson, M.A. Animal models and programming of the metabolic syndrome, *Br. Med. Bull.* 60, 103, 2001.

32. Van Assche, F.A., Holemans, K., and Aerts, L. Long-term consequences for offspring of diabetes during pregnancy, *Br. Med. Bull.* 60, 173, 2001.

33. Holemans, K., Aerts, L., and Van Assche, F.A. Lifetime consequences of abnormal fetal pancreatic development, *J. Physiol.* 547, 11, 2003.

34. Snoeck, A. et al. Effect of a low protein diet during pregnancy on the fetal rat endocrine pancreas, *Biol. Neonate* 57, 107, 1990.

35. Dahri, S. et al. Islet function in offspring of mothers on low-protein diet during gestation, *Diabetes* 40, Suppl. 2, 115, 1991.

36. Plagemann, A. et al. Hypothalamic nuclei are malformed in weanling offspring of low protein malnourished rat dams, *J. Nutr.* 130, 2582, 2000.

37. Berney, D.M. et al. The effects of maternal protein deprivation on the fetal rat pancreas: major structural changes and their recuperation, *J. Pathol.* 183, 109, 1997.

38. Latorraca, M.Q. et al. Protein deficiency and nutritional recovery modulate insulin secretion and the early steps of insulin action in rats, *J. Nutr.* 128, 1643, 1998.

39. Ozanne, S.E. and Hales, C.N. The long-term consequences of intra-uterine protein malnutrition for glucose metabolism, *Proc. Nutr. Soc.* 58, 615, 1999.

40. Bennis-Taleb, N. et al. A low-protein isocaloric diet during gestation affects brain development and alters permanently cerebral cortex blood vessels in rat offspring, *J. Nutr.* 129, 1613, 1999.

41. Desai, M. et al. Organ-selective growth in the offspring of protein-restricted mothers, *Br. J. Nutr.* 76, 591, 1996.

42. Malandro, M.S. et al. Effect of low-protein diet-induced intrauterine growth retardation on rat placental amino acid transport, *Am. J. Physiol.* 271, C295, 1996.

43. Rees, W.D. et al. The effects of maternal protein restriction on the growth of the rat fetus and its amino acid supply, *Br. J. Nutr.* 81, 243, 1999.

44. Hoet, J.J. and Hanson, M.A. Intrauterine nutrition: its importance during critical periods for cardiovascular and endocrine development, *J. Physiol.* 514, Pt. 3, 617, 1999.

45. Petrik, J. et al. A low protein diet alters the balance of islet cell replication and apoptosis in the fetal and neonatal rat and is associated with a reduced pancreatic expression of insulin-like growth factor-II, *Endocrinology* 140, 4861, 1999.

46. Cherif, H. et al. Stimulatory effects of taurine on insulin secretion by fetal rat islets cultured *in vitro*, *J. Endocrinol.* 151, 501, 1996.

47. Langley-Evans, S.C. Hypertension induced by foetal exposure to a maternal low-protein diet in the rat is prevented by pharmacological blockade of maternal glucocorticoid synthesis, *J. Hypertens.* 15, 537, 1997.

48. Perkins, M.N. et al. Activation of brown adipose tissue thermogenesis by the ventromedial hypothalamus, *Nature* 289, 401, 1981.

49. Krukoff, T.L. et al. Expression of c-fos protein in rat brain elicited by electrical and chemical stimulation of the hypothalamic paraventricular nucleus, *Neuroendocrinology* 59, 590, 1994.

50. Garofano, A., Czernichow, P., and Breant, B. Beta-cell mass and proliferation following late fetal and early post-natal malnutrition in the rat, *Diabetologia* 41, 1114, 1998.

51. Garofano, A., Czernichow, P., and Breant, B. Effect of ageing on beta-cell mass and function in rats malnourished during the perinatal period, *Diabetologia* 42, 711, 1999.
52. Aerts, L., Holemans, K., and Van Assche, F.A. Maternal diabetes during pregnancy: consequences for the offspring, *Diabetes Metab. Rev.* 6, 147, 1990.
53. Reusens-Billen, B. et al. Cell proliferation in pancreatic islets of rat fetuses and neonates from normal and diabetic mothers: an *in vitro* and *in vivo* study, *Horm. Metab. Res.* 16, 565, 1984.
54. Eriksson, U. et al. Diabetes in pregnancy: effects on the foetal and newborn rat with particular regard to body weight, serum insulin concentration and pancreatic contents of insulin, glucagon and somatostatin, *Acta Endocrinol.* 94, 354, 1980.
55. Kervran, A., Guillaume, M., and Jost, A. The endocrine pancreas of the fetus from diabetic pregnant rat, *Diabetologia* 15, 387, 1978.
56. Bihoreau, M.T., Ktorza, A., and Picon, L. Gestational hyperglycaemia and insulin release by the fetal rat pancreas *in vitro*: effect of amino acids and glyceraldehyde, *Diabetologia* 29, 434, 1986.
57. Plagemann, A. et al. Perinatal elevation of hypothalamic insulin, acquired malformation of hypothalamic galaninergic neurons, and syndrome x-like alterations in adulthood of neonatally overfed rats, *Brain Res.* 836, 146, 1999.
58. Gallou-Kabani, C. and Junien, C. Nutritional epigenomics of metabolic syndrome: new perspective against the epidemic, *Diabetes* 54, 1899, 2005.
59. Srinivasan, M. et al. Maternal high fat diet consumption results in fetal malprogramming predisposing to the onset of metabolic syndrome-like phenotype in their adulthood, *Am. J. Physiol. Endocrinol. Metab.* 2006.
60. Cerf, M.E. et al. Islet cell response in the neonatal rat after exposure to a high-fat diet during pregnancy, *Am. J. Physiol. Regul. Integr. Comp. Physiol.* 288, R1122, 2005.
61. Guo, F. and Jen, K.L. High-fat feeding during pregnancy and lactation affects offspring metabolism in rats, *Physiol. Behav.* 57, 681, 1995.
62. Taylor, P.D. et al. Impaired glucose homeostasis and mitochondrial abnormalities in offspring of rats fed a fat-rich diet in pregnancy, *Am. J. Physiol. Regul. Integr. Comp. Physiol.* 288, R134, 2005.
63. Khan, I.Y. et al. A high-fat diet during rat pregnancy or suckling induces cardiovascular dysfunction in adult offspring, *Am. J. Physiol. Regul. Integr. Comp. Physiol.* 288, R127, 2005.
64. Khan, I.Y. et al. Gender-linked hypertension in offspring of lard-fed pregnant rats, *Hypertension* 41, 168, 2003.
65. Armitage, J.A., Taylor, P.D., Poston, L., Experimental models of developmental programming: consequences of exposure to an energy rich diet during development, *J. Physiol.* 565, 3, 2005.
66. Palinski, W. et al. Maternal hypercholesterolemia and treatment during pregnancy influence the long-term progression of atherosclerosis in offspring of rabbits, *Circ. Res.* 89, 991, 2001.
67. Kozak, R., Richy, S., and Beck, B., Persistent alterations in neuropeptide Y release in the paraventricular nucleus of rats subjected to dietary manipulation during early life, *Eur. J. Neurosci.* 21, 2887, 2005.
68. Mortensen, E.L. et al. The association between duration of breastfeeding and adult intelligence, *JAMA* 287, 2365, 2002,

69. Haisma, H. et al. Breast milk and energy intake in exclusively, predominantly, and partially breast-fed infants, *Eur. J. Clin. Nutr.* 57, 1633, 2003.

70. Baker, J.L. et al. Maternal prepregnant body mass index, duration of breastfeeding, and timing of complementary food introduction are associated with infant weight gain, *Am. J. Clin. Nutr.* 80, 1579, 2004.

71. Owen, C.G. et al. The effect of breastfeeding on mean body mass index throughout life: a quantitative review of published and unpublished observational evidence, *Am. J. Clin. Nutr.* 82, 1298, 2005.

72. Owen, C.G. et al. Effect of infant feeding on the risk of obesity across the life course: a quantitative review of published evidence, *Pediatrics* 115, 1367, 2005.

73. Armstrong, J. and Reilly, J.J. Breastfeeding and lowering the risk of childhood obesity, *Lancet* 359, 2003, 2002.

74. Heinig, M.J. et al. Energy and protein intakes of breast-fed and formula-fed infants during the first year of life and their association with growth velocity: the DARLING Study, *Am. J. Clin. Nutr.* 58, 152, 1993.

75. Lucas, A. et al. Breast versus bottle: endocrine responses are different with formula feeding, *Lancet* 1, 1267, 1980.

76. McCance, R.A. Food, growth, and time, *Lancet* 2, 671, 1962.

77. Faust, I.M., Johnson, P.R., and Hirsch, J. Long-term effects of early nutritional experience on the development of obesity in the rat, *J. Nutr.* 110, 2027, 1980,

78. Plagemann, A. et al. Obesity and enhanced diabetes and cardiovascular risk in adult rats due to early post-natal overfeeding, *Exp. Clin. Endocrinol.* 99, 154, 1992.

79. Waterland, R.A. and Garza, C. Early post-natal nutrition determines adult pancreatic glucose-responsive insulin secretion and islet gene expression in rats, *J. Nutr.* 132, 357, 2002.

80. Davidowa, H. and Plagemann, A. Hypothalamic neurons of post-natally overfed, overweight rats respond differentially to corticotropin-releasing hormones, *Neurosci. Lett.* 371, 64, 2004.

81. Plagemann, A. et al. Elevation of hypothalamic neuropeptide Y-neurons in adult offspring of diabetic mother rats, *Neuroreport* 10, 3211, 1999.

82. Plagemann, A. et al. Increased number of galanin-neurons in the paraventricular hypothalamic nucleus of neonatally overfed weanling rats, *Brain Res.* 818, 160, 1999.

83. Walker, C.D. Nutritional aspects modulating brain development and the responses to stress in early neonatal life, *Progr. Neuropsychopharmacol. Biol. Psychiatr.* 29, 1249, 2005.

84. Li, Y., Plagemann, A., and Davidowa, H. Increased inhibition by agouti-related peptide of ventromedial hypothalamic neurons in rats overweight due to early post-natal overfeeding, *Neurosci. Lett.* 330, 33, 2002.

85. Cryer, A. and Jones, H.M. The development of white adipose tissue: effect of litter size on the lipoprotein lipase activity of four adipose-tissue depots, serum immunoreactive insulin and tissue cellularity during the first year of life in male and female rats, *Biochem. J.* 186, 805, 1980.

86. Kaung, H.L. Growth dynamics of pancreatic islet cell populations during fetal and neonatal development of the rat, *Dev. Dyn.* 200, 163, 1994.

87. Hall, W.G. Weaning and growth of artificially reared rats, *Science* 190, 1313, 1975.

88. Srinivasan, M. et al. Neonatal nutrition: metabolic programming of pancreatic islets and obesity, *Exp. Biol. Med.* 228, 15, 2003.

89. Haney, P.M. et al. Precocious induction of hepatic glucokinase and malic enzyme in artificially reared rat pups fed a high-carbohydrate diet, *Arch. Biochem. Biophys.* 244, 787, 1986.

90. Hiremagular, B.K., Johanning, G.L., and Patel, M.S. Long-term effects of feeding of high carbohydrate diet in preweaning by gastrosomy: a new rat model for obesity, *Int. J. Obes.* 17, 495, 1993.

91. Vadlamudi, S. et al. Long-term effects on pancreatic function of feeding a HC formula to rats during the preweaning period, *Am. J. Physiol.* 265, E565, 1993.

92. Patel, M.S. and Srinivasan, M. Metabolic programming: causes and consequences, *J. Biol. Chem.* 277, 1629, 2002.

93. Patel, M.S. and Srinivasan, M. *Metabolic Programming as a Consequence of the Nutritional Environment during Fetal and the Immediate Post-natal Periods*, Cambridge University Press, New York, 2006.

94. Aalinkeel, R. et al. A dietary intervention (high carbohydrate) during the neonatal period causes islet dysfunction in rats, *Am. J. Physiol.* 277, E1061, 1999.

95. Bratanova-Tochkova, T.K. et al. Triggering and augmentation mechanisms, granule pools, and biphasic insulin secretion, *Diabetes* 51, Suppl. 1, S83, 2002.

96. Kieffer, T.J. and Habener, J.F. The glucagon-like peptides, *Endocr. Rev.* 20, 876, 1999.

97. Srinivasan, M. et al. Adaptive changes in insulin secretion by islets from neonatal rats raised on a high-carbohydrate formula, *Am. J. Physiol. Endocrinol. Metab.* 279, E1347, 2000.

98. Sharp, G.W. Mechanisms of inhibition of insulin release, *Am. J. Physiol.* 271, C1781, 1996.

99. Srinivasan, M. et al. Molecular adaptations in islets from neonatal rats reared artificially on a high carbohydrate milk formula, *J. Nutr. Biochem.* 12, 575, 2001.

100. Song, F. et al. Use of a cDNA array for the identification of genes induced in islets of suckling rats by a high-carbohydrate nutritional intervention, *Diabetes* 50, 2053, 2001.

101. Petrik, J. et al. A long-term high-carbohydrate diet causes an altered ontogeny of pancreatic islets of Langerhans in the neonatal rat, *Pediatr. Res.* 49, 84, 2001.

102. Hiremagalur, B.K., Johanning, G.L., and Patel, M.S. Alterations in hepatic lipogenic capacity in rat pups artificially reared on a milk-substitute formula high in carbohydrate or medium-chain triglycerides, *J. Nutr. Biochem.* 3, 474, 1992.

103. Schwartz, M.W. et al. Central nervous system control of food intake, *Nature* 404, 661, 2000.

104. Seeley, R.J. and Woods, S.C. Monitoring of stored and available fuel by the CNS: implications for obesity, *Nat. Rev. Neurosci.* 4, 901, 2003.

105. Aalinkeel, R. et al. Programming into adulthood of islet adaptations induced by early nutritional intervention in the rat, *Am. J. Physiol. Endocrinol. Metab.* 281, E640, 2001.

106. Vadlamudi, S., Kalhan, S.C., and Patel, M.S. Persistence of metabolic consequences in the progeny of rats fed a HC formula in their early post-natal life, *Am. J. Physiol.* 269, E731, 1995.

107. Srinivasan, M. et al. Programming of islet functions in the progeny of hyperinsulinemic/obese rats, *Diabetes* 52, 984, 2003.

108. Srinivasan, M. et al. Maternal hyperinsulinemia predisposes rat fetuses for hyperinsulinemia, and adult-onset obesity and maternal mild food restriction reverses this phenotype, *Am. J. Physiol. Endocrinol. Metab.* 290, E129, 2006.

109. Srinivasan, M., Vadlamudi, S., and Patel, M.S. Glycogen synthase regulation in hyperinsulinemic/obese progeny of rats fed a high carbohydrate formula in their infancy, *Int. J. Obes. Relat. Metab. Disord.* 20, 981, 1996.
110. Srinivasan, M. et al. Glycogen synthase activation in the epididymal adipose tissue from chronic hyperinsulinemic/obese rats, *Nutr. Biochem.* 9, 81, 1998.
111. Waterland, R.A. and Garza, C. Potential mechanisms of metabolic imprinting that lead to chronic disease, *Am. J. Clin. Nutr.* 69, 179, 1999.

5 Oxidative Stress and Antioxidants in the Perinatal Period

Hiromichi Shoji, Yuichiro Yamashiro, and Berthold Koletzko

CONTENTS

INTRODUCTION

Oxidative stress is thought to be implicated in many pathologic processes of human disorders.[1] Pregnancy is a physiological state of oxidative stress accompanied by high metabolic demands and elevated requirements for tissue oxygen.[2] Childbirth is also accompanied by increased oxidative stress. Numerous disorders in infancy have been linked with oxidative stress, while the role of oxidative stress in the pathogenesis and progression of these diseases is only partially defined.[3] In 1988, Saugstad advocated "oxygen radical disease in neonatology" as a unifying hypothesis for the role of oxidative stress in a wide range of neonatal morbidities, implying different manifestations of the same condition.[4] Thus it has become important for clinicians who treat patients in the perinatal period to understand oxidative stress. The purpose of this chapter is to review information about the role of oxidative stress in pregnancy and post-partum.

OXIDATIVE STRESS

Free radicals are highly reactive chemical molecules containing one or more unpaired electrons. They donate or take electrons from other molecules in an attempt to pair their elections and generate a more stable species. Reactive oxygen species (ROS) is a collective term that includes both the oxygen free radicals (the $O_2^{.-}$ superoxide and the OH· hydroxyl radical) and some non-radical (hydrogen peroxide, H_2O_2) derivatives of oxygen. ROS are normally produced in living organisms with the potential of reacting with almost all types of molecules in living cells. ROS may be generated by different mechanisms such as ischemia–reperfusion, neutrophil and macrophage activation, Fenton chemistry, endothelial cell xanthine oxidase, free fatty acid and prostaglandin metabolism, and hypoxia (Figure 5.1)[.5,6]

Mitochondrial respiration is also one of the main physiological sources for the generation of ROS. Superoxide is formed when electrons leak from the electron transport chain.[7] ROS are capable of damaging all biologic macromolecules including lipids, proteins, polysaccharides, and DNA. When ROS are overproduced, they are important mediators of cell and tissue injury[8,9] and may cause cell death by apoptotic and necrotic mechanisms.[10]

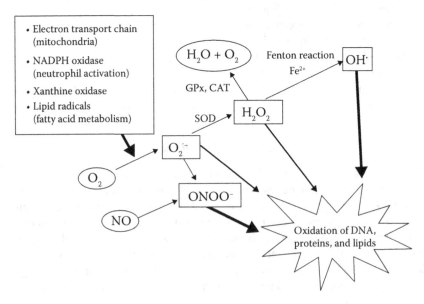

FIGURE 5.1 Reactive oxygen species produced in tissues.

TABLE 5.1
Major Antioxidants

Enzymes		Superoxide dismutase
		Catalase
		Glutathione peroxidase
		Glutathione reductase
Non-enzymatic antioxidants	Vitamins	Vitamin E
		Vitamin A
		Vitamin C
		Coenzyme Q
		β-carotene
	Reducing agents	Glutathione
		Cysteine
		Thioredoxin
	Binding proteins	Albumin
		Ceruloplasmin
		Lactoferrin
		Transferrin
	Enzyme constituents	Zinc
		Selenium
	Others	Uric acid
		Bilirubin
		Erythropoietin

A more recently appreciated group of molecules known as reactive nitrogen species are derived from reactions involving another free radical, nitric oxide (NO). NO can act either as a pro-oxidant or as an antioxidant.[11] $O_2^{\cdot-}$ reacts with NO to form a cytotoxic oxidant peroxynitrite (ONOO⁻) (Figure 5.1).

Oxidative stress can be defined as an imbalance between the amount of ROS and the intracellular and extracellular antioxidant protection systems.[12] An antioxidant may be classified as "any substance that can delay or prevent oxidation of a particular substrate" (Table 5.1)[13] and it may be broadly classified as enzymatic (superoxide dismutases; SOD, catalase; CAT, glutathione peroxidase; GPx and glutathione; GSH and their precursors) or non-enzymatic.[14] The antioxidant defense can be divided into extracellular and intracellular defenses that protect against ROS-induced cell damage. Transferrin, ceruloplasmin, vitamin C, vitamin E, uric acid, bilirubin, sulfhydryl groups, and other unidentified antioxidants contribute to the total antioxidant capacities of extracellular fluids.[15]

The extent of oxidative stress has been variably determined by measurement of a decrease in total antioxidant capacity, through depletion of individual antioxidants such as vitamin E, vitamin C, or GPx. Otherwise it is defined as the product (marker) of oxidative stress to lipids, proteins, and DNA.

PREGNANCY: OXIDATIVE STRESS

Pregnancy accompanied by a high metabolic demand and elevated requirements for tissue oxygen represents a physiological state of oxidative stress.[2] The placenta has been identified as an important source of lipid peroxidation because of its enrichment with polyunsaturated fatty acids (PUFAs).[16] Falkay et al. suggested that the increase in the lipid peroxidation levels is due to increased prostaglandin synthesis in the placenta.[17] Levels of peroxidation markers such as lipid hydroperoxide and malondialdehyde (MDA) are higher in pregnant than in non-pregnant women.[18] Lipid peroxidation is enhanced in the second trimester, tapers off later in gestation, and decreases after delivery.[19]

A certain amount of ROS promotes embryonic development.[20–22] ROS seem to play a role in signal transduction by modulating transcription factors including hypoxia-inducible factor (HIF-1)[23] and activated protein-1 (AP-1)[24] that control the expression of cell growth mediators. ROS may also affect activation and release of the nuclear factor NFκB — an important regulator of cytokine and anti-apoptotic gene expression.[23,25,26] However, over-production of ROS may also induce early embryonic developmental block and retardation (Figure 5.2).[20,23,27]

The placenta is also a source of antioxidative enzymes to control placental lipid peroxidation during uncomplicated pregnancies. All the major antioxidant systems including SOD, CAT, GPx, glutathione S-transferase, and GSH, and vitamins C and E are present in the placenta.[28–31] The activities of SOD and CAT increased as gestation progressed, while the activity of GPx and vitamin E concentration did not significantly change.[29,32,33] One study found that placental concentration of lipid peroxidation decreased as gestation advanced, with increased activities of SOD and CAT in the placenta.[34] Placental antioxidant defense systems may be sufficient to control the lipid peroxidation in normal pregnancies.[35,36]

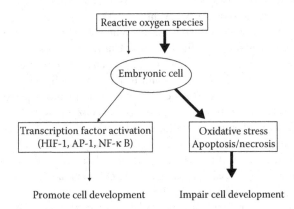

FIGURE 5.2 Role of reactive oxygen species in fetal development. (*Source:* Modified from Dennery, P.A. *Antioxid. Redox Signal.* 6, 147, 2004.)

Abnormal placentation (inadequate placental invasion of the maternal spiral arteries) leads to reduced uteroplacental blood flow and placental ischemia.[37] The ischemia–reperfusion to the placenta leads to abnormal generation of placental oxidative stress. Excessive placental oxidative stress was assumed to play a role in the pathogenesis of preeclampsia (PE)[16,38,39] and intrauterine fetal growth retardation (IUGR).[39–42]

PREECLAMPSIA

Preeclampsia (PE) is a pregnancy-specific disorder that complicates 5% of all pregnancies and 11% of all first pregnancies.[42] Clinically, PE is usually diagnosed in late pregnancy by increased blood pressure and proteinuria, and the symptoms of PE typically disappear shortly after delivery of the placenta. The placenta plays a major role in the pathogenesis of PE, characterized by abnormal placentation and reduced placental perfusion.[43] Recent evidence suggests that a disturbance of normal endothelial cell function may be a primary cause in the pathogenesis of PE.[44,45] PE is associated with severe maternal and fetal mortality[46] and is a major risk factor for preterm delivery and IUGR.[47]

Placental oxidative stress resulting from ischemia–reperfusion injury is involved in the pathogenesis of PE. A significant increase in placental lipid peroxidation levels in the placenta of PE has been demonstrated.[48–50] Several reports confirm that circulating levels of lipid peroxides are significantly elevated in women with PE compared to normal pregnant women.[18,51–53] However, other studies showed no evidence of increased circulating lipid peroxidation in PE.[54–56]

Lluba et al. speculated that circulating lipid peroxides generated in preeclamptic women are neutralized by the increasing concentration of antioxidant enzymes and vitamin E.[55] On the other hand, several important antioxidants are significantly decreased in pregnant women with PE. The levels of vitamin C, vitamin E, vitamin A, β-carotene, coenzyme Q10, and GSH are all significantly lower than in normal pregnancy.[52,53,57,58] Although most studies report decreased levels of vitamin E in preeclamptic women, some do not.[54,55]

Vitamin C and vitamin E have been studied related to prevention of PE. Early intervention at 16 to 22 weeks of pregnancy (283 women) with vitamin C (1000 mg/day) and E (400 IU/day) supplementation resulted in significant reductions of PE in the supplement group[59] but a recent report of a randomized trial (100 women) failed to reveal the benefits of PE prevention after the same amounts of vitamins C and E were supplemented at 14 to 20 weeks of gestation.[60]

OXIDATIVE STRESS AT CHILDBIRTH

Childbirth is an oxidative challenge for newborns. The transition from fetal to neonatal life at birth implies acute and complex physiologic changes. The fetus transfers from an intrauterine hypoxic environment with a pO_2 of 20 to 25 mm Hg to an extrauterine normoxic environment with a pO_2 of 100 mm Hg.[61] This four- to five-fold increase is believed to induce increased production of ROS.

Neonatal plasma has relatively poor antioxidative defenses.[62] At birth, neo-natal plasma concentrations of vitamins A, E and β-carotene were significantly lower than maternal plasma levels, while neonatal levels of vitamin C were significantly higher.[63] Uric acid and vitamin C constitute most of the extracellular antioxidant capacity, totaling 75%. At 2 weeks of age, these two components represent only 35% of the extracellular capacity. This change is caused by the rapid decline of vitamin C levels during the first few days of life and the increasing concentration of bilirubin which acts as an antioxidant in the first 1 to 2 weeks post-partum.[64]

Term labor is associated with oxidative stress for the neonate. Reported MDA levels were higher for term infants born by cesarean section than for those born by spontaneous vaginal delivery.[65,66] In a case controlled study, the serum levels of lipid peroxidation products were significantly higher (110%) in 20 women during labor compared to the 20 controls (pregnant women not in labor and matched for maternal gestational age).[67] On the other hand, women during term labor showed up-regulations of red blood cell GSH in cord blood. Fetal (difference in GSH concentration in umbilical vein and artery) and maternal (pre- and post-delivery) GSH concentrations were significantly lower during labor at term than in elective cesarean section.[68]

PERINATAL ASPHYXIA

Perinatal asphyxia is characterized by transient hypoxia during the ischemic phase followed by reperfusion. One definition of birth asphyxia is based on the finding of three of the following four criteria: (1) pH of umbilical arterial cord blood <7.00, (2) Apgar score <4 for more than 5 min, (3) multiple organ damage, and (4) hypoxic ischemic encephalopathy (HIE).[69] HIE in newborns remains a major cause of both mortality and long-term morbidity and leaves significant handicaps in 100% of survivors of severe HIE and 20% of survivors of moderate HIE.[70]

Hypoxia and ischemia during perinatal asphyxia give rise to an inadequate substrate supply to brain tissue, resulting in damage of neuronal cells. During tissue ischemia, an intracellular accumulation of a large amount of hypoxanthine and conversion of xanthine dehydrogenase to xanthine oxidase occur.[71] Upon reperfusion, xanthine oxidase promotes formation of large quantities of $O_2^{\cdot-}$ and H_2O_2, and hypoxanthine is reconverted to xanthine (Figure 5.3). The amount of ROS produced is directly dependent on oxygen as well as on hypoxanthine and other purine concentrations in tissues. Production of ROS in the early reperfusion phase plays a substantial role in the resulting brain cell damage.[35]

The brain of a term fetus is at higher risk of oxidative stress than the brain of a preterm fetus. This appears to result from the term brain's higher concentration of PUFAs and the greater maturity of the N-methyl-D-aspartate receptor system — a system of heterotetrameric amino acid receptors that functions as a membrane calcium channel.[72] Mishra et al. demonstrated in animal studies that term brains showed higher levels of lipid peroxidation than preterm brains[73] and increased activation of NMDA receptors during brain development as gestation

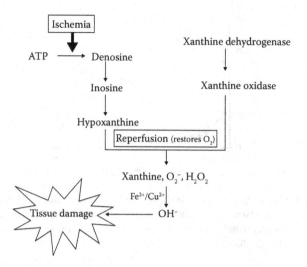

FIGURE 5.3 Suggested mechanism for tissue damage upon reperfusion of hypoxic and ischemic tissues. (*Source:* Modified from Halliwell, B. and Gutteridge, J.M., *Free Radicals in Biology and Medicine*, Oxford University Press, New York, 1999.)

approached term.[40] ROS cause endothelial cell damage and abnormalities in N-methyl-D-aspartate receptors, synaptosome structures, and astrocyte functions, thus contributing to the development of brain injury after a hypoxic–ischemic episode.[74–76]

USE OF 100% OXYGEN IN RESUSCITATION

The optimal concentration of oxygen for neonatal resuscitation is uncertain. Traditionally, neonatal resuscitation has been performed with 100% oxygen. Many textbooks and the advisory statement of the International Liaison Committee on Resuscitation[77] recommend that resuscitation of newborn infants should be performed with 100% oxygen. Resuscitation of depressed newborns with presumed hypoxia and/or asphyxia in the delivery room is currently the only remaining clinical indication for the use of unregulated 100% oxygen in infants.

Exposure to high levels of oxygen during reoxygenation may promote the formation of excessive levels of ROS in tissue, causing tissue injury. Hyperoxemia has been associated with numerous negative side effects including delayed initiation of spontaneous respiration, increased oxygen consumption, and irregularities in cerebral circulation.[78] Therefore, several groups have investigated the safety and efficacy of using room air for neonatal resuscitation.

Evidence from animal studies proves that room air is as effective as 100% oxygen in resuscitation. No significant differences were noted between two groups (hypoxemic newborn pigs resuscitated with 21% O_2 or 100% O_2 for 20 to 25 min followed by 21% O_2) in arterial blood pressure, base deficit, plasma

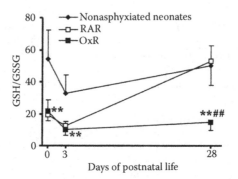

FIGURE 5.4 Reduced gluthatione (GSH) and oxidized gluthathione (GSSG) ratio in asphyxiated neonates resuscitated with 100% oxygen (OxR) or room air (RAR). ** p <0.01 versus non-asphyxiated neonates. ## p <0.01 versus RAR. (*Source:* Venton, M. et al., *Pediatrics* 107, 642, 2001.)

hypoxanthine, cerebral blood flow, and forebrain O_2 uptake (determined by radio-active microspheres).[79,80]

Human infants have been studied. Vento et al.[81] showed that neonates resuscitated with 100% oxygen revealed persistent oxidative stress. The reduced GSH:oxidized glutathione (GSSG) ratios (determined in whole blood) of the 100% oxygen-resuscitated infants were significantly different from those of room-air-resuscitated infants and the non-asphyxiated control infants at 28 days of life (Figure 5.4). One multicenter study enrolled approximately 600 infants (inclusion criteria: apnea with heart rate <80 beats per min at birth and birth weights >1000 g) to receive either 21 or 100% oxygen during resuscitation. The follow-up study showed no significant differences between groups in the primary outcomes, early neonatal deaths, and/or neurological handicap (cerebral palsy and/or mental or other delay) at ages of 18 to 24 months.[82,83]

OXIDATIVE STRESS IN PREMATURE INFANTS

Premature infants are probably more prone to oxidative stress than term infants because premature infants very often are exposed to high oxygen concentrations as a result of pulmonary surfactant deficiency. They have lower and less efficient antioxidant defenses and more often are exposed to infections and inflammation resulting in the release of pro-inflammatory cytokines activating the production of ROS. Moreover, free iron is found in the plasma and tissues of premature infants at higher concentrations than in term infants.[84]

Several clinical studies in premature infants noted deficiencies in antioxidant defenses including lower cord blood SOD activities,[85] lower concentrations of GSH in bronchoalveolar lavage fluid,[86] and lower plasma GSH concentrations.[87] Antioxidants such as vitamins E, A, and C, ceruloplasmin, transferrin, and trace metals such as copper, zinc, and selenium (metal cofactors required for antioxidant synthesis and activities) are low in preterm infants compared to term infants.[88,89] Induction of antioxidative enzymes following oxidative stress was found in term but not in preterm infants.[90] These data strongly suggest increased susceptibilities of premature infants to oxidative stress.[62] Clinical conditions in which oxidative stress may be particularly relevant in premature infants are discussed below.

BRONCHOPULMONARY DYSPLASIA

Ventilated premature infants are at risk of developing bronchopulmonary dysplasia (BPD), a chronic lung disease of prematurity. When BPD was first described, exposure to a high oxygen level was identified as a risk factor for its development[91,92] and it was soon related to ROS.[93] It is widely assumed that BPD is due to ROS-related injury to the immature lung.[94] Two studies found a close relationship between high oxygen exposure and the development of BPD.[95,96] Van Marter et al. found that inspired oxygen concentrations were higher in infants who developed BPD compared with those who did not. A fractional inspired oxygen level of 1.0 in the first day of life almost doubled the risk of BPD relative to an FiO_2.[96] Oxidative stress detected by the augmentation of lipid peroxidation products in the very first days of life is associated with the subsequent development of BPD.[97,98]

Antioxidants present in the alveolar epithelial lining fluid are well positioned to protect against tissue damage caused by ROS generated by these mechanisms.[99] Infants who develop BPD have decreased concentrations of glutathione in bronchoalveolar fluid compared to those who did not require supplemental oxygen at 36 weeks post-conceptional age on the first day of life.[86] Robbins et al. reported that the administration of recombinant human SOD (rhSOD) in newborn piglets mitigated the lung inflammatory changes (analyzed by BAL neutrophil chemotactic activity and cell count), MDA concentration in lung tissue, and acute lung injury induced by 100 ppm of NO and 90% O_2.[100] In a multicenter controlled study of 26 preterm infants, significant reductions in radiological evidence of BPD were noted in infants treated with rhSOD compared to infants treated with placebo.[101] However, in a later and larger study of 302 infants, no differences in the development of BPD in infants treated with rhSOD and those treated with placebo were noted.[102]

NO is a free radical that may be oxidized or reduced, depending on its concentration and the presence of other oxidants such as oxygen. Hyperoxic exposure of rat pups up-regulated both inducible and endothelial NO synthase and therefore increased the concentrations of NO and subsequently peroxynitrite as well.[103] NO is highly toxic for preterm infants. However, a clinical trial with inhaled NO did not report higher occurrences of BPD in NO-treated premature infants compared with controls.[104] Hamon et al. recently reported that low doses (5 ppm) of inhaled NO administered soon after birth in hypoxemic premature infants were associated with significant decreases in their oxidative stress after 24 hr as assessed by plasma MDA.[105]

NEONATAL NECROTIZING ENTEROCOLITIS

Neonatal necrotizing enterocolitis (NEC) is the most common life-threatening disease of the gastrointestinal tract in the neonatal period; it primarily affects premature infants.[106] NEC is characterized by various degrees of mucosal or transmural necrosis of the intestine. Its causes remain unclear but are most likely multifactorial. Intestinal ischemia and colonization of the intestinal lumen by several viral and bacterial organisms were hypothesized to cause NEC.[107] Immunologic immaturity of the preterm gut[108] and neonatal hypoxia were also included as causative factors. Upon reperfusion and reoxygenation of the damaged intestine, a flood of ROS is generated by the xanthine oxidase system. This burst of ROS can cause severe tissue damage and may be the final pathway for tissue injury of the gut mucosa noted in NEC.[71]

ROS can alter inflammation in the gut and lead to activation of platelet activating factor (PAF), an endogenous phospholipid mediator[109] that results in increased coagulation and subsequent microthrombus formation in small vessels. A model of acute inflammatory intestinal injury induced by PAF was significantly attenuated by administration of allopurinol (a xanthine inhibitor).[110]

PERIVENTRICULAR LEUKOMALACIA

Periventricular leukomalacia (PVL) in premature infants is a distinctive lesion of cerebral white matter (Figure 5.5) associated with severe adverse neurologic outcomes. The pathogenesis of cerebral white matter injury in premature infants

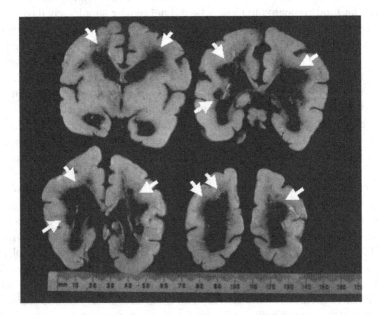

FIGURE 5.5 Coronal sections of brain with periventricular leukomalacia.

is not entirely clear, although ischemia-reperfusion and infection-inflammation appear important.[70] The brain is at special risk due to its high concentration of PUFAs and deficiency of SOD and GPx.[111] Immature oligodendrocytes are more prone to oxidative stress than mature ones.[112] These immature forms, the so-called preoligodendrocytes, have been shown recently to account for 90% of the total population of oligodendrocytes in cerebral white matter of infants under the gestational age of 31 weeks[113] who represent a high risk group for the development of PVL.[114]

Houdou et al. also reported that CAT-positive glia did not appear in the deep white matter before 31 to 32 weeks of gestation.[115] These very premature infants are most at risk for white matter injury.[116] In two model systems of free radical accumulation, the early differentiating oligodendrocyte was shown to be exquisitely vulnerable to free radical attack.[113,117]

Direct support for an association between products of oxidation and PVL in premature infants is limited. However, in a recent study, premature infants with subsequent evidence of PVL on magnetic resonance imaging at term had higher levels of cerebrospinal fluid protein carbonyls (markers of oxidized protein) than healthy premature or term infants.[116]

RETINOPATHY OF PREMATURITY

Retinopathy of prematurity (ROP) is a vasoproliferative retinal disorder of premature infants. Its basic pathogenesis is not fully understood but exposure to the extrauterine environment including necessarily high inspired oxygen concentrations produces cellular damage mediated by ROS. Hyperoxygenation favors peroxidation of vasoactive isoprostanes, resulting in vasoconstriction and vascular cytotoxicity leading to ischemia, which predisposes to the development of vasoproliferative retinopathy.[118] Severe ROP can lead to lifelong visual impairment or blindness.[119]

Early animal models of ROP first suggested that oxygen was involved in the normal developing retinal vasculature, probably via a putative angiogenic factor[120] that was subsequently identified as vascular endothelial growth factor (VEGF).[121] VEGF plays an important role in endothelial cell proliferation, migration, and blood vessel formation[122,123] and in the development of ROP.[124]

Several studies have shown that greater variability of transcutaneous oxygen during the first 2 weeks of life is associated with the development of severe ROP.[125,126] The so-called STOP-ROP (supplemental therapeutic oxygen for pre-threshold retinopathy of prematurity) trial enrolled about 600 premature infants with confirmed threshold ROP in at least one eye. The risks of adverse pulmonary events including BPD increased with the use of supplemental oxygen at pulse oximetry saturation of 96 to 99% compared with pulse oximetry targeting 89 to 94%. However, this therapy had no significant effect on the progression of ROP.[127]

Supplementation of antioxidants has been reported to protect against ROP. A meta-analysis evaluated data from six randomized clinical trials of vitamin E prophylaxis (15 to 100 mg/kg/day from day 1 until discharge) that included a total of 704 premature infants in the vitamin E prophylaxis groups and 714 control infants. The overall incidence of any stage ROP was similar between the groups, but incidence of severe (stage +3) ROP was lower in the vitamin E-treated group (2.4% in the vitamin E group versus 5.3% of controls). The authors concluded that the role of the vitamin E antioxidant in reducing severe ROP must be re-evaluated.[128]

HUMAN MILK AND ANTIOXIDATIVE PROTECTION

Human milk (HM) is considered the ideal food for healthy infants[129] and is known to contain various bioactive substances, some of which are reported to be antioxidants.[130] HM contains many antioxidants such as enzymes (CAT, GPx, SOD), vitamins (A, C, E), binding proteins such as lactoferrin, and constituents of antioxidative enzymes (Cu, Zn).[131–133] In contrast, antioxidative enzymes are absent from infant formula.[134] Most infant formula has higher amounts of vitamins added than are present in HM to make up for the reduced bioavailability. Thus, the overall antioxidant capacity of HM versus infant formula is difficult to assess, although assessment would probably favor HM.[14]

We previously reported *in vitro* results showing that HM alleviated H_2O_2-induced oxidative damage in intestinal epithelial cell lines, whereas bovine milk or infant formula did not.[135] Confluent intestinal epithelial (IEC-6) cells were preincubated with defatted HM, bovine milk, or three artificial milks for 24 hr followed by H_2O_2 challenge. HM-treated cells showed the highest survival rates (50%) compared with bovine milk-treated (6%) or infant formula-treated (13 to 16%) cells (Figure 5.6).[135] Buescher and McIlherhan[131] reported that human colostrum manifested antioxidant properties, proving capable of spontaneous reduction of cytochrome C, depletion of polymorphonuclear leukocyte-produced H_2O_2, and protection of epithelial cells from polymorphonuclear leukocyte-mediated detachment.

Several clinical studies demonstrated antioxidative properties of HM. HM-fed infants had higher plasma trapping ability (a measure of resistance to oxidative stress *in vitro*) than did control infants fed formula.[136] We compared the oxidative stress levels in 41 healthy 1-month-old infants and 29 premature infants by measuring urinary 8-hydroxy-2-deoxyguanosine (8-OHdG — a marker of oxidative DNA damage). In the 1-month-old group, urinary 8-OHdG excretions of the breast-fed infants were significantly lower than those of the artificial formula-fed infants (Figure 5.7).[137] In the premature group, urinary 8-OHdG excretions of the breast-fed infants at 14 and 28 days of age were significantly lower than those of the formula-fed infants (Figure 5.8).[138] These data indicate that HM provides more antioxidant properties than infant formula during early infancy. This may be due to the presence of antioxidants in HM that may exhibit antioxidant effects in the gut and may pass through the relatively porous neonatal intestine early in infancy.[131,132] In fact, feeding with HM has been associated with a lower incidence of a variety of illnesses including NEC,[139] respiratory disease,[140] and ROP[141] in premature infants.

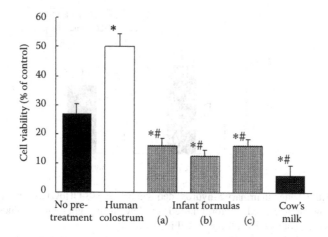

FIGURE 5.6 Antioxidative properties of human colostrum, bovine milk, and artificial formulas and H_2O_2-induced oxidative damage in IEC-6 cells. Survival cell rates are expressed as percentages of control (non-H_2O_2-challenged) cells. Values are means + SD. * p <0.05 compared with values of non-pretreatment cells. # p <0.05 compared with values of cells preincubated with colostrum.

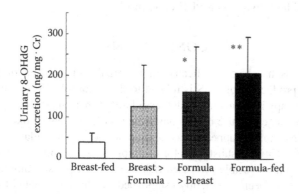

FIGURE 5.7 Urinary 8-hydroxydeoxyguanosine (8-OHdG) in one-month-old term infants. Breast-fed group received 90% of their intake as breast milk. Breast>formula group received 50 to 90% of their intake as breast milk. Formula>breast group received 50 to 90% of their intake as formula. Formula-fed group received 90% of their intake as formula. Values = means + SD. * p <0.05 compared with values of breast-fed group. ** p <0.01 compared with values of breast-fed group.

HM also contains long-chain PUFAs such as docosahexaenoic acid (DHA) and arachidonic acid (AA) that have been shown to enhance visual and cognitive functions in infants. Recently, infant formulas have been fortified with DHA and AA. However, long-chain PUFAs are subject to oxidative damage and concerns have been raised that these nutrients may increase the vulnerability of infants to

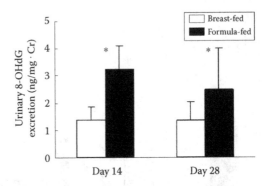

FIGURE 5.8 Change in urinary 8-hydroxydeoxyguanosine (8-OHdG) excretion at 14 and 28 days of age in breast-fed and formula-fed very low birthweight infants. Values = means + SD. *p <0.01 compared with breast-fed infants.

be damaged by ROS.[142] Lipid peroxidation of PUFAs may also enhance oxidative damage of essential amino acids.[143] Previous studies reported that the addition of long-chain PUFAs to infant formula did not affect plasma vitamin E and A concentrations[144] or values of essential amino acids[145] in term infants. In contrast, vitamin E:total lipid ratios in erythrocyte membranes significantly decreased in low birth weight infants fed long-chain PUFA-fortified formula compared with infants fed without long-chain PUFAs or HM.[142]

CONCLUSIONS

Oxidative stress has been implicated in maternal and neonatal diseases, and uncontrolled production of ROS is related to their morbidity. Oxygen is usually required as therapy in premature or depressed infants but excessive oxygen causes physiologic damage because infants have less protection against oxidative stress. Breast feeding of infants appears beneficial in providing antioxidative protection. The supply of nutritional substances for the improvement of infant formula should take into account a possible increased risk of oxidative stress. Antioxidant therapies have not yet been sufficiently established for diseases related to oxidative stress in the perinatal period. Further biochemical and clinical investigations are needed to define effective antioxidative therapeutic approaches.

ACKNOWLEDGMENTS

The studies reported in this chapter were carried out with partial financial support from the Commission of the European Committees, specifically the RTD programme titled "Quality of Life and Management of Living Resources" within the Sixth Framework Programme, Contract FOOD-CT-2005-007036 EARNEST (Early Nutrition Programming Project, www.metabolic-programming.org). This manuscript does not necessarily reflect the view of the commission and in no way anticipates future policies in this area. BK is the recipient of a Freedom to Discover Award of the Bristol Myers Squibb Foundation, New York, NY, U.S.A.

REFERENCES

1. Halliwell, B. and Gutteridge, J.M. *Free Radicals in Biology and Medicine*, Oxford University Press, New York, 1999.
2. Spatling, L., Fallenstein, F., Huch, A., Huch, R., and Rooth, G. The variability of cardiopulmonary adaptation to pregnancy at rest and during exercise. *Br. J. Obstet. Gynaecol.* 99, Suppl. 8, 1, 1992.
3. Granot, E. and Kohen, R. Oxidative stress in childhood — in health and disease states. *Clin. Nutr.* 23, 3, 2004.
4. Saugstad, O.D. Hypoxanthine as an indicator of hypoxia: its role in health and disease through free radical production. *Pediatr. Res.* 23, 143, 1988.
5. McCord, J.M. Oxygen-derived free radicals in postischemic tissue injury. *New Engl. J. Med.* 312, 159, 1985.
6. Mishra, O.P. and Delivoria-Papadopoulos, M. Cellular mechanisms of hypoxic injury in the developing brain. *Brain Res. Bull.* 48, 233, 1999.
7. Halliwell, B., Gutteridge, J.M., and Cross, C.D. Free radicals, antioxidants, and human disease: where are we now? *J. Lab. Clin. Med.* 119, 598, 1992.
8. Fridovich, I. Oxygen toxicity: a radical explanation. *J. Exp. Biol.* 201, 1203, 1998.
9. Halliwell, B. Free radicals, antioxidants, and human disease: curiosity, cause, or consequence? *Lancet* 344, 721, 1994.
10. Wiseman, H. and Halliwell, B. Damage to DNA by reactive oxygen and nitrogen species: role in inflammatory disease and progression to cancer. *Biochem. J.* 313, Pt. 1, 17, 1996.
11. Patel, R.P. et al. Biological aspects of reactive nitrogen species. *Biochim. Biophys. Acta* 1411, 385, 1999.
12. Sies, H. Oxidative stress: oxidants and antioxidants. *Exp. Physiol.* 82, 291, 1997.
13. Halliwell, B. The antioxidant paradox. *Lancet* 355, 1179, 2000.
14. Rassin, D.K. and Smith, K.E. Nutritional approaches to improve cognitive development during infancy: antioxidant compounds. *Acta Paediatr. Suppl.* 92, 34, 2003.
15. Jankov, R.P., Negus, A., and Tanswell, A.K. Antioxidants as therapy in the newborn: some words of caution. *Pediatr. Res.* 50, 681, 2001.
16. Mutlu-Turkoglu, U. et al. Imbalance between lipid peroxidation and antioxidant status in preeclampsia. *Gynecol. Obstet. Invest.* 46, 37, 1998.
17. Falkay, G., Herczeg, J., and Sas, M. Microsomal lipid peroxidation in human pregnant uterus and placenta. *Biochem. Biophys. Res. Commun.* 79, 843, 1977.
18. Morris, J.M. et al. Circulating markers of oxidative stress are raised in normal pregnancy and pre-eclampsia. *Br. J. Obstet. Gynaecol.* 105, 1195, 1998.
19. Little, R.E. and Gladen, B.C. Levels of lipid peroxides in uncomplicated pregnancy: a review of the literature. *Reprod. Toxicol.* 13, 347, 1999.
20. Dennery, P.A. Role of redox in fetal development and neonatal diseases. *Antioxid. Redox. Signal.* 6, 147, 2004.
21. Fisher, J.C. et al. Oxidation–reduction (redox) controls fetal hypoplastic lung growth. *J. Surg. Res.* 106, 287, 2002.

22. Harvey, A.J., Kind, K.L., and Thompson, J.G.. Redox regulation of early embryo development. *Reproduction* 123, 479, 2002.
23. Haddad, J.J., Olver, R.E., and Land, S.C. Antioxidant/pro-oxidant equilibrium regulates HIF-1 alpha and NF-kappa B redox sensitivity: evidence for inhibition by glutathione oxidation in alveolar epithelial cells. *J. Biol. Chem.* 275, 21130, 2000.
24. Gomez, D.A. et al. Antioxidants and AP-1 activation: a brief overview. *Immunobiology* 198, 273, 1997.
25. Burdon, R.H. Superoxide and hydrogen peroxide in relation to mammalian cell proliferation. *Free Radical Biol. Med.* 18, 775, 1995.
26. Mattson, M.P. et al. Roles of nuclear factor kappa-B in neuronal survival and plasticity. *J. Neurochem.* 74, 443, 2000.
27. Bedaiwy, M.A. et al. Differential growth of human embryos *in vitro*: role of reactive oxygen species. *Fertil. Steril.* 82, 593, 2004.
28. Guthenberg, C. and Mannervik, B. Glutathione S-transferase (transferase pi) from human placenta is identical or closely related to glutathione S-transferase (transferase rho) from erythrocytes. *Biochim. Biophys. Acta* 661, 255, 1981.
29. Myatt, L. and Cui, X. Oxidative stress in the placenta. *Histochem. Cell. Biol.* 122, 369, 2004.
30. Van Hien, P., Kovacs, K., and Matkovics, B. Properties of enzymes. I. Study of superoxide dismutase activity change in human placenta of different ages. *Enzyme* 18, 341, 1974.
31. Wang, Y. and Walsh, S.W. Antioxidant activities and mRNA expression of superoxide dismutase, catalase, and glutathione peroxidase in normal and preeclamptic placentas. *J. Soc. Gynecol. Investig.* 3, 179, 1996.
32. Watson, A.L. et al. Variations in expression of copper/zinc superoxide dismutase in villous trophoblast of the human placenta with gestational age. *Placenta* 18, 295, 1997.
33. Spinnato, J.A. and Livingston, J.C. Prevention of preeclampsia with antioxidants: evidence from randomized trials. *Clin. Obstet. Gynecol.* 48, 416, 2005.
34. Qanungo, S. and Mukherjea, M. Ontogenic profile of some antioxidants and lipid peroxidation in human placental and fetal tissues. *Mol. Cell Biochem.* 215, 11, 2000.
35. Gitto, E. et al. Causes of oxidative stress in the pre- and perinatal period. *Biol. Neonate* 81, 146, 2002.
36. Mueller, A. et al. Placental defence is considered sufficient to control lipid peroxidation in pregnancy. *Med. Hypotheses* 64, 553, 2005.
37. van Beck, E. and Peters, L.L. Pathogenesis of preeclampsia: a comprehensive model. *Obstet. Gynecol. Surv.* 53, 233, 1998.
38. Gupta, S., Agarwal, A., and Sharma, R.K. The role of placental oxidative stress and lipid peroxidation in preeclampsia. *Obstet. Gynecol. Surv.* 60, 807, 2005.
39. Takagi, Y. et al. Levels of oxidative stress and redox-related molecules in the placenta in preeclampsia and fetal growth restriction. *Virchows Arch.* 444, 49, 2004.
40. Mishra, O.P. and Delivoria-Papadopoulos, M. Modification of modulatory sites of NMDA receptor in the fetal guinea pig brain during development. *Neurochem. Res.* 17, 1223, 1992.
41. Scholl, T.O. and Stein, T.P. Oxidant damage to DNA and pregnancy outcome. *J. Matern. Fetal Med.* 10, 182, 2001.

42. Agarwal, A., Gupta, S., and Sharma, R.K. Role of oxidative stress in female reproduction. *Reprod. Biol. Endocrinol.* 3, 28, 2005.
43. Roberts, J.M. Preeclampsia: what we know and what we do not know. *Semin. Perinatol.* 24, 24, 2000.
44. Hubel, C.A. Oxidative stress in the pathogenesis of preeclampsia. *Proc. Soc. Exp. Biol. Med.* 222, 222, 1999.
45. VanWijk, M.J. et al. Vascular function in preeclampsia. *Cardiovasc. Res.* 47, 38, 2000.
46. Villar, J. et al. Methodological and technical issues related to the diagnosis, screening, prevention, and treatment of pre-eclampsia and eclampsia. *Int. J. Gynaecol. Obstet.* 85, Suppl. 1, S28, 2004.
47. Lim, K.H. and Friedman, S.A. Hypertension in pregnancy. *Curr. Opin. Obstet. Gynecol.* 5, 40, 1993.
48. Atamer, Y. et al. Lipid peroxidation, antioxidant defense, status of trace metals and leptin levels in preeclampsia. *Eur. J. Obstet. Gynecol. Reprod. Biol.* 119, 60, 2005.
49. Vanderlelie, J. et al. Increased biological oxidation and reduced anti-oxidant enzyme activity in pre-eclamptic placentae. *Placenta* 26, 53, 2005.
50. Walsh, S.W. et al. Placental isoprostane is significantly increased in preeclampsia. *FASEB J.* 14, 1289, 2000.
51. Maseki, M. et al. Lipid peroxide levels and lipids content of serum lipoprotein fractions of pregnant subjects with or without pre-eclampsia. *Clin. Chim. Acta* 115, 155, 1981.
52. Wang, Y.P. et al. Imbalance between thromboxane and prostacyclin in preeclampsia is associated with an imbalance between lipid peroxides and vitamin E in maternal blood. *Am. J. Obstet. Gynecol.* 165, 1695, 1991.
53. Yanik, F.F. et al. Pre-eclampsia associated with increased lipid peroxidation and decreased serum vitamin E levels. *Int. J. Gynaecol. Obstet.* 64, 27, 1999.
54. Bowen, R.S. et al. Oxidative stress in pre-eclampsia. *Acta Obstet. Gynecol. Scand.* 80, 719, 2001.
55. Llurba, E. et al. A comprehensive study of oxidative stress and antioxidant status in preeclampsia and normal pregnancy. *Free Radic. Biol. Med.* 37, 557, 2004.
56. Regan, C.L. et al. No evidence for lipid peroxidation in severe preeclampsia. *Am. J. Obstet. Gynecol.* 185, 572, 2001.
57. Mikhail, M.S. et al. Preeclampsia and antioxidant nutrients: decreased plasma levels of reduced ascorbic acid, alpha-tocopherol, and beta-carotene in women with preeclampsia. *Am. J. Obstet. Gynecol.* 171, 150, 1994.
58. Palan, P.R. et al. Lipid-soluble antioxidants and pregnancy: maternal serum levels of coenzyme Q10, alpha-tocopherol and gamma-tocopherol in preeclampsia and normal pregnancy. *Gynecol. Obstet. Invest.* 58, 8, 2004.
59. Chappell, L.C. et al. Effect of antioxidants on the occurrence of pre-eclampsia in women at increased risk: a randomised trial. *Lancet* 354, 810, 1999.
60. Beazley, D. et al. Vitamin C and E supplementation in women at high risk for preeclampsia: a double-blind, placebo-controlled trial. *Am. J. Obstet. Gynecol.* 192, 520, 2005.
61. Muller, D.P. Free radical problems of the newborn. *Proc. Nutr. Soc.* 46, 69, 1987.
62. Saugstad, O.D. Oxidative stress in the newborn: a 30-year perspective. *Biol. Neonate* 88, 228, 2005.

63. Scaife, A.R. et al. Maternal intake of antioxidant vitamins in pregnancy in relation to maternal and fetal plasma levels at delivery. *Br. J. Nutr.* 95, 771, 2006.

64. Berger, H.M. et al. Extracellular defence against oxidative stress in the newborn. *Semin. Neonatol.* 3, 183, 1998.

65. Rogers, M.S. et al. Lipid peroxidation in cord blood at birth: the effect of labour. *Br. J. Obstet. Gynaecol.* 105, 739, 1998.

66. Yaacobi, N., Ohel, G., and Hochman, A. Reactive oxygen species in the process of labor. *Arch. Gynecol. Obstet.* 263, 23, 1999.

67. Fainaru, O. et al. Active labour is associated with increased oxidisibility of serum lipids *ex vivo. Br. J. Obstet. Gynaecol.* 109, 938, 2002.

68. Buhimschi, I.A. et al. Beneficial impact of term labor: nonenzymatic antioxidant reserve in the human fetus. *Am. J. Obstet. Gynecol.* 189, 181, 2003.

69. Committee on Fetus and Newborn, American Academy of Pediatrics, and Committee on Obstetric Practice, American College of Obstetricians and Gynecologists. Use and abuse of the Apgar score. *Pediatrics* 98, 141, 1996.

70. Volpe, J.J. *Neurology of the Newborn*, W.B. Saunders, Philadelphia, 2001, p. 314.

71. Jassem, W. and Roake, J. The molecular and cellular basis of reperfusion injury following organ transplantation. *Transplantation Rev.* 12, 14, 1998.

72. Buonocore, G., Perrone, S., and Bracci, R. Free radicals and brain damage in the newborn. *Biol. Neonate* 79, 180, 2001.

73. Mishra, O.P. and Delivoria-Papadopoulos, M. Lipid peroxidation in developing fetal guinea pig brain during normoxia and hypoxia. *Brain Res. Dev. Brain Res.* 45, 129, 1989.

74. Berger, R. and Garnier, Y. Perinatal brain injury. *J. Perinat. Med.* 28, 261, 2000.

75. Bracci, R., Perrone, S., and Buonocore, G. Red blood cell involvement in fetal/neonatal hypoxia. *Biol. Neonate* 79, 210, 2001.

76. Fellman, V. and Raivio, K.O. Reperfusion injury as the mechanism of brain damage after perinatal asphyxia. *Pediatr. Res.* 41, 599, 1997.

77. Kattwinkel, J. et al. Resuscitation of the newly born infant: an advisory statement from the Pediatric Working Group of the International Liaison Committee on Resuscitation. *Resuscitation* 40, 71, 1999.

78. Saugstad, O.D. Resuscitation with room-air or oxygen supplementation. *Clin. Perinatol.* 25, 741, 1998.

79. Rootwelt, T. et al. Hypoxemia and reoxygenation with 21% or 100% oxygen in newborn pigs: changes in blood pressure, base deficit, and hypoxanthine and brain morphology. *Pediatr. Res.* 32, 107, 1992.

80. Rootwelt, T. et al. Cerebral blood flow and evoked potentials during reoxygenation with 21 or 100% O2 in newborn pigs. *J. Appl. Physiol.* 75, 2054, 1993.

81. Vento, M. et al. Resuscitation with room air instead of 100% oxygen prevents oxidative stress in moderately asphyxiated term neonates. *Pediatrics* 107, 642, 2001.

82. Saugstad, O.D., Rootwelt, T., and Aalen, O. Resuscitation of asphyxiated newborn infants with room air or oxygen: an international controlled trial. *Pediatrics* 102, E1, 1998.

83. Saugstad, O.D. et al. Resuscitation of newborn infants with 21% or 100% oxygen: follow-up at 18 to 24 months. *Pediatrics* 112, 296, 2003.

84. Saugstad, O.D. Bronchopulmonary dysplasia, oxidative stress and antioxidants. *Semin. Neonatol.* 8, 39, 2003.

85. Phylactos, A.C. et al. Erythrocyte cupric/zinc superoxide dismutase exhibits reduced activity in preterm and low-birthweight infants at birth. *Acta Paediatr.* 84, 1421, 1995.

86. Grigg, J., Barber, A., and Silverman, M. Bronchoalveolar lavage fluid glutathione in intubated premature infants. *Arch. Dis. Child.* 69, 49, 1993.

87. Jain, A. et al. Glutathione metabolism in newborns: evidence for glutathione deficiency in plasma, bronchoalveolar lavage fluid, and lymphocytes in prematures. *Pediatr. Pulmonol.* 20, 160, 1995.

88. Baydas, G. et al. Antioxidant vitamin levels in term and preterm infants and their relation to maternal vitamin status. *Arch. Med. Res.* 33, 276, 2002.

89. Galinier, A. et al. Reference range for micronutrients and nutritional marker proteins in cord blood of neonates appropriated for gestational ages. *Early Hum. Dev.* 81, 583, 2005.

90. Frank, L. and Sosenko, I.R. Failure of premature rabbits to increase antioxidant enzymes during hyperoxic exposure: increased susceptibility to pulmonary oxygen toxicity compared with term rabbits. *Pediatr. Res.* 29, 292, 1991.

91. Bancalari, E. and Gerhardt, T. Bronchopulmonary dysplasia. *Pediatr. Clin. North Am.* 33, 1, 1986.

92. Northway, W.H., Jr. and Rosan, R.C. Radiographic features of pulmonary oxygen toxicity in the newborn: bronchopulmonary dysplasia. *Radiology* 91, 49, 1968.

93. Bonta, V.W., Gawron, E.R., and Warshaw, J.B. Neonatal red cell superoxide dismutase enzyme levels: possible role as a cellular defense mechanism against pulmonary oxygen toxicity. *Pediatr. Res.* 11, 754, 1977.

94. Abman, S.H. and Groothius, J.R. Pathophysiology and treatment of bronchopulmonary dysplasia: current issues. *Pediatr. Clin. North Am.* 41, 277, 1994.

95. Gaynon, M.W. and Stevenson, D.K. What can we learn from STOP-ROP and earlier studies? *Pediatrics* 105, 420, 2000.

96. Van Marter, L.J. et al. Do clinical markers of barotrauma and oxygen toxicity explain interhospital variation in rates of chronic lung disease? *Pediatrics* 105, 1194, 2000.

97. Ogihara, T. et al. Raised concentrations of aldehyde lipid peroxidation products in premature infants with chronic lung disease. *Arch. Dis. Child. Fetal Neonatal Ed.* 80, F21, 1999.

98. Saugstad, O.D. Chronic lung disease: the role of oxidative stress. *Biol. Neonate* 74, Suppl. 1, 21, 1998.

99. van Klaveren, R.J., Demedts, M., and Nemery, B. Cellular glutathione turnover *in vitro*, with emphasis on type II pneumocytes. *Eur. Resp. J.* 10, 1392, 1997.

100. Robbins, C.G. et al. Recombinant human superoxide dismutase reduces lung injury caused by inhaled nitric oxide and hyperoxia. *Am. J. Physiol.* 272, L903, 1997.

101. Rosenfeld, W.N. et al. Safety and pharmacokinetics of recombinant human superoxide dismutase administered intratracheally to premature neonates with respiratory distress syndrome. *Pediatrics* 97, 811, 1996.

102. Davis, J.M. et al. Pulmonary outcome at 1 year corrected age in premature infants treated at birth with recombinant human CuZn superoxide dismutase. *Pediatrics* 111, 469, 2003.

103. Potter, C.F. et al. Effects of hyperoxia on nitric oxide synthase expression, nitric oxide activity, and lung injury in rat pups. *Pediatr. Res.* 45, 8, 1999.

104. Lipkin, P.H. et al. Neurodevelopmental and medical outcomes of persistent pulmonary hypertension in term newborns treated with nitric oxide. *J. Pediatr.* 140, 306, 2002.

105. Hamon, I. et al. Early inhaled nitric oxide improves oxidative balance in very preterm infants. *Pediatr. Res.* 57, 637, 2005.

106. Behrman, R.E. et al. *Nelson's Textbook of Pediatrics*, W.B. Saunders, Philadelphia, 2003.

107. Kosloske, A.M. Pathogenesis and prevention of necrotizing enterocolitis: a hypothesis based on personal observation and a review of the literature. *Pediatrics* 74, 1086, 1984.

108. Kosloske, A.M. Epidemiology of necrotizing enterocolitis. *Acta Paediatr. Suppl.* 396, 2, 1994.

109. Bhatia, A.M., Feddersen, R.M., and Musemeche, C.A. Role of luminal nutrients in intestinal injury from mesenteric reperfusion and platelet-activating factor in the developing rat. *J. Surg. Res.* 63, 152, 1996.

110. Cueva, J.P. and Hsueh, W. Role of oxygen derived free radicals in platelet activating factor induced bowel necrosis. *Gut* 29, 1207, 1988.

111. Inder, T.E. et al. Lipid peroxidation as a measure of oxygen free radical damage in the very low birthweight infant. *Arch. Dis. Child. Fetal Neonatal Ed.* 70, F107, 1994.

112. Baud, O. et al. Glutathione peroxidase-catalase cooperativity is required for resistance to hydrogen peroxide by mature rat oligodendrocytes. *J. Neurosci.* 24, 1531, 2004.

113. Back, S.A. et al. Maturation-dependent vulnerability of oligodendrocytes to oxidative stress-induced death caused by glutathione depletion. *J. Neurosci.* 18, 6241, 1998.

114. Back, S.A. et al. Late oligodendrocyte progenitors coincide with the developmental window of vulnerability for human perinatal white matter injury. *J. Neurosci.* 21, 1302, 2001.

115. Houdou, S. et al. Developmental immunohistochemistry of catalase in the human brain. *Brain Res.* 556, 267, 1991.

116. Inder, T. et al. Elevated free radical products in cerebrospinal fluid of VLBW infants with cerebral white matter injury. *Pediatr. Res.* 52, 213, 2002.

117. Oka, A. et al. Vulnerability of oligodendroglia to glutamate: pharmacology, mechanisms, and prevention. *J. Neurosci.* 13, 1441, 1993.

118. Hardy, P. et al. Oxidants, nitric oxide and prostanoids in the developing ocular vasculature: a basis for ischemic retinopathy. *Cardiovasc. Res.* 47, 489, 2000.

119. Multicenter trial of cryotherapy for retinopathy of prematurity: natural history ROP: ocular outcome at 5(1/2) years in premature infants with birth weights less than 1251 g. *Arch. Ophthalmol.* 120, 595, 2002.

120. Ashton, N., Ward, B., and Serpell, G.. Effect of oxygen on developing retinal vessels with particular reference to the problem of retrolental fibroplasia. *Br. J. Ophthalmol.* 38, 397, 1954.

121. Stone, J. et al. Development of retinal vasculature is mediated by hypoxia-induced vascular endothelial growth factor (VEGF) expression by neuroglia. *J. Neurosci.* 15, 4738, 1995.

122. Carmeliet, P. Mechanisms of angiogenesis and arteriogenesis. *Nat. Med.* 6, 389, 2000.

123. Helmlinger, G. et al. Formation of endothelial cell networks. *Nature* 405, 139, 2000.

124. Smith, L.E. Pathogenesis of retinopathy of prematurity. *Semin. Neonatol.* 8, 469, 2003.

125. Cunningham, S. et al. Transcutaneous oxygen levels in retinopathy of prematurity. *Lancet* 346, 1464, 1995.

126. Saito, Y. et al. The progression of retinopathy of prematurity and fluctuation in blood gas tension. *Graefes Arch. Clin. Exp. Ophthalmol.* 231, 151, 1993.

127. Supplemental therapeutic oxygen for prethreshold retinopathy of prematurity (STOP-ROP): a randomized, controlled trial. I. Primary outcomes. *Pediatrics* 105, 295, 2000.

128. Raju, T.N. et al. Vitamin E prophylaxis to reduce retinopathy of prematurity: a reappraisal of published trials. *J. Pediatr.* 131, 844, 1997.

129. Anderson, G.H. Human milk feeding. *Pediatr. Clin. North Am.* 32, 335, 1985.

130. Goldman, A.S., Goldblum, R.M., and Ganson, L.A. Anti-inflammatory systems in human milk. *Adv. Exp. Med. Biol.* 262, 69, 1990.

131. Buescher, E.S. and McIlheran, S.M. Antioxidant properties of human colostrum. *Pediatr. Res.* 24, 14, 1988.

132. L'Abbe, M.R. and Friel, J.K. Superoxide dismutase and glutathione peroxidase content of human milk from mothers of premature and full-term infants during the first 3 months of lactation. *J. Pediatr. Gastroenterol. Nutr.* 31, 270, 2000.

133. Hamosh, M. Bioactive factors in human milk. *Pediatr. Clin. North Am.* 48, 69, 2001.

134. Friel, J.K. et al. Milk from mothers of both premature and full-term infants provides better antioxidant protection than does infant formula. *Pediatr. Res.* 51, 612, 2002.

135. Shoji, H. et al. Effects of human milk and spermine on hydrogen peroxide-induced oxidative damage in IEC-6 cells. *J. Pediatr. Gastroenterol. Nutr.* 41, 460, 2005.

136. Zoeren-Grobben, D. et al. Postnatal changes in plasma chain-breaking antioxidants in healthy preterm infants fed formula and/or human milk. *Am. J. Clin. Nutr.* 60, 900, 1994.

137. Shoji, H. et al. Effect of human breast milk on urinary 8-hydroxy-2'-deoxyguanosine excretion in infants. *Pediatr. Res.* 53, 850, 2003.

138. Shoji, H. et al. Suppressive effects of breast milk on oxidative DNA damage in very low birthweight infants. *Arch. Dis. Child. Fetal Neonatal Ed.* 89, F136, 2004.

139. Lucas, A. and Cole, T.J. Breast milk and neonatal necrotising enterocolitis. *Lancet* 336, 1519, 1990.

140. Watkins, C.J., Leeder, S.R., and Corkhill, R.T. Relationship between breast and bottle feeding and respiratory illness in the first year of life. *J. Epidemiol. Commun. Health* 33, 180, 1979.

141. Hylander, M.A. et al. Association of human milk feedings with a reduction in retinopathy of prematurity among very low birthweight infants. *J. Perinatol.* 21, 356, 2001.

142. Koletzko, B., Decsi, T., and Sawatzki, G.. Vitamin E status of low birthweight infants fed formula enriched with long-chain polyunsaturated fatty acids. *Int. J. Vitam. Nutr. Res.* 65, 101, 1995.

143. Nair, V. et al. The chemistry of lipid peroxidation metabolites: crosslinking reactions of malondialdehyde. *Lipids* 21, 6, 1986.

144. Decsi, T. and Koletzko, B. Growth, fatty acid composition of plasma lipid classes, and plasma retinol and alpha-tocopherol concentrations in full-term infants fed formula enriched with omega-6 and omega-3 long-chain polyunsaturated fatty acids. *Acta Paediatr.* 84, 725, 1995.

145. Decsi, T., Burus, I., and Koletzko, B. Effects of dietary long-chain polyunsaturated fatty acids on plasma amino acids and indices of protein metabolism in infants: results from a randomized clinical trial. *Ann. Nutr. Metab.* 42, 195, 1998.

146. Decsi, T., Burus, I., and Koletzko, B. Effects of dietary long-chain polyunsaturated fatty acids on plasma amino acids and indices of protein metabolism in infants: results from a randomized clinical trial. *Ann. Nutr. Metab.* 42, 195, 1998.

6 Maternal Obesity, Glucose Intolerance, and Inflammation in Pregnancy

Janet C. King

CONTENTS

ABSTRACT

The prevalence of obesity among pregnant women is at an all-time high. Maternal obesity increases the risks of mortality and morbidity in the mother and baby. The risk of gestational diabetes, one of the most prevalent metabolic complications in pregnancy, is significantly greater among obese women. Emerging research suggests an association of obesity, inflammation, and insulin resistance in non-pregnant and pregnant individuals. Cytokines secreted by the placenta, such as tumor necrosis factor-α and leptin, may mediate that link in pregnancy. Studies of maternal body mass index (BMI), insulin resistance, and circulating levels of cytokines, i.e., tumor necrosis factor-α and C-reactive protein, show that maternal obesity is associated with increased levels of inflammatory markers and that elevated levels of these markers are linked to glucose intolerance in women. These metabolic adjustments may exacerbate fetal overgrowth and excessive deposition of fat stores. Currently, obese pregnant women do not receive any guidance for reducing the risk of developing a pro-inflammatory, insulin-resistant state. Preliminary studies suggest that moderate physical activity and consuming a diet high in fiber and a higher proportion of polyunsaturated fatty acids may be beneficial.

INTRODUCTION

The prevalence of obesity among women of reproductive age has reached an all-time high. National survey data show that the number of women with body mass indexes (BMIs) greater than 30 averaged 31% in white women, 40% in Hispanic women, and 51% in black women in 1999 and 2000.[1] Obese women encounter more health problems during pregnancy.[1] Common disorders include pregnancy-induced hypertension, pre-eclampsia, large-for-gestational age (LGA) babies, need for cesarean or assisted deliveries, and post-delivery infections. The risk of gestational diabetes mellitus (GDM) is about three- to four-fold higher among obese compared to lean women.[2,3] The incidence of impaired glucose tolerance (IGT), i.e., one abnormal blood glucose value after an oral glucose load, is likely to be more frequent than GDM in obese women.

The prevalence of GDM varies widely within and between populations. Although the U.S. national average is about 4%,[4] the prevalence is as high as 16% in some high-risk populations. As with type 2 diabetes, the prevalence varies by race, ethnicity, and BMI. For example, in a study of 28,330 women from the Northern California Kaiser Permanente Medical Care Program, 7.4% of the Asian women, 5.6% of the Hispanic women, 4% of the African-American women, and 3.9% of the white women developed GDM.[5] The rates among American Indian mothers were considerably higher: 15.1% among Zuni women and 10.4% among Navajo women.[4] Some of the differences due to race or ethnicity reflect the higher prevalence of obesity in certain populations. A study of 552 African-American women and 653 Latina women in Detroit showed that very obese Latina women (BMIs >35) demonstrated a 6.5-fold increased risk for developing GDM compared to normal weight Latina women; very obese African-American women had nearly a four-fold increased risk compared to normal weight women.[6] Although the risk of GDM increased with maternal body weights in both ethnic groups, the increase among Latina women was about 2.5 times greater than that in African-American women.

OBESITY, INFLAMMATION, AND INSULIN RESISTANCE

Obesity in non-pregnant adults is associated with subclinical inflammation and insulin resistance.[7] The inflammatory and insulin-resistant states arise from changes in cellular and molecular functions and metabolism when adipocytes become enlarged in obese individuals. Perlipin, a phosphoprotein on the surfaces of triglyceride droplets that acts as a gatekeeper preventing lipases from

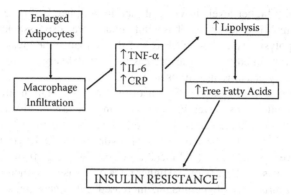

FIGURE 6.1 Relationship of obesity, inflammation, and insulin resistance. Enlarged adipocytes in the adipose tissues of obese individuals cause enhanced production of pro-inflammatory cytokines that increase rates of lipolysis, circulating levels of free fatty acids, and subsequent tissue insulin resistance.

hydrolyzing triglycerides and releasing free fatty acids, declines when adipocytes become enlarged, leading to an increased release of free fatty acids.[7] In turn, free fatty acids promote insulin resistance by stimulating the phosphorylation of insulin-receptor substrate-1 (IRS-1) on its serine residues instead of the usual phosphorylation of tyrosine residues.[8] Once serine is phosphorylated, IRS-1 becomes a poor substrate for the insulin receptor and insulin sensitivity declines. Cytokines, such as tumor necrosis factor-alpha (TNF-α) and interleukin-6 (IL-6) also stimulate IRS-1 phosphorylation on the serine residues.

Obesity appears to predispose individuals to a pro-inflammatory state because enlarged fat stores promote the invasion of macrophages into adipose tissue to scavenge moribund adipocytes that tend to increase with obesity[7] (Figure 6.1). Thus, obese individuals often have increased circulating levels of cytokines such as TNF-α and IL-6. The cause of the increased number of moribund adipocytes in obese individuals is unknown, but one hypothesis is that clusters of adipocytes distant from capillary networks experience hypoxia before angiogenesis occurs.[9] Cytokine secretion by macrophages also appears to be potently stimulated by acute-phase serum amyloid A; the decline in circulating levels of this protein after weight loss by obese individuals is associated with a reduction in circulating cytokine levels.[10]

The mild inflammatory state resulting from the increased secretion of cytokines by enlarged adipocytes contributes to the metabolic adjustments and insulin resistance associated with obesity. Some cytokines are thought to reduce adiponectin expression,[7] and serum adiponectin levels tend to fall with obesity.[11] Adiponectin is a potent inhibitor of TNF-α-induced monocyte adhesion, which may explain in part the link between obesity and cardiovascular disease.

TNF-α also increases adipocyte lipolysis, possibly by reducing perlipin[7] and thereby promoting the release of free fatty acids into circulation and reducing insulin sensitivity. IL-6 expression is also increased in the adipocytes of obese individuals with higher levels in visceral than in peripheral adipose tissue.[7] The elevated plasma IL-6 levels in obese individuals are also associated with an increase in lipolysis. Thus, both TNF-α and IL-6 enhance the breakdown of fats to free fatty acids in adipose tissue, increasing their delivery to liver and muscle and subsequently causing insulin resistance.

Endogenous overproduction of cortisol or exogenous administration of cortisteroids generally causes weight gain with an increase in visceral fat stores compared with peripheral stores. However, obese people usually do not have higher levels of cortisol.[7] The activity of 11β-hydroxysteroid dehydrogenase type 1, which converts inactive cortisol metabolites into active cortisol, is elevated in the adipose tissues of obese people. Higher rates of cortisol production in adipose tissue may contribute to hyperphagia, increased cytokine expression, gain in visceral adipose tissue, hyperlipidemia, and insulin resistance in obese individuals without any increases in circulating cortisol concentrations.

Clearly, adipocytes are not merely storage reservoirs for fat, but they are also endocrine organs with multiple functions. Their metabolic function changes as they enlarge with increasing obesity. Enlarged adipocytes recruit macrophages, promote inflammation, and enhance the secretion of a variety of metabolites that predispose toward insulin resistance. An active area of research is identification of therapies that alter adipose tissue metabolic pathways leading to inflammation and insulin resistance.

PREGNANCY: AN INFLAMMATORY, INSULIN-RESISTANT STATE

Profound metabolic adjustments occur during pregnancy to assure an adequate nutrient supply is available to support fetal growth. Since glucose is the preferred fuel of the fetus, maternal metabolism is shifted toward a hyperglycemic state. This ensures facilitated glucose diffusion from maternal circulation across the placenta to the fetus. Maternal hyperglycemia is created by establishing an insulin-resistant state. Maternal insulin resistance increases throughout gestation, reaching a peak in late gestation when fetal fuel demands are the highest.[12] The rise in insulin resistance is ascribed to alterations in maternal cortisol levels and placental hormones (human placental lactogen, progesterone, and estrogen).[12] However, the changes in insulin resistance have never been correlated with these hormonal changes in a prospective, longitudinal study.[13]

The recent evidence that adipokines such as TNF-α and leptin affect insulin sensitivity in non-pregnant individuals has led investigators to propose that similar mechanisms may occur in pregnancy. Pro-inflammatory cytokines may influence insulin metabolism in pregnant as well as non-pregnant women for several reasons. First, pregnant women with adequate food supplies gain fat during the first two trimesters.[1] The amounts gained vary from 0 to 10 kg and the average is about 3.5 kg. Expansion of the fat tissue and enlarged adipocytes may increase adipose tissue cytokine secretion and subsequent insulin resistance in pregnant women similar to that seen in non-pregnant individuals. Second, the placenta expresses all known cytokines including TNF-α, IL-6, IL-10, leptin, resistin, and PAI-1.[14]

The physiological role of these placental cytokines is uncertain, but many of the same cytokines are produced by the placenta as are secreted by adipose tissue. Possibly, the high placental cytokine secretion assists in creating the maternal insulin-resistant state. Since maternal fat gain is very limited or non-existent in pregnant women with limited energy intakes, the placenta is likely to be more important for creating a maternal insulin-resistant state than is an increase in maternal adipose tissue.

To test the hypothesis that placental cytokines play a role in modifying insulin sensitivity in pregnancy, Kirwan and co-workers measured circulating hormones, cytokines, and insulin resistance in 15 women (5 with GDM and 10 with normal glycemia) before pregnancy, at 12 to 14 weeks of gestation, and at 34 to 36 weeks of gestation.[13] Changes in insulin sensitivity were compared to placental hormones, cortisol, leptin, and TNF-α. TNF-α was more strongly associated with insulin sensitivity than any other hormone or cytokine measured; higher TNF-α levels were associated with lower insulin sensitivity ($r = -0.69$, $p < 0.006$). Furthermore, the *change* in TNF-α from pregravid to late pregnancy was the only factor significantly predicting the decline in insulin sensitivity ($r = -0.60$, $p < 0.02$). After adjusting for differences in body fat, TNF-α explained 21% of the variance in insulin sensitivity. None of the placental reproductive hormones was correlated with insulin sensitivity in late pregnancy. These data suggest that placental TNF-α plays a major role in causing the decline of insulin sensitivity during pregnancy.

To further evaluate the effects of pro-inflammatory cytokines on insulin resistance during pregnancy, Radaelli and co-workers measured the effect of GDM on the expression of placental genes. Placentas were collected from 15 women at the time of cesarean section (7 GDM women and 8 controls). A total of 22,823 gene sequences were surveyed; 435 genes were significantly modified in GDM placenta. Of those 435 genes modified in the placentas of GDM women, 18% or the largest cluster were related to inflammatory responses. Interleukins, leptin, TNF-α, and their downstream molecular adaptors were up-regulated. These findings confirm that placentas from diabetic mothers create an inflammatory milieu that enhances insulin resistance. These placental adjustments in gene expression may also cause adverse fetal programming.

Studies in non-pregnant individuals show that proinflammatory cytokines promote lipolysis in adipose tissue, enhance the delivery of free fatty acids to liver and muscle, and thereby contribute to insulin resistance. In pregnancy, plasma free fatty acid levels increased, reaching their highest levels in late pregnancy when insulin resistance also peaked.[15] This rise in free fatty acids may be linked to the rise in placental cytokines in maternal circulation. However, lipolysis and the release of free fatty acids from maternal adipose tissue were also enhanced by placental hormones, specifically human placental lactogen.[16]

In fact, artificially elevating plasma free fatty acids concentrations by infusing lipid during euglycemic clamp studies of pregnant women at 14 to 17 weeks of gestation inhibited total body glucose uptake and oxidation to rates similar to those observed in late pregnancy.[16] This confirmed that the levels of free fatty acids play an important role in regulating insulin sensitivity in pregnancy. Recent research suggests that the circulating levels of cytokines secreted by the placenta may mediate shifts in free fatty acid concentrations and subsequent insulin resistance.

MATERNAL OBESITY AND INFLAMMATION

Studies of inflammation and insulin resistance in pregnancy were performed in non-obese women.[13,17] Since obesity precipitates inflammatory responses, excessive free fatty acid release, and subsequent insulin-resistant states in non-pregnant individuals, it is reasonable to assume that inflammatory responses and insulin resistance would be enhanced in obese compared to lean pregnant women. Comparisons of metabolic adjustments in lean and obese pregnant women are limited, but the few studies done show that obese women rely more on fat oxidation as a source of energy in late pregnancy than do lean women.[18,19]

The increase in fat oxidation among the obese women was significantly correlated with serum leptin concentrations (r = 0.76, p <0.005).[19,20] Longitudinal changes in insulin sensitivity were also compared among women with BMIs <25, between 25 and 30, and ≥30.[21] Although 50 to 60% decreases in insulin sensitivity were noted in all groups from before conception through late pregnancy (p <0.0001), the obese subjects were significantly less insulin-sensitive or more insulin-resistant than the lean women (p <0.0001) and overweight women (p <0.004), particularly prior to conception or at 12 to 14 weeks of gestation.

We studied changes in insulin sensitivity in 8 lean women (body fat <30%) and 12 obese women (body fat ≥30%) followed from 12 to 34 weeks of gestation[22] (Figure 6.2). Insulin sensitivity was estimated from serum glucose and insulin

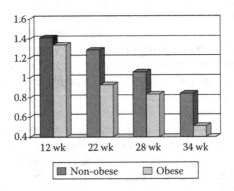

FIGURE 6.2 Decline in insulin sensitivity in lean and obese pregnant women. Insulin sensitivity was estimated from post-oral glucose tolerance test serum glucose and insulin levels[23] in 8 lean and 12 obese (body fat >30%) women at 12, 22, 28, and 34 weeks of gestation. The decline in insulin sensitivity was greater in obese than in lean women between 12 and 34 weeks, p <0.005.

concentrations following oral glucose tolerance tests.[23] Insulin sensitivity declined by 34% in the non-obese group from 12 to 34 weeks and by 60% in the obese group; the change in the two groups did not achieve statistical significance (p = 0.09). Insulin sensitivity did not differ between the two groups at 12 weeks of gestation, but it was significantly lower in the obese than in the non-obese at 34 weeks (p <0.005), reflecting the larger drop in insulin sensitivity in obese compared to lean women during gestation.

Since these preliminary studies showed that insulin sensitivity was lower in obese than in non-obese pregnant women, we studied the relationship between BMI and maternal cytokine levels in 51 women at 28 weeks' gestation.[24] At the beginning of the third trimester, the BMI values of these women averaged 32 ± 9 kg/m^2; the range was 21 to 54 kg/m^2. BMI correlated with C-reactive protein (CRP) levels (r = 0.53; p <0.001) and leptin (r = 0.79; p <0.001). The relationship with TNF-α was borderline (r = 0.26; p <0.07). Thus, obese pregnant women, like obese non-pregnant women, tend to have higher circulating levels of pro-inflammatory cytokines.

The women with higher BMIs in this study also had higher levels of fasting glucose (r = 0.53; p <0.001), fasting insulin (r = 0.62, p <0.001), and homeostasis model assessment (HOMA) — an estimate of insulin resistance calculated from the fasting glucose:insulin ratio — (r = 0.64, p <0.001). Serum adiponectin levels were negatively associated with BMI (r = –0.32, p = 0.02). Women with higher BMIs at the beginning of the third trimester of pregnancy showed higher circulating levels of inflammatory markers, hyperglycemia, hyperinsulinemia, hyperleptinemia, hypoadiponectinemia, and insulin resistance. These findings are consistent with studies of pro-inflammatory cytokines, adipocyte hormones, and glucose tolerance in non-pregnant, obese women.[7]

In a study of 24 lean women (BMI = 22.1 kg/m^2) and 23 obese women (BMI = 31.0 kg/m^2), Ramsay and co-workers found that both CRP and leptin concentrations were about twice as high in the obese compared to lean women.[25] The logs of CRP and of leptin were strongly related with the log of fasting insulin (CRP: r = 0.47, p <0.001; leptin: r = 0.74, p <0.001). Furthermore, a step-wise regression model showed that both leptin and BMI were independent predictors of insulin; together, they explained 57% of the insulin variability, suggesting that body fatness and adipocyte hormones work together to create a maternal insulin-resistant state in pregnancy.

Our studies and that of Ramsay[25] are limited by the fact that they are cross-sectional in nature. Longitudinal, prospective data are needed to comprehensively evaluate the relationship of maternal obesity, inflammation, and glycemic control during pregnancy. Nevertheless, preliminary data demonstrate that obesity in pregnancy is associated with marked hyperinsulinemia (in advance of gestational diabetes) and inflammatory up-regulation. Such metabolic perturbations may not only contribute to GDM and other maternal complications such as pre-eclampsia but also affect placental function and fetal growth and development.

MATERNAL INFLAMMATION, INSULIN RESISTANCE, AND FETAL GROWTH

Obese women tend to have big babies irrespective of the amount of weight they gain.[26] Also, maternal glucose metabolism and adiposity are highly correlated with fetal growth and body composition.[21,27] Although gestational age at birth is the strongest predictor of both birth weight and infant fat-free mass, maternal pregravid BMI is the strongest predictor of infant fat mass ($r^2 = 0.066$), explaining about 7% of the variance in body fat of newborns.

Maternal diabetes is a strong predictor of fetal overgrowth, with diabetic women facing a three-fold higher risk of having an LGA baby than obese women. Four times as many LGA infants are born to obese women than to diabetic women because there are many more obese women than diabetic women. Also, diabetic women usually are carefully monitored by their health care providers during pregnancy whereas no special care is provided to obese women. The additional care given to diabetic women undoubtedly reduces the number of LGA babies born to them. Recent studies from both North America and Europe report increases in mean birth weight over the past 30 years.[21] This rise in birth weight is likely related to the increasing incidence of maternal obesity. In fact, Catalano and his co-workers in Cleveland, Ohio found that a mean increase in birth weight of 116 g over the past 30 years was more strongly correlated with maternal weight at delivery than any other maternal characteristic.

Is this rise in birth weight linked to an increased maternal subclinical inflammation and insulin resistance associated with obesity? Preliminary data suggest that it may be.[28] For example, Radaelli and co-workers[29] measured circulating maternal and fetal cytokines and growth factors in three mother–infant cohorts divided into tertiles according to neonatal body fat. The only fetal factor associated with neonatal body fat was leptin (p <0.006), whereas the only maternal factor associated with neonatal body fat was IL-6 (p <0.01). None of the other inflammatory cytokines was elevated in the mothers at delivery. Possibly, the placenta releases some of the mediators of inflammation such as IL-6 into maternal circulation, which in turn blunts maternal insulin action and enhances the fetal fuel supply.

Studies of placenta from lean and obese (≥16% body fat) babies showed that the main placental phospholipases were increased in the obese compared to lean infants (p <0.05). Higher levels of phospholipases were associated with increased amounts of omega-3-derived metabolites in the placenta which may in turn modify the membrane lipid biolayers, membrane fluidity, and the efficiency of nutrient transfer. Studies in animals supported the conclusion that phospholipase influences adipose tissue metabolism and glucose tolerance. Mice with phospholipase A2 mutations that reduced their adipose tissues had higher levels of insulin sensitivity.[30]

INTERVENTIONS TO REDUCE MATERNAL INFLAMMATION AND INSULIN RESISTANCE

Studies in non-pregnant obese individuals provided evidence that changes in lifestyle (weight reduction, changes in dietary fat and fiber, and increased physical activity) reduced the risks of type 2 diabetes mellitus in obese individuals.[31] Weight loss is a very effective way to reduce circulating cytokine levels and fasting insulin concentrations in obese women[32] but weight loss is not indicated for pregnant women. However, modest reductions in the total amounts gained by obese women are appropriate. The Institute of Medicine recommends that obese women gain at least 15 lb during gestation; normal weight women are advised to gain 25 to 35 lb.[33]

Physical activity is an effective intervention for reducing the risk of type 2 diabetes and associated metabolic anomalies such as insulin resistance, oxidative stress, and dyslipidemia.[34] Physical activity activates the AMP-activated protein kinase (AMPK) enzyme, which increases glucose transport into the muscle, enhances fat oxidation, and reduces insulin resistance.[7]. Exercise, even intermittently, reduces the risk of GDM among obese women with BMIs >33 by nearly two-fold.[35] Women who exercise throughout pregnancy (i.e., perform endurance exercises ≥4 times/week) gain significantly less fat and had significantly lower increases in TNF-α and leptin during gestation.[36] The changes in leptin, but not TNF-α, were correlated with reduced fat mass in physically active women. Possibly, the differences in TNF-α levels reflect the exercise-induced reductions in insulin resistance whereas the leptin changes are more closely linked to fat accretion. Nevertheless, moderate physical activity during pregnancy may be an effective way to reduce subclinical inflammation and insulin resistance during pregnancy.

Changes in the types of dietary fat and carbohydrates consumed may be other ways to reduce maternal inflammation and insulin resistance. Decreases in total fat and saturated fat intakes and increases in dietary fiber are recommended for reducing the risk of type 2 diabetes mellitus in non-pregnant adults.[37] Results from the limited number of studies of pregnant women suggest that similar changes in dietary fats and carbohydrates also effectively reduced the risks of glucose intolerance. Clapp randomized 12 pregnant women to a low or high glycemic index diet prior to conception.[38] The amounts of carbohydrate consumed were similar in the two groups: 56% of the total energy.

During pregnancy, the women on the low glycemic diet showed no significant changes in their glycemic responses to mixed meals whereas the women on the high glycemic diet experienced 190% increases in their responses compared to pre-pregnancy values. These findings are similar to those reported by Fraser et al.[39] who showed that a high fiber diet reduced the post-prandial response to a meal in comparison to low fiber intake. Bronstein and co-workers have also shown that reducing the glycemic index of a test meal lowered post-prandial glucose and insulin responses in *both* lean and obese women studied in the third trimester[40] (Figure 6.3). These findings suggest that the maternal insulin-resistant state and

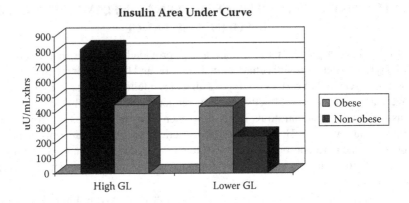

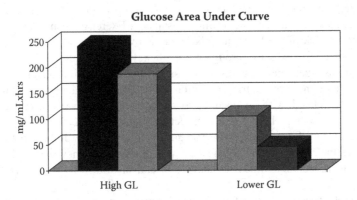

FIGURE 6.3 Serum insulin and glucose responses to lower glycemic test meals in lean and obese pregnant women. The areas under the curve for serum insulin and glucose were measured over a 2-hour period in 6 lean and 8 obese pregnant women studied at 32 to 36 weeks of gestation.[40]

hyperglycemic responses to meals reflect the intake of a Westernized, low fiber, high glycemic diet rather than a typical metabolic response to pregnancy.

Studies of non-pregnant, obese individuals also suggest that increasing the ratio of polyunsaturated to saturated fats in the diet may reduce the risk of developing metabolic syndrome and its complications as early as adolescence.[41] Similar findings have been reported for pregnant women. In a study of 171 pregnant Chinese women with or without impaired glucose tolerance, the type of dietary fat predicted impaired glucose tolerance and GDM.[42] In a logistic regression analysis, increased body weight, decreased polyunsaturated fat intake, and a low dietary polyunsaturated-to-saturated fat ratio independently predicted glucose intolerance. Bo and co-workers also found that glucose intolerance in pregnant women without conventional risk factors (i.e., family history, age, and

BMI) was related to the percent of saturated and polyunsaturated fats in the diet with high intakes of saturated fat increasing the risk and high intakes of poly-unsaturated fat decreasing the risk.[43]

The conventional dietary treatment of women with GDM is to reduce the amount of dietary carbohydrate and increase slightly the amount of fat. The preliminary data reviewed here suggest that it is more important to consider the *types* of carbohydrates and fats rather than the amounts. Increasing dietary fiber (or lowering the glycemic index) and the proportion of polyunsaturated fatty acids may be effective interventions for reducing inflammation and insulin resistance in pregnancy.

CONCLUSIONS

Placental cytokine secretion induces a pro-inflammatory state during pregnancy. This metabolic state is exacerbated by additional cytokines secreted by enlarged adipocytes in obese pregnant women. The inflammatory state leads to an increase in circulating free fatty acids and an insulin-resistant state. This pro-inflammatory, insulin-resistant state in a mother may affect placental function and fetal growth and development. Accumulating evidence suggests that the fetal fuel supply is enhanced, leading to accelerated growth and excessive fetal fat stores. Currently, obese pregnant women are not given any specific advice for reducing the pro-inflammatory response and insulin resistance. However, studies in both non-pregnant and pregnant adults suggest that moderate physical activity and increased intakes of dietary fiber and the proportion of polyunsaturated fatty acids attenuate these metabolic abnormalities and thereby potentially improve preg-nancy outcomes.

ACKNOWLEDGMENTS

The author acknowledges the contributions of her research collaborators to the original, unpublished research findings included in this paper: Jessica DeHaene, Dina El Kady, James L. Graham, Peter J. Havel, Liza Kunz, Meredith Milet, Ratna Mukherjea, Kimber L. Stanhope, and Leslie R. Woodhouse.

REFERENCES

1. J.C. King. Maternal obesity, metabolism, and pregnancy outcomes. *Annu Rev Nutr* 2006.
2. L.C. Castro and Avina R.L. Maternal obesity and pregnancy outcomes. *Curr Opin Obstet Gynecol* 2002; 14: 601.
3. K.R. Andreasen, Andersen M.L., and Schantz A.L. Obesity and pregnancy. *Acta Obstet Gynecol Scand* 2004; 83: 1022.

4. G.L.A. Beckles and Thompson-Reid P.E. Diabetes and women's health across life states: a public health perspective (translation). Centers for Disease Control and Prevention, National Center for Chronic Disease Prevention and Health Promotion, Atlanta, 2001.

5. A. Ferrara et al. Prevalence of gestational diabetes mellitus detected by the National Diabetes Data Group or the Carpenter and Coustan plasma glucose thresholds. *Diabetes Care* 2002; 25: 1625.

6. E.C. Kieffer et al. Obesity and gestational diabetes among African-American women and Latinas in Detroit: implications for disparities in women's health. *J Am Med Womens Assn* 2001; 56: 181.

7. A.S. Greenberg and Obin M.S. Obesity and the role of adipose tissue in inflammation and metabolism. *Am J Clin Nutr* 2006; 83: 461S.

8. G.S. Hotamisligil. Inflammatory pathways and insulin action. *Int J Obes* 2003; 27: 553.

9. P. Trayhurn. Endocrine and signalling role of adipose tissue: new perspectives on fat. *Acta Physiol Scand* 2005; 184: 285.

10. R.Z. Yang et al. Acute-phase serum amyloid A: an inflammatory adipokine and potential link between obesity and its metabolic complications. *Plos Med* 2006; 3: 884.

11. P.J. Havel. Control of energy homeostasis and insulin action by adipocyte hormones: leptin, acylation stimulating protein, and adiponectin production. *Curr Opin Lipidol* 2002; 13: 51.

12. J.L. Kitzmiller. The endocrine pancreas and maternal metabolism, in D. Tulchinsky and Ryan K.J., Eds, *Maternal–Fetal Endocrinology*, W.B. Saunders, Philadelphia, 1980, p. 58.

13. J.P. Kirwan et al. TNF-α is a predictor of insulin resistance in human pregnancy. *Diabetes* 2002; 51: 2207.

14. S. Hauguel-de Mouzon and Guerre-Millo M. The placenta cytokine network and inflammatory signals. *Placenta* 2005.

15. R.L. Phelps, Metzger B.E., and Freinkel N. Carbohydrate metabolism in pregnancy. XVII. Diurnal profiles of plasma glucose, insulin, free fatty acids, triglycerides, cholesterol, and individual amino acids in late normal pregnancy. *Am J Obstet Gynecol* 1981; 140: 730.

16. E. Sivan et al. Free fatty acids and insulin resistance during pregnancy. *J Clin Endocrinol Metab* 1998; 83: 2338.

17. T. Radaelli et al. Gestational diabetes induces placental genes for chronic stress and inflammatory pathways. *Diabetes* 2003; 52: 2951.

18. P.M. Catalano et al. Carbohydrate metabolism during pregnancy in control subjects and women with gestational diabetes. *Am J Physiol* 1993; 264: E60.

19. N.C. Okereke et al. Longitudinal changes in energy expenditure and body composition in obese women with normal and impaired glucose tolerance. *Am J Physiol Endocrinol Metab* 2004; 287: E472.

20. N.C. Okereke et al. Longitudinal changes in energy expenditure and body composition in obese women with normal and impaired glucose tolerance. *Am J Physiol* 2004, publication.

21. P. Catalano and Ehrenberg H. The short- and long-term implications of maternal obesity on the mother and her offspring. *Bjog* 2006.

22. K.E. Lang. Maternal metabolic and pregnancy outcomes in obese and non-obese women. Unpublished data, University of California, Davis, 2006.

23. M. Matsuda and DeFronzo R.A. Insulin sensitivity indices obtained from oral glucose tolerance testing: comparison with the euglycemic insulin clamp. *Diabetes Care* 1999; 22: 1462.

24. L. Kunz. Circulating levels of cytokines during the third trimester of pregnancy. Unpublished data, University of California, Davis, 2006.

25. J.E. Ramsay et al. Maternal obesity is associated with dysregulation of metabolic, vascular, and inflammatory pathways. *J Clin Endocrinol Metab* 2002; 87: 4231.

26. B.F. Abrams and Laros R.K., Jr. Pre-pregnancy weight, weight gain, and birth weight. *Am J Obstet Gynecol* 1986; 154: 503.

27. P.M. Catalano et al. Increased fetal adiposity: a very sensitivie marker of abnormal *in utero* development. *Am J Obstet Gynecol* 2003; 189: 1698.

28. P.M. Catalano. Obesity and pregnancy: the propagation of a viscous cycle? (editorial). *J Clin Endocrinol Metab* 2003; 88: 3505.

29. T. Radaelli et al. Maternal interleukin-6: marker of fetal growth and adiposity. *J Soc Gynecol Investig* 2006; 13: 53.

30. K.W. Huggins, Boileau A.C., and Hui D.Y. Portection against diet-induced obesity and obesity-related insulin resistance in group 1B PLA-2-deficient mice. *Am J Physiol Endocrinol Metab* 2002; 283: E994.

31. J. Tuomilehto et al. Prevention of type 2 diabetes mellitus by changes in lifestyle among subjects with impaired glucose tolerance. *New Engl J Med* 2001; 344: 1343.

32. P Ziccardi et al. Reduction of inflammatory cytokine concentrations and improvement of endothelial functions in obese women after weight loss over one year. *Circulation* 2002; 105: 804.

33. Food and Nutrition Board Institute of Medicine. Nutrition during Pregnancy. Part I. Weight Gain. Part II. Nutrient Supplements. National Academy Press, Washington, 1990.

34. J.C. Dempsey et al. Prospective study of gestational diabetes mellitus risk in relation to maternal recreational physical activity before and during pregnancy. *Am J Epidemiol* 2004; 159: 663.

35. T.D. Dye et al. Physical activity, obesity, and diabetes in pregnancy. *Am J Epidemiol* 1997; 146: 961.

36. J.F. Clapp, 3rd and Kiess W. Effects of pregnancy and exercise on concentrations of the metabolic markers tumor necrosis factor alpha and leptin. *Am J Obstet Gynecol* 2000; 182: 300.

37. U.S. Department of Health and Human Services and U.S. Department of Agriculture, *Dietary Guidelines for Americans*, U.S. Government Printing Office, Washington, 2005.

38. J.F. Clapp, 3rd. Effect of dietary carbohydrate on the glucose and insulin response to mixed caloric intake and exercise in both nonpregnant and pregnant women. *Diab Care* 1998; 21: B107.

39. R.B. Fraser, Ford F.A., and Lawrence G.F. Insulin sensitivity in third trimester pregnancy: a randomized study of dietary effects. *Br J Obstet Gynaecol* 1988; 95: 223.

40. M.N. Bronstein, Mak R.P., and King J.C. The thermic effect of food in normal weight and overweight pregnant women. *Br J Nutr* 1995; 75: 261.

41. C. Klein-Platat et al. Plasma fatty acid composition is associated with the metabolic syndrome and low-grade inflammation in overweight adolescents. *Am J Clin Nutr* 2005; 82: 1178.

42. Y. Wang et al. Dietary variables and glucose tolerance in pregnancy. *Diabetes Care* 2000; 23: 460.
43. S. Bo et al. Dietary fat and gestational hyperglycaemia. *Diabetologia* 2001; 44: 972.

7 Obesity, Nutrigenomics, Metabolic Syndrome, and Type 2 Diabetes

David Heber

CONTENTS

INTRODUCTION

The traditional medical paradigm for understanding diabetes mellitus has undergone a major shift in the past two decades with the discovery of the many immunological functions of fat cells, especially those located in the mesenteric or visceral fat of the abdomen. This shift in thinking occurred in the molecular–genetic era of the last decade, beginning with the discovery of leptin.

Once it was realized that leptin is also a cytokine that stimulates angiogenesis, the many connections between obesity, inflammation, and diabetes began to emerge. A large number of cytokines and chemokines originating in fat cells were subsequently discovered. Recently, our group has demonstrated that a small but significant 5% weight loss could lead to a much larger decrease in the levels of circulating C-reactive protein, a biomarker of inflammation. This connection of inflammation and obesity, leading in genetically susceptible individuals to diabetes mellitus type 2, provides a unifying pathophysiological mechanism for many of the co-morbid diseases associated with diabetes and the metabolic syndrome including cardiovascular disease, renal disease, liver disease, and hypertension.

OBESITY EPIDEMIC AND ITS SOLUTIONS

Obesity is defined as excess body fat, but the difficulties inherent in measuring body fat in tens of thousands of individuals have resulted in the substitution of a surrogate measure called body mass index (BMI), defined as weight in kilograms over height in meters squared. This is a regression equation that approximates excess body fat in large populations, but can be significantly erroneous in individuals such as athletes, whose excess weight is due to muscle, or in young sedentary women consuming inadequate dietary protein who may have low body weights with abnormally high percentages of body fat.

Practical methods such as bioelectrical impedance can be used to assess body fat in individuals in clinical settings, but the BMI has proven invaluable in population studies. Overweight, defined as a BMI of at least 25 but less than 30 kg/m², and obesity (used here to denote greater overweight), defined as a BMI of 30 kg/m² or greater, are major contributors to morbidity and mortality in the United States today. The risk for some of our most devastating diseases, especially cardiovascular disease and diabetes mellitus, is significantly correlated with an individual's BMI[1] and a huge portion of the country's healthcare resources are used to treat the consequences of overweight and obesity. The United States is in the midst of an explosive epidemic of obesity. Although the early maps from 1985 to 1990 present all states in shades of light blue, indicating obesity prevalences of less than 15%, the latest maps are almost entirely orange (20 to 24% prevalence), with an ominous streak of red (25% or more) arcing from Texas to Michigan through Appalachia.[2] (Maps are available at http://www.cdc.gov/nccdphp/dnpa/obesity/trend/maps/.)

Weight gain is primarily a matter of energy imbalance — energy intake that is greater than energy output. This imbalance is influenced by genetic, metabolic, behavioral, and environmental factors. Only the latter two factors could have changed enough in the last 20 years to cause the obesity epidemic, resulting in both an increase in average energy intake and a decrease in average energy expenditure. Energy intake has increased in the United States due to a greater availability of highly palatable and energy-dense foods, larger portions becoming the norm, more food consumed outside the home (where less heed is paid to nutrition), and greater consumption of high-sugar beverages and high-fat processed and fast foods.

Over the past 20 years, nutrition researchers and food scientists over-emphasized reducing fat consumption with less regard for total calories, resulting in an ineffective low-fat food campaign while the incidence of obesity continued to increase. A significant decrease in energy output has been associated with increased hours of television viewing, increased numbers of individuals in sedentary jobs, and the increased prevalence of labor-saving devices such as elevators, automobiles, and remote controls. Some experts hold that the current obesity epidemic is caused by relatively small but chronic energy imbalances. Because a pound of body fat represents 3500 stored calories, a chronic energy imbalance — intake over output — of only 100 calories per day will cause a weight gain

TABLE 7.1
Patient Assessment: Factors to Consider in Developing Weight Loss Strategies

Factor	Guidelines
Body mass index (BMI)	BMI 25.0 to 29.9 kg/m² (overweight)
	BMI ≥30 kg/m² (obese)
Waist circumference	Men: >102 cm (40 inches)
	Women: >88 cm (35 inches)
Risk status	High:
	Presence of coronary heart disease, atherosclerosis, type 2 diabetes, sleep apnea, OR any three of the following:
	Cigarette smoking
	Hypertension (systolic blood pressure ≥140 mm Hg or diastolic blood pressure ≥90 mm Hg)
	High LDL cholesterol (≥160 mg/dL)
	Low HDL cholesterol (<35 mg/dL)
	Impaired fasting glucose (110–125 mg/dL)
	Family history of premature coronary heart disease (definite myocardial infarction or sudden death before age 55 in male relative or before age 65 in female relative), age (≥45 years for men; ≥55 years or post-menopausal for women)
	Other risk factors:
	Physical inactivity, high serum triglycerides (>200 mg/dL)
Patient motivation	Reasons and motivation for weight reduction
	History of successful and unsuccessful weight loss attempts
	Support (family, friends, co-workers)
	Patient's understanding of causes and health risks of obesity
	Attitude toward and capacity for physical activity
	Time available for weight loss intervention
	Financial considerations

Adapted from NIH/NHLBI clinical guidelines on overweight and obesity. Body mass index (kg/m²) can be calculated from pounds and inches using the formula: BMI (kg/m²) = [weight (pounds)/height (inches)²] × 704.5. An interactive BMI calculator and BMI chart are included in the NIH/NHLBI report.

of 10 pounds a year. The only way to restore energy balance is to reduce energy intake and/or increase energy output.

The assessment process, as described in the National Institutes of Health/National Heart, Lung, and Blood Institute report titled *Clinical Guidelines on the Identification, Evaluation, and Treatment of Overweight and Obesity in Adults*, involves considering a patient's BMI, waist circumference, motivation to lose weight, and overall risk status (Table 7.1).

Weight reduction has well documented health benefits and can usually be achieved with a loss of 1 to 2 pounds per week (0.5 to 1.0 pounds for lower starting weights), requiring an energy deficit of 500 to 1000 calories per day for the obese or 300 to 500 calories per day for the overweight. Achieving this energy deficit requires a three-part strategy that includes dietary therapy, physical activity, and behavior therapy.

Dietary therapy may be the most controversial of these elements. Patients tend to diet to achieve weight loss and then stop when their goals are reached. However, obesity is a chronic disease that cannot be cured and must be controlled through permanent lifestyle changes. Indications of what comprises successful diet strategies have been gleaned from the National Weight Control Registry (NWCR), which is following more than 4500 adults who were able to maintain weight losses of at least 30 pounds for at least a year. The typical participant (with an average weight loss of 60 pounds maintained for 5 years) made a permanent change to a low-calorie diet, what is called "chronic restrained eating."[4]

There are many possible means to attaining a state of chronic restrained eating. Thus, it makes little difference which form of diet one chooses to follow, whether it involves restricted carbohydrates (e.g., Atkins), macronutrient balance (e.g., Zone), restricted fat (e.g., Ornish), or simply restricted portion sizes and calories (e.g., Weight Watchers); what matters primarily is whether one is able to adhere to the diet.[5] Once a weight goal is achieved, there is general agreement that an optimally healthy diet consisting of seven servings per day of colorful fruits and vegetables, whole grains, low fat protein sources, and limited amounts of refined carbohydrates should be consumed. Significant amounts of research also suggest that structured diet plans using meal replacements such as high protein shakes or frozen meals are helpful adjuncts to weight loss. The most important requirements for an effective diet are that it (1) establishes a calorie deficit, (2) is healthy, and (3) fits one's lifestyle.

The next component of a successful weight loss strategy is physical activity. In fact, a daily expenditure of 300 or more calories through physical activity may be the most important factor for maintaining weight loss according to the NWCR. To accomplish this, a formerly obese person would need to perform 60 to 90 minutes per day of moderate-intensity physical activity, equivalent to walking at 3.5 to 4.0 miles per hour. Lesser amounts are recommended to prevent excess weight gain or reduce the risk of chronic disease.

A popular and simple program that started in Japan is the 10,000 steps per day idea, which provides a concrete goal. Since the goal can be worked on throughout the day, this strategy encourages cumulative episodes of physical activity. A goal counted in daily steps has been shown to induce more total walking than a goal measured in, say, hours of brisk walking.[6]

Abdominal obesity due to excess visceral fat is associated with an increased risk of developing cardiovascular disease.[1,7] Moreover, excess visceral fat is linked to an increased risk of metabolic syndrome, which includes a greater risk of developing type 2 diabetes mellitus[3] with its associated cardiometabolic disorders.[4]

TYPE 2 DIABETES MELLITUS AND OBESITY: "DIABESITY"

The focus of physicians treating diabetes has been strictly on glucose control for much of the last century. In fact, the term *diabetes mellitus* derives from the Latin meaning "sweet urine." In Ayurvedic medicine, the term for diabetes (madhumeda) translates as "one whose urine attracts ants." This glucocentric model of diabetes evolved from observations in the 1920s that surgical removal of the pancreas in a dog led to diabetes mellitus. Insulin injections could then restore glucose metabolism.

In juvenile or type 1 diabetes mellitus patients with autoimmune destruction of the pancreas early in life, insulin treatment could delay or prevent many of the complications of this terrible disease such as blindness, renal failure, and need for limb amputations. However, this form of diabetes is much less common today, accounting for less than 5% of all diabetes cases. Today, over 95% of all diabetes is type 2 and it occurs in children and adolescents as well as adults. In an individual with a BMI of 30, the risk of diabetes type 2 is increased 60- to 80-fold in comparison to lean individuals. In contrast, the risk for heart disease is only 4- to 6-fold increased at a BMI of 30 (see Figure 7.1). The association of diabetes type 2 with obesity goes beyond typical risk factors and justifies naming the disease *diabesity*. While 10% of type 2 diabetes patients are said to be lean, this assessment is based on weight and not body composition; many of these individuals may have excess abdominal fat.

As type 2 diabetes mellitus gained recognition in the 1970s, its etiology remained poorly understood. The idea that insulin, by controlling complications, was central to therapy was simply applied to type 2 diabetes mellitus as it had been to type 1, with the assumption that the results would be the same. However, insulin treatment failed to reduce cardiovascular mortality.

The late onset of diabetes mellitus type 2, usually in the fourth to sixth decade of life, was attributed to aging of the beta cell within the pancreas which secretes insulin. The pathophysiologic progression of type 2 diabetes mellitus was first described by Drenick and Johnson. Over a 2- to 10-year period, their model predicted evolution from hyperinsulinemia with euglycemia to a condition of insulin deficiency and hyperglycemia as the beta cell was exhausted. Dr. Peter Butler recently described a likely cause of beta cell exhaustion by uncovering the role of an amyloid protein called insulin-associated polypeptide or IAPP.

METABOLIC SYNDROME: PRE-DIABETES OR CARDIOMETABOLIC RISK?

A particular cluster of risk factors that seems especially coherent and predictive of atherosclerotic cardiovascular disease (ASCVD) has been called the metabolic syndrome. It comprises abdominal obesity, atherogenic dyslipidemia, hypertension, and elevated plasma glucose, along with the pro-thrombotic and pro-inflammatory states. Identification of this syndrome has proven useful in unifying

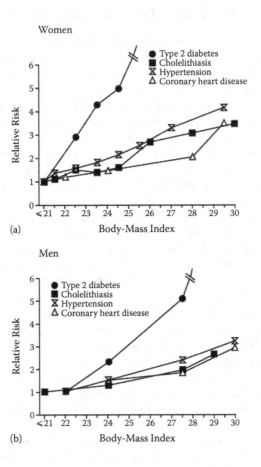

FIGURE 7.1 Relation of body mass index (BMI) and relative risk of diabetes, cholelithiasis, hypertension, and coronary heart disease. (Source: From Willett, W.C. et al. *New Engl J Med* 341, 427, 1999.)

medical approaches to the previously disparate realms of diabetes and cardiovascular disease.[8]

The National Heart, Lung, and Blood Institute in conjunction with the American Heart Association recently published updated guidelines for the diagnosis of metabolic syndrome[9] (Table 7.2). Note that elevated low-density lipoprotein cholesterol (LDL-C) is not listed as a component; this is because LDL-C levels are not generally divergent from normal levels in metabolic syndrome. However, LDL particles are seen to be smaller and denser in the syndrome, so that small, dense LDL (sdLDL) particle size is an experimental observation in metabolic syndrome but is not practical to assess clinically. The unifying framework for the relationships of obesity, metabolic syndrome, type 2 diabetes, and cardiovascular disease is hypothesized to be the release of free fatty acids (FFAs) preferentially

TABLE 7.2
Criteria for Diagnosing Metabolic Syndrome

Three or more of the following characteristics define metabolic syndrome:
1. Abdominal obesity: waist circumference >102 cm in men and >88 cm in women
2. Hypertriglyceridemia: ≥150 mg/dL (1.69 mmol/L)
3. Low high-density lipoprotein (HDL) cholesterol: <40 mg/dL (1.04 mmol/L) in men and <50 mg/dL (1.29 mmol/L) in women
4. High blood pressure: ≥130/85 mm Hg
5. High fasting glucose: ≥110 mg/dL (≥6.1 mmol/L)

According to Third Report of Expert Panel on Detection, Evaluation, and Treatment of High Blood Cholesterol in Adults (ATP III), National Cholesterol Education Program, National Heart, Lung, and Blood Institute, National Institutes of Health. NIH Publication 01-3670, May 2001.

from abdominal (visceral, as opposed to subcutaneous) fat stores, which then leads to insulin resistance at the liver and muscle as well as atherogenesis.[10]

Insulin resistance plays a fundamental role in the metabolic syndrome and its sequelae, but the ultimate failure of the ability of the beta cell to secrete adequate insulin leads to the decompensation of glucose control late in the course of the disease. The natural history of type 2 diabetes is a preexisting long-term insulin resistance for 2 to 10 years while within the failing beta cell there is an underlying progression from impaired glucose tolerance (IGT) to uncontrolled hyperglycemia. As indicated above, insulin resistance is thought to be induced by the excess FFAs associated with abdominal obesity, and is believed to underlie the constellation of CVD risk factors of metabolic syndrome including — besides hyperglycemia — dyslipidemia, hypertension, abnormal vascular function, vascular inflammation, and a prothrombotic state.[11]

NUTRIGENETICS OF TYPE 2 DIABETES MELLITUS

The metabolic syndrome and type 2 diabetes have underlying genetic predispositions that remain poorly understood. One candidate is the SEPS1 gene, first discovered in the Israeli sand rat which is highly adapted to starvation.[12] This gene codes for a protein that channels misfolded proteins out of the endoplasmic reticulum. When present in a non-functional polymorphic form, misfolded proteins accumulate in the endoplasmic reticulum, triggering apoptosis. The non-functional genetic polymorphisms of the SEPS1 gene are associated with increased expression of tumor necrosis factor (TNF) and interleukin-6 (IL-6) inflammatory cytokines in Mexican-American families.

The discovery of the human homolog based on a genetic polymorphism present in the Israeli sand rat points out the relationship of a predisposition to diabetes type 2 and being adapted to a brisk inflammatory response as well as starvation. Both the adaptation to starvation and the adaptation to its primary lethal complication — infection — have been essential to human survival over the past 100,000 years. It is likely that many forms of diabetes mellitus type 2 are associated with numerous genetic polymorphisms concerning both inflammation and the adaptation to starvation.

As of October 2005, 176 human obesity cases due to single-gene mutations in 11 different genes have been reported, 50 loci related to Mendelian syndromes relevant to human obesity have been mapped to a genomic region, and causal genes or strong candidates have been identified for most of these syndromes.[13]

There are 244 genes that, when mutated or expressed as transgenes in mice, result in phenotypes that affect body weight and adiposity. The number of studies reporting associations between DNA sequence variation in specific genes and obesity phenotypes has also increased considerably, with 426 findings of positive associations with 127 candidate genes. The nutrigenetics of diabetes mellitus type 2 and obesity will most likely be closely related, if not identical, in many individuals with the difference being that environmental influences uncover underlying type 2 diabetes mellitus.

The famous example of the Pima Indians in the Southwest United States points out the influence of environmental factors. Living only a few hundred miles apart, members of the same tribe were separated by the United States–Mexican border but were genetically similar. Members of the tribe living in the United States were forced onto a reservation in Arizona where they could not pursue their agricultural lifestyle. As a result, the age- and sex-adjusted prevalence of type 2 diabetes in the Mexican Pima Indians (6.9%) was less than one-fifth that in the U.S. Pima Indians (38%) and similar to that of non-Pima Mexicans (2.6%). The prevalence of obesity was similar in the Mexican Pima Indians (7% in men and 20% in women) and non-Pima Mexicans (9% in men and 27% in women) but was much lower than in the U.S. Pima Indians.[14] Other examples can be drawn both from immigrants to the U.S. or internationally in countries such as China, Japan, Korea, and Thailand where Western diets are rapidly replacing traditional Asian diets, leading to increases in obesity and diabetes.

While the treatment strategies for this vast complex of medical problems traditionally includes treatment of the complications (hypertension, dyslipidemia, and type 2 diabetes), or management of the ensuing coronary heart disease risk, it is a challenge to incorporate prevention and reduction of the "hypertriglyceridemic waist" into standard medical practice.

In patients who have already developed type 2 diabetes mellitus, a knowledge of proper blood glucose monitoring along with the ability to assess emotional status and garner adequate social support augment the efforts to increase physical activity and modify nutritional habits. Among the drugs used that can complement lifestyle change, metformin, which is off patent currently, is the drug of choice. Metformin blocks hepatic gluconeogenesis by an unknown mechanism, but has minimal effect on insulin resistance. It also reduces the prothrombotic state by an effect on plasminogen activator inhibitor type 1 (PAI-1), and its early use can prevent the progression from prediabetes to diabetes. Metformin's effect on CVD risk is uncertain, although it does cause a modest weight loss.

Drugs aimed at increasing insulin secretion (e.g., the sulfonylureas) have shown no significant impacts on CVD risk and may even increase risk after a myocardial infarction (MI).

Insulin limits markers of inflammation, reduces intimal medial thickness (IMT, a surrogate marker of CVD risk), and has been shown to lower risk after coronary artery bypass and to reduce post-MI mortality. On the whole, intensive treatment of hyperglycemia *per se* has been seen to have minimal effect on CVD risk.[15]

Besides insulin, metformin, and sulfonylureas, a new class of drug has emerged in the last decade, the thiazolidinediones (TZDs) or glitazones. These drugs are uniquely specific in reversing insulin resistance as agonists of the peroxisome proliferator-activated receptor gamma (PPAR-gamma). This is a nuclear receptor found in adipocytes and elsewhere that regulates the expression of genes involved in lipid and glucose metabolism, vascular function, thrombotic control, and inflammation. However, glitazones can also stimulate adipogenesis and so are used as second-line agents after metformin which encourages weight loss.

Exenatide is a new injectable treatment developed after observation of the "incretin effect" in which insulin levels were seen to rise significantly more after oral glucose than after an IV injection of glucose due to the action of a then-unidentified gut peptide.[16] Exenatide is a synthetic version of a salivary peptide in the Gila monster, and it mimics an endogenous incretin, glucagon-like peptide 1 in stimulating glucose-dependent insulin secretion, regulating gastric emptying, and inhibiting glucagon secretion, food intake, and acute plasma glucose. Exenatide is indicated as an adjunct to metformin and has been associated with weight loss as well.

Another new injectable treatment is pramlintide, a more soluble analog of amylin, which is co-secreted by the pancreatic gamma cell with insulin. Amylin, and thus pramlintide, promotes satiety, inhibits glucagon release, and delays gastric emptying, all through an effect via the central nervous system.[16]

BETA CELL FAILURE AND TYPE 2 DIABETES MELLITUS

Type 2 diabetes involves a progressive defect of insulin secretion that precedes the development of hyperglycemia.[17] This defect appears to be at least in part due to a deficit in beta cell mass.[18–20] Several therapeutic strategies now being proposed may reverse the defect in beta cell mass in people with type 2 diabetes, for example, glucagon-like peptide 1 or glucagon-like peptide 1–like surrogates.[21]

In humans there is a curvilinear relationship between the relative beta cell volume (and presumably the beta cell mass) and fasting blood glucose concentration. The present data reveal a narrow range of blood glucose over a wide range of fractional beta cell volume (up to ~10%) and then a much wider range of blood glucose values over a narrow range of volumes at low beta cell volumes, with the threshold set by the curve at ~1.1% defining that difference. These findings imply a much greater tolerance for variance in insulin sensitivity above this threshold and that below-the-threshold variance in insulin sensitivity and functional defects in insulin secretion have a much greater impact on blood glucose.

Autopsy studies involve some important limitations. The numbers of cases are frequently relatively small. The studies are inevitably cross-sectional and retrospective. Butler and others have presented data suggesting that the decline in beta cell mass in type 2 diabetes is caused by increased beta cell apoptosis.[19,22] Therefore, these data suggest that inhibition of beta cell apoptosis to avoid a beta cell deficit may be effective to delay and/or avoid the onset of diabetes. Indeed, both metformin and TZDs have been reported to inhibit beta cell apoptosis *in vitro*[22,23] and delay onset of type 2 diabetes in clinical studies.[24,25]

The potential mechanisms underlying increased beta cell apoptosis in type 2 diabetes include toxicity from islet amyloid polypeptide oligomer formation, free fatty acids (lipotoxicity), free oxygen radical toxicity, and, once hyperglycemia supervenes, glucose-induced apoptosis (glucotoxicity).[26,27] The steep increase in blood glucose concentration with beta cell deficiency is consistent with the well known deleterious effects of hyperglycemia *per se* on beta cell function. These include defective glucose sensing due to reduced glucokinase activity,[28] impaired glucose-induced insulin secretion due to increased uncoupling protein 2 activity,[29] and depletion of immediately secretable insulin stores.[30] Also, hyperglycemia reduces insulin sensitivity, further compounding the effects of decreased insulin secretion.[31] While these observations suggest that relatively small increases in beta cell mass may have useful actions in restoring blood glucose control, it is likely that aggressive normalization of blood glucose concentrations is required to accompany any strategy to increase beta cell mass to overcome the deleterious effects of hyperglycemia as well as glucose-induced beta cell apoptosis.

ADIPOKINES

Adipose tissue secretes bioactive peptides, termed *adipokines*, which act locally and distally through autocrine, paracrine and endocrine effects. In obesity, increased production of most adipokines impacts on multiple functions such as appetite and energy balance, immunity, insulin sensitivity, angiogenesis, blood pressure, lipid metabolism and hemostasis, all of which are linked with cardiovascular disease. Enhanced activities of TNF and IL-6 are involved in the development of obesity-related insulin resistance. Angiotensinogen has been implicated in hypertension and PAI-1 in impaired fibrinolysis.

Other adipokines like adiponectin and leptin, at least in physiological concentrations, are insulin sparing as they stimulate beta oxidation of fatty acids in skeletal muscle. The role of resistin is less understood. It is implicated in insulin resistance in rats, but probably not in humans. Reducing adipose tissue mass through weight loss in association with exercise can lower TNF-α and IL-6 levels and increase adiponectin concentrations, whereas drugs such as TZDs increase endogenous adiponectin production. In-depth understanding of the pathophysiology and molecular actions of adipokines may in the coming years lead to effective therapeutic strategies designed to protect against atherosclerosis in obese patients.

Obesity associated with unfavorable changes in adipokine expression such as increased levels of TNF-α, IL-6, resistin, PAI-1 and leptin, and reduced levels of adiponectin affects glycemic homeostasis, vascular endothelial function, and the coagulation system, thus accelerating atherosclerosis. Adipokines and a low-grade inflammatory state may be the link between the metabolic syndrome with its cluster of obesity and insulin resistance and cardiovascular disease.

In fact, atherosclerosis is now recognized as an inflammatory process of the arterial wall. Monocytes adhere to the endothelium and then migrate into the subendothelial space where they become foam cells loaded with oxidized lipoproteins. Foam cell production of metalloproteinases leads to rupture of the atherosclerotic plaque's fibrous cap and then to rupture of the plaque itself.[32] Thus, an inflammatory process accounts for both the development and evolution of atherosclerosis.

In this inflammatory process, adipokines play multiple roles. TNF-α activates the transcription factor nuclear factor-κβ, with subsequent inflammatory changes in vascular tissue. These include increased expression of intracellular adhesion molecule (ICAM)-1 and vascular cell adhesion molecule (VCAM)-1,[33,34] which enhances monocyte adhesion to the vessel wall, greater production of MCP-1 and M-CSF from endothelial cells and vascular smooth muscle cells,[35,36] and up-regulated macrophage expression of inducible nitric oxide (NO) synthase, interleukins, superoxide dismutase, etc.[37,38] Leptin, especially in the presence of high glucose, stimulates macrophages to accumulate cholesterol.[39] IL-6 exerts pro-inflammatory activity in itself and by increasing IL-1 and TNF-α.[40] Importantly, IL-6 also stimulates liver production of C-reactive protein which is considered a predictor of atherosclerosis.[41] IL-6 may also influence glucose tolerance by regulation of visfatin. Visfatin, a newly discovered adipocytokine in human visceral fat, exerts insulin-mimetic effects in cultured cells and lowers plasma glucose levels in mice through activation of the insulin receptor.[42]

PAI-1 concentrations regulated by transcription factor nuclear factor-κβ are abnormally high in hyperglycemia, obesity, and hypertriglyceridemia,[43] because of increased PAI-1 gene expression.[44] PAI-1 inhibits fibrin clot breakdown, thereby favoring thrombus formation upon ruptured atherosclerotic plaques.[45] In humans, circulating PAI-1 levels correlate with atherosclerotic events and mortality, and some studies suggest PAI-1 is an independent risk factor for coronary artery disease.[46] Angiotensinogen is a precursor of angiotensin II (AngII), which stimulates ICAM-1, VCAM-1, MCP-1, and M-CSF expression in vessel wall cells.[47] AngII also reduces NO bioavailability[48] with loss of vasodilator capacity and increased platelet adhesion to vessel walls.

In humans, endothelial dysfunction is indicative of the preclinical stages of atherosclerosis and is prognostic of future cardiovascular events.[49,50] High concentrations of pro-inflammatory adipokines may contribute to development of endothelial dysfunction. At this stage of disease, the role of resistin is particularly interesting. *In vitro* studies show resistin activates endothelial cells which, when incubated with recombinant human resistin, release more endothelin-1 and VCAM-1.[51] Recombinant human resistin is also reported to induce higher expression of mRNA of VCAM, ICAM-1, and pentraxin-3 from endothelial cells,[52] thus expressing a biochemical pattern of dysfunctional endothelium. Finally, resistin also induces proliferation of aortic smooth muscle cells.[53] In asymptomatic patients with family histories of coronary heart disease, plasma resistin levels are predictive of coronary atherosclerosis even after control for other established risk factors.[54,55]

CONCLUSION

The molecular effects of adipokines represent a challenging area of research and in-depth understanding of their pathophysiology and molecular actions will undoubtedly lead to the discovery of effective therapeutic interventions. Reducing adipose tissue mass and consequently adipokine concentrations will prevent the metabolic syndrome and, if the hypothesis of adipokine-related linkage with atherosclerosis is proven, help prevent the development of atherosclerosis. Despite the new findings in the field of adipokines, researchers are still led to focus back on obesity as an essential primary target in the continued effort to reduce the risk of developing the metabolic syndrome and type 2 diabetes with its associated cardiovascular complications.

REFERENCES

1. Wilson PWF et al. Overweight and obesity as determinants of cardiovascular risk: the Framingham experience. *Arch Intern Med* 2002; 162: 1867.
2. Centers for Disease Control and Prevention, Department of Health and Human Services. Overweight and obesity: U.S. obesity trends, 1985–2004. Available at: http://www.cdc.gov/nccdphp/dnpa/obesity/trend/maps/. Accessed April 13, 2006.
3. National Heart, Lung, and Blood Institute, North American Association for the Study of Obesity. The practical guide: identification, evaluation, and treatment of overweight and obesity in adults. Available at http://www.nhlbi.nih.gov/guidelines/obesity/practgde.htm. Accessed April 13, 2006.
4. Shick SM et al. Persons successful at long-term weight loss and maintenance continue to consume a low calorie, low fat diet. *J Am Dietetic Assoc* 1998; 98: 408.
5. Dansinger ML et al. Comparison of the Atkins, Ornish, Weight Watchers, and Zone diets for weight loss and heart disease risk reduction: a randomized trial. *JAMA* 2005; 293: 43.
6. Hultquist CN, Albright C, and Thompson DL. Comparison of walking recommendations in previously inactive women. *Med Sci Sports Exerc* 2005; 37: 676.
7. Thomas RA and Dlugosz CK. Tools and techniques for fighting the obesity epidemic. Program and abstracts of the American Pharmacists Association Annual Meeting and Exposition; March 17–21, 2006, San Francisco.
8. Grundy S. Metabolic syndrome: connecting and reconciling cardiovascular and diabetes worlds. *J Am Coll Cardiol* 2006; 47: 1093. Available at Medscape Cardiology, http://www.medscape.com/viewarticle/524081. Accessed April 13, 2006.

9. Grundy SM et al. Diagnosis and management of metabolic syndrome: an American Heart Association/National Heart, Lung, and Blood Institute scientific statement. *Circulation* 2005; 112: e285.

10. Expert Panel on Detection, Evaluation, and Treatment of High Blood Cholesterol in Adults. Executive Summary of the Third Report of National Cholesterol Education Program (NCEP) Expert Panel on Detection, Evaluation, and Treatment of High Blood Cholesterol in Adults (Adult Treatment Panel III). *JAMA* 2001; 285: 2486.

11. Kendall DM and Harmel AP. The metabolic syndrome, type 2 diabetes, and cardiovascular disease: understanding the role of insulin resistance. *Am J Manag Care* 2002; 8: S635.

12. Curran JE et al. Genetic variation in selenoprotein S influences inflammatory response. *Nat Genet* 2005; 37: 1234.

13. Rankinen T et al. The human obesity gene map: the 2005 update. *Obesity* 2006; 14: 529.

14. Schulz LO et al. Effects of traditional and Western environments on prevalence of type 2 diabetes in Pima indians in Mexico and the U.S. *Diabetes Care* 2006; 29: 1866.

15. United Kingdom Prospective Diabetes Study 24: a 6-year, randomized, controlled trial comparing sulfonylurea, insulin, and metformin therapy in patients with newly diagnosed type 2 diabetes that could not be controlled with diet therapy. *Ann Intern Med* 1998; 128: 165.

16. Aronoff SL et al. Glucose metabolism and regulation: beyond insulin and glucagons. *Diabetes Spectrum* 2004; 17: 183.

17. United Kingdom Prospective Diabetes Study Group. Study 16: overview of 6 years' therapy of type II diabetes: a progressive disease. *Diabetes* 1995; 44: 1249.

18. Kloppel G et al. Islet pathology and the pathogenesis of type 1 and type 2 diabetes mellitus revisited. *Surv Synth Pathol Res* 1985; 4: 110.

19. Butler AE et al. Beta cell deficit and increased beta cell apoptosis in humans with type 2 diabetes. *Diabetes* 2003; 52: 102.

20. Yoon KH et al. Selective beta-cell loss and alpha-cell expansion in patients with type 2 diabetes mellitus in Korea. *J Clin Endocrinol Metab* 2003; 88: 2300.

21. Nauck MA. Glucagon-like peptide 1 (GLP-1) in the treatment of diabetes. *Horm Metab Res* 2004; 36: 852.

22. Marchetti P et al. Pancreatic islets from type 2 diabetic patients have functional defects and increased apoptosis that are ameliorated by metformin. *J Clin Endocrinol Metab* 2004; 89: 5535.

23. Zeender E et al. Pioglitazone and sodium salicylate protect human beta-cells against apoptosis and impaired function induced by glucose and interleukin-1-beta. *J Clin Endocrinol Metab* 2004; 89: 5059.

24. Knowler WC et al. Reduction in the incidence of type 2 diabetes with lifestyle intervention or metformin. *New Engl J Med* 2002; 346: 393.

25. Buchanan TA et al. Preservation of pancreatic ß-cell function and prevention of type 2 diabetes by pharmacological treatment of insulin resistance in high-risk hispanic women. *Diabetes* 2002; 51: 2796.

26. Butler AE et al. Diabetes due to a progressive defect in ß-cell mass in rats transgenic for human islet amyloid polypeptide (HIP Rat): a new model for type 2 diabetes. *Diabetes* 2004; 53: 1509.

27. Donath MY and Halban PA. Decreased beta-cell mass in diabetes: significance, mechanisms and therapeutic implications (review). *Diabetologia* 2004; 47: 581.

28. Matschinsky FM, Glaser B, and Magnuson MA. Pancreatic ß-cell glucokinase: closing the gap between theoretical concepts and experimental realities. *Diabetes* 1998; 47: 307.

29. Krauss S et al. Superoxide-mediated activation of uncoupling protein 2 causes pancreatic beta cell dysfunction. *J Clin Invest* 2003; 112: 1831.

30. Ritzel RA et al. Induction of beta-cell rest by a Kir6.2/SUR1-selective K(ATP)-channel opener preserves beta-cell insulin stores and insulin secretion in human islets cultured at high (11 mM) glucose. *J Clin Endocrinol Metab* 2004; 89: 795.

31. Tomas E et al. Hyperglycemia and insulin resistance: possible mechanisms. *Ann NY Acad Sci* 2002; 967: 43.

32. Libby, P. Changing concepts of atherogenesis. *J Int Med* 2000; 247, 349.

33. Landry DB et al. Activation of the NF- B and I-B system in smooth muscle cells after rat arterial injury. Induction of vascular cell adhesion molecule-1 and monocyte chemoattractant protein-1. *Am J Pathol* 1997; 151, 1085.

34. Iademarco MF, McQuillan JJ, and Dean DC. Vascular cell adhesion molecule 1: contrasting transcriptional control mechanisms in muscle and endothelium. *Proc Natl Acad Sci USA* 1993; 90, 3943.

35. Eck SL et al. Inhibition of phorbol ester-induced cellular adhesion by competitive binding of NF-B *in vivo. Mol Cell Biol* 1993; 13, 6530.

36. Clesham GJ et al. High adenoviral loads stimulate NFB-dependent gene expression in human vascular smooth muscle cells. *Gene Ther* 1998; 174.

37. Xie QW, Kashiwabara Y, and Nathan C. Role of transcription factor NF-B/Rel in induction of nitric oxide synthase. *J Biol Chem* 1994; 269, 4705.

38. Goto M et al. Involvement of NF-B p50/p65 heterodimer in activation of the human prointerleukin-1 gene at two subregions of the upstream enhancer element. *Cytokine* 1999; 11, 16.

39. O'Rourke L et al. Glucose-dependent regulation of cholesterol ester metabolism in macrophages by insulin and leptin. *J Biol Chem* 2002; 277, 42557.

40. Yudkin JS et al. Inflammation, obesity, stress and coronary heart disease: is interleukin-6 the link? *Atherosclerosis* 2000; 148, 209.

41. Ridker PM et al. Comparison of C-reactive protein and low-density lipoprotein cholesterol levels in the prediction of first cardiovascular events. *New Engl J Med* 2002; 347, 1557.

42. Fukuhara A et al. Visfatin: a protein secreted by visceral fat that mimics the effects of insulin. *Science* 2005; 307, 426.

43. Stentz FB et al. Proinflammatory cytokines, markers of cardiovascular risks, oxidative stress, and lipid peroxidation in patients with hyperglycemic crises. *Diabetes* 2004; 53, 2079.

44. Gabriely I et al. Hyperglycemia induces PAI-1 gene expression in adipose tissue by activation of the hexosamine biosynthetic pathway. *Atherosclerosis* 2002; 160, 115.

45. Sobel BE. Increased plasminogen activator inhibitor-1 and vasculopathy: a reconcilable paradox. *Circulation* 1999; 99, 2496.

46. Thogersen AM et al. High plasminogen activator inhibitor and tissue plasminogen activator levels in plasma precede a first acute myocardial infarction in both men and women: evidence for the fibrinolytic system as an independent primary risk factor. *Circulation* 1998; 98, 2241.

47. Tham DM et al. Angiotensin II is associated with activation of NF-B-mediated genes and downregulation of PPARs. *Physiol Genomics* 2002; 11, 21.
48. Cai H et al. NAD (P) H oxidase-derived hydrogen peroxide mediates endothelial nitric oxide production in response to angiotensin II. *J Biol Chem* 2002; 277, 48311.
49. Widlansky ME et al. The clinical implications of endothelial dysfunction. *J Am Coll Cardiol* 2003; 42, 1149.
50. Jambrik Z et al. Peripheral vascular endothelial function testing for the diagnosis of coronary artery disease. *Am Heart J* 2004; 148, 684.
51. Verma S et al. Resistin promotes endothelial cell activation: further evidence of adipokine–endothelial interaction. *Circulation* 2003; 108, 736.
52. Kawanami D et al. Direct reciprocal effects of resistin and adiponectin on vascular endothelial cells: a new insight into adipocytokine–endothelial cell interactions. *Biochem Biophys Res Commun* 2004; 314, 415.
53. Calabro P et al. Resistin promotes smooth muscle cell proliferation through activation of extracellular signal-regulated kinase 1/2 and phosphatidylinositol 3-kinase pathways. *Circulation* 2004; 110, 3335.
54. Pinkney JH et al. Endothelial dysfunction: cause of the insulin resistance syndrome. *Diabetes* 1997; 46, S9.
55. Reilly MP et al. Resistin is an inflammatory marker of atherosclerosis in humans. *Circulation* 2005; 111, 932.

8 Post-Prandial Endothelial Dysfunction, Oxidative Stress, and Inflammation in Type 2 Diabetes

Antonio Ceriello

CONTENTS

ABSTRACT

Increasing evidence suggests that the post-prandial state is a contributing factor to the development of atherosclerosis. In diabetes, the post-prandial phase is characterized by a rapid and large increase in blood glucose levels. The possibility that the post-prandial "hyperglycemic spikes" may be relevant to the onset of cardiovascular complications recently received much attention. Epidemiological studies and preliminary intervention studies have shown that post-prandial hyperglycemia is a direct and independent risk factor for cardiovascular disease. Most of the cardiovascular risk factors are modified in the post-prandial phase in diabetic subjects and directly affected by an acute increase of glycemia. The mechanisms through which acute hyperglycemia exerts its effects may be identified in the production of free radicals. This alarmingly suggestive body of evidence for a harmful effect of post-prandial hyperglycemia on diabetic complications has been sufficient to influence guidelines from key professional scientific societies. Correcting post-prandial hyperglycemia may form part of a strategy for the prevention and management of cardiovascular diseases in diabetes.

INTRODUCTION

Diabetes mellitus is characterized by a high incidence of cardiovascular disease[1] and poor control of hyperglycemia appears to play a significant role in the development of cardiovascular disease in diabetes.[2] Increasing evidence indicates that the post-prandial state is an important contributing factor to the development of atherosclerosis.[3] In diabetes, the post-prandial phase is characterized by a rapid and large increase in blood glucose levels. The possibility that these post-prandial "hyperglycemic spikes" may be relevant to the pathophysiology of late diabetic complications is recently receiving much attention.

In this chapter, epidemiological data and preliminary results of intervention studies indicating that post-prandial hyperglycemia represents an increased risk for cardiovascular disease are surveyed. The proposed mechanisms involved in this effect are summarized.

POSSIBLE ROLE OF HYPERGLYCEMIC SPIKES IN CARDIOVASCULAR DISEASES

FASTING HYPERGLYCEMIA AND CARDIOVASCULAR DISEASE

Over the last 10 years, many studies showed an independent relationship between cardiovascular diseases and glycemic control in patients with type 2 diabetes.[2] These studies involved thousands of subjects, often newly diagnosed, who were followed up for periods ranging from 3.5 to 11 years and were evaluated on the basis of various cardiovascular end-points.[2] It is necessary to underline that the majority of these studies used a single baseline fasting glycemic value or a single value of glycated hemoglobin A1c (HbA1c) to predict cardiovascular events occurring many years later.

For instance, the observational version of the United Kingdom Prospective Diabetes Study (UKPDS) showed that the mean HbA1c value was a good predictor of ischemic heart disease.[4] In particular, multivariate analysis showed that for each 1% increment in HbA1c there was an approximate 10% increase in the risk of coronary heart disease.[4] This evidence is not substantially different compared with the results of the interventional version of the UKPDS. In this trial, even though the result was not significant (p <0.052), intensive treatment leading to an approximate 1% reduction in HbA1c levels led to a 16% reduction in the occurrence of myocardial infarction.[5]

Interestingly, the UKPDS noted significant impacts on cardiovascular events in the metformin-treated group.[6] However, it is reasonable that metformin, while improving insulin resistance, may have significantly improved the "cluster" of cardiovascular risk factors associated with insulin resistance.

The relationship existing between macroangiopathy and fasting plasma glucose or HbA1c is weaker than that observed with microangiopathy.[2] This was found in cross-sectional or longitudinal studies. These data support the hypothesis that fasting plasma glucose or HbA1c alone is unable to describe thoroughly the glycemic disorders occurring in diabetes and its impact on cardiovascular disease. In addition to fasting glycemia and HbA1c, emphasis has recently been given to the relationship between post-prandial hyperglycemia and cardiovascular diseases.

POST-PRANDIAL HYPERGLYCEMIA AND CARDIOVASCULAR DISEASE: EPIDEMIOLOGICAL EVIDENCE

The oral glucose tolerance test (OGTT) has been used mostly in epidemiological studies that attempt to evaluate the risk of cardiovascular disease. The main advantage of the OGTT is its simplicity: a single plasma glucose measurement 2 hours after a glucose load determines whether glucose tolerance is normal, impaired, or indicates overt diabetes.

The caveats of the OGTT are numerous because 75 or 100 g glucose is almost never ingested during a meal and, more importantly, many events associated with ingesting a pure glucose solution do not incorporate the numerous metabolic events associated with eating a mixed meal. Moreover, the relationship between glycemia and meal content is contingent upon the contents of the meal.[7] However, it has recently been demonstrated that the level of glycemia reached 2 hours after an OGTT was closely related to the level of glycemia after a standardized meal (mixed meal in the form of wafers containing oat fractionation products, soy protein, and canola oil sweetened with honey: 345 kcal, 10.7 g fat, 12.1 g protein, 8.9 g simple sugars, 41.1 g starch, and 3.8 g dietary fiber), suggesting that the OGTT may represent a valid tool to reveal altered carbohydrate metabolism during a meal.[8] Interestingly, the correlation is more consistent for the values of glycemia in the impaired glucose tolerance range.[8]

From an epidemiological point of view, the Hoorn Study,[9] the Honolulu Heart Study,[10] the Chicago Heart Study,[11] and the more recent DECODE Study[12] have clearly shown that glucose serum levels 2 hours after oral challenge with glucose are powerful predictors of cardiovascular risk. This evidence is confirmed also by two important meta-analyses: the first by Coutinho et al. examined studies of 95,783 subjects.[13] The second, covering more than 20,000 subjects, pooled the data of the Whitehall Study, the Paris Prospective Study, and the Helsinki Policemen Study.[14]

The possible role of post-prandial hyperglycemia as an independent risk factor has also been supported by the Diabetes Intervention Study showing how in type 2 diabetics post-prandial hyperglycemia predicts infarction[15] and by another study associating post-prandial hyperglycemia levels with medio-intimal carotid thickening.[16]

TABLE 8.1
Epidemiological Studies Showing Association of Post-Prandial Hyperglycemia with Risk of CVD and Mortality

Study	Result	Ref. No.
Hoorn Study	2-hour glucose better predictor of mortality than HbA1c	9
Honolulu Heart Program	1-hour glucose predicts coronary heart disease	10
Chicago Heart Study	2-hour post-challenge glucose predicts all cause mortality	11
DECODE	High 2-hour post-load blood glucose associated with increased risk of death independent of fasting glucose	12
Coutinho M et al.	2-hour glucose associated with CHD	13
Whitehall, Paris, and Helsinki Studies	2-hour post-challenge glucose predicts all cause and CHD mortality	14
Diabetes Intervention Study	Post-meal, but not fasting glucose, associated with CHD	15

Intriguing evidence comes from a study that demonstrates how medio-intimal carotid thickening is correlated not only with post-prandial glucose serum level but particularly with glycemic spikes during the OGTT.[17] A post-challenge glucose spike was defined as the difference between the maximal post-challenge glucose level during OGTT, irrespective of the time after glucose challenge, and the level of fasting plasma glucose.[17] Epidemiological studies are summarized in Table 8.1.

Indirect evidence of the unfavorable role of acute hyperglycemia on cardiovascular diseases is also available. Hyperglycemia during a cardiovascular acute event is unfavorable from a prognostic view in cases both of myocardial infarction[18,19] and stroke.[20,21] A worst prognosis has been demonstrated for both cases in diabetic and non-diabetic subjects.[18–21]

Regarding infarction, it has been recently demonstrated by a meta-analysis that there is a continuous correlation between glucose serum levels and the seriousness of prognosis even in non-diabetic subjects,[22] while intensive insulin treatment during acute myocardial infarction reduced long-term mortality in diabetic patients.[23] This is consistent with the evidence that an acute increase of glycemia significantly prolongs the QT in normal subjects[24] and that during myocardial infarction increased glucose level is capable of inducing such electrophysiological alterations as to favor the occurrence of arrhythmias whose outcomes could even be fatal.[25]

POST-PRANDIAL HYPERGLYCEMIA AND CARDIOVASCULAR DISEASE: INTERVENTION STUDIES

One of the major concerns about the role of post-prandial hyperglycemia in cardiovascular disease has been, until now, the absence of intervention studies. That situation is changing.

The STOP-NIDDM trial has presented data indicating that treatment of patients who have IGT with acarbose, an α-glucosidase inhibitor that specifically reduces post-prandial hyperglycemia, is associated not only with a 36% reduction in the risk of progression to diabetes[26] but also with a 34% risk reduction in the development of new cases of hypertension and a 49% risk reduction in cardiovascular events.[27] In addition, in a subgroup of patients, carotid intima media thickness (CIMT) was measured before randomization and at the end of the study.[28] Acarbose treatment was associated with a significant decrease in the progression of intima media thickness, an accepted surrogate for atherosclerosis.[28]

In a recent meta-analysis of type 2 diabetic patients, acarbose treatment was associated with significant reductions in cardiovascular events, even after adjusting for other risk factors.[29] Finally, very recently, the effects of two insulin secretagogues, repaglinide and glyburide, known to have different efficacies for post-prandial hyperglycemia, CIMT, and markers of systemic vascular inflammation in type 2 diabetic patients, have been evaluated.[30] After 12 months, post-prandial glucose peak measurements were 148 ± 28 mg/dL in the repaglinide group and 180 ± 32 mg/dL in the glyburide group (p <0.01). HbA1c showed similar decreases in both groups (–0.9%). CIMT regression, defined as a decrease of >0.020 mm, was observed in 52% of diabetics receiving repaglinide and in 18% of those receiving glyburide (p <0.01).

Interleukin-6 (p = 0.04) and C-reactive protein (p = 0.02) decreased more in the repaglinide group than in the glyburide group. The reductions in CIMT were associated with changes in post-prandial but not fasting hyperglycemia.[30] Emerging evidence suggests that treating post-prandial hyperglycemia may positively affect the development of cardiovascular disease.

MECHANISMS INVOLVED

Acceptance of the hypothesis that post-prandial hyperglycemia has a direct and harmful effect on the cardiovascular system requires, at the very least, a link between acute hyperglycemia and one or more risk factors for cardiovascular disease. Most cardiovascular risk factors are affected directly by acute increases of glycemia in individuals with diabetes and are modified in the post-prandial phase. LDL oxidation in diabetes is related to metabolic control,[31,32] and it has been shown in type 2 diabetic patients that after meals LDL oxidation increases,[33] and that this phenomenon is in strict relationship with the degree of hyperglycemia.[34]

Endothelial function is altered early in diabetes. It has been demonstrated that the vasodilating response to stimuli is diminished in diabetics and that this anomaly is related to glycemic control.[35] *In vivo* studies have demonstrated that hyperglycemic spikes induce endothelial dysfunction in both diabetic and normal subjects.[36-38] This effect of hyperglycemia is probably linked with reduced production or bioavailability of NO since the hyperglycemia-induced endothelial dysfunction is counterbalanced by arginine.[38] It is interesting that a rapid decrease of flow-mediated vasodilation has been shown in the post-prandial phase in type

2 diabetic patients and that the decrease correlated inversely with the magnitude of post-prandial hyperglycemia.[39]

The possible role of hyperglycemia in the activation of blood coagulation has previously been reviewed.[40] It emerges that acute glycemic variations are matched with a series of alterations of coagulation that are likely to cause a thrombosis. This tendency is documented by studies that demonstrate how inducing hyperglycemia produces a shortening of fibrinogen half-life[41] and increases in fibrinopeptide A,[42,43] fragments of prothrombin,[44] factor VII,[45] and platelet aggregation[46] in both normal and diabetic subjects. These data indicate that coagulation is activated during experimental hyperglycemia. It is interesting that the over-production of thrombin by post-prandial hyperglycemia in diabetics has already been already documented. The phenomenon is strictly dependent on the glycemic levels reached.[47]

Adhesion molecules regulate the interactions of endothelium and leukocytes.[48] They participate in the process of atherogenesis because their greater expression would imply an increase in the adhesion of leukocytes (monocytes in particular) to the endothelium.[49] It is well known that this is considered one of the early stages of the process leading to atheromatous lesions. Among the various pro-adhesive molecules, ICAM-1 has received particular interest. Increases in the circulating form of this molecule have been demonstrated in subjects with vascular disease[50] and with diabetes mellitus with or without vascular disease.[51,52] These increases have been considered indications of the activation of the atherogenic process.

The soluble form of ICAM-1 is stored in cells and can be quickly expressed outside them as a consequence of various stimuli. It has been demonstrated that acute hyperglycemia in both normal and diabetic subjects is a sufficient stimulus for the circulating level of ICAM-1 to increase, thus activating one of the first stages of the atherogenic process.[53,54]

The concept of atherosclerosis as an inflammatory disease even in diabetes is now well established.[55] Studies support the evidence that acute hyperglycemia during a hyperglycemic clamp[56] or in a post-prandial state[57] can increase the production of plasma IL-6, TNF-α, and IL-18.

POST-PRANDIAL HYPERGLYCEMIA AND OXIDATIVE AND NITROSATIVE STRESS

Recent studies demonstrated that hyperglycemia induces an over-production of superoxide by the mitochondrial electron-transport chain.[58] Superoxide over-production is accompanied by increased nitric oxide generation due to eNOS and iNOS in an uncoupled state — a phenomenon favoring the formation of peroxynitrite, a strong oxidant that in turn damages DNA.[59] DNA damage is an obligatory stimulus for the activation of the poly(ADP-ribose) polymerase nuclear enzyme. Poly(ADP-ribose) polymerase activation in turn depletes the intracellular concentration of its substrate NAD^+, slowing the rate of glycolysis, electron

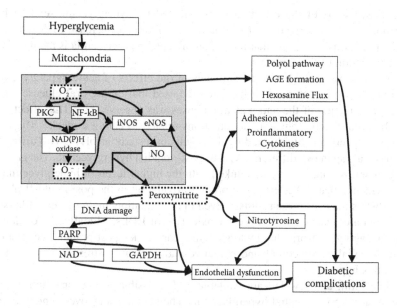

FIGURE 8.1 In endothelial cells, glucose can pass freely in an insulin-independent manner through the cell membrane. Intracellular hyperglycemia induces over-production of superoxide at the mitochondrial level. Over-production of superoxide is the first and key event in the activation of all other pathways involved in the pathogenesis of diabetic complications such as polyol pathway flux, increased AGE formation, activation of protein kinase C and NF-kB, and increased hexosamine pathway flux. O_2^- reacting with NO produces peroxynitrite ($ONOO^-$). Superoxide over-production reduces eNOS activity but through NF-kB and PKC activates NAD(P)H and increases iNOS expression: the final effect is an increased nitric oxide generation. This condition favors the formation of the strong peroxynitrite oxidant that in turn produces in iNOS and eNOS an uncoupled state, resulting in the production of superoxide rather than NO, and damages DNA. DNA damage is an obligatory stimulus for the activation of the poly(ADP-ribose) polymerase nuclear enzyme. Poly(ADP-ribose) polymerase activation in turn depletes the intracellular concentration of its substrate NAD^+, slowing the rate of glycolysis, electron transport, and ATP formation, and also produces an ADP ribosylation of the GAPDH. This process results in acute endothelial dysfunction in diabetic blood vessels that contributes to the development of diabetic complications. NF-kB activation also induces a pro-inflammatory condition and adhesion molecule over-expression. All these alterations produce the final picture of diabetic complications.

transport, and ATP formation and also produces an ADP ribosylation of the GAPDH.[59] These processes result in acute endothelial dysfunction in diabetic blood vessels that, convincingly, contributes to the development of CVD.[59] These pathways are summarized in Figure 8.1.

Several indirect and direct approaches support the concept that acute hyperglycemia works through the production of oxidative and nitrosative stress. Indirect evidence is obtained through the use of antioxidants. The fact that antioxidants can hinder some of the effects acutely induced by hyperglycemia–endothelial dysfunction,[36,60,61] activation of coagulation,[44] plasmatic increase of ICAM-1,[53] and interleukins[57] suggests that the action of acute hyperglycemia is mediated by the production of free radicals.

Direct evidence is linked to estimates of the effects of acute hyperglycemia on oxidative stress markers. It has been reported that during oral glucose challenge, a reduction of the antioxidant defenses was observed.[62–64] This effect can be observed in other physiologic situations related to eating a meal.[65] The role of hyperglycemia is highlighted by the results of a study involving two different meals resulting in two different levels of post-prandial hyperglycemia. The greater drop in antioxidant activity was linked with the higher levels of hyperglycemia.[34] The evidence that LDLs are more prone to oxidation in the post-prandial phase in diabetics matches these results.[33] Even in this situation, higher levels of hyperglycemia are matched with a greater oxidation of LDLs.[34] Finally, the evidence that managing post-prandial hyperglycemia can reduce post-prandial generation of endothelial dysfunction[66] and oxidative and nitrosative stress[67] strongly supports this hypothesis.

Interesting and new data are available on the possible generation of nitrosative stress during post-prandial hyperglycemia. The simultaneous over-generation of NO and superoxide favors the production of a toxic reaction product, the peroxynitrite anion.[68] The peroxynitrite anion is cytotoxic because it oxidizes sulfhydryl groups in proteins, initiates lipid peroxidation, and nitrates amino acids such as tyrosine which affects many signal transduction pathways.[68] The production of peroxynitrite can be indirectly inferred by the presence of nitrotyrosine,[68] and it has recently been reported that nitrotyrosine is an independent predictor of CVD.[69]

Several pieces of evidence support a direct role of hyperglycemia in favoring nitrotyrosine over-generation. Nitrotyrosine formation was detected in the artery walls of monkeys during hyperglycemia[70] and also in the plasma of healthy subjects during hyperglycemic clamp[71] or OGTT.[72,73] Hyperglycemia was also accompanied by nitrotyrosine deposition in perfused working hearts from rats, and was reasonably related to unbalanced production of NO and superoxide through iNOS over-expression.[74] Nitrotyrosine formation is followed by the development of endothelial dysfunction in both healthy subjects[71,72] and in coronaries of perfused hearts.[74] This effect is not surprising because it has been shown that nitrotyrosine can also be directly harmful to endothelial cells.[75]

Dyslipidemia also is a recognized risk factor for cardiovascular disease in diabetes[76] and post-prandial hyperlipidemia contributes to this risk.[77] In non-obese type 2 diabetes patients with moderate fasting hypertriglyceridemia, atherogenic lipoprotein profiles were amplified in the post-prandial state.[78] Such observations have raised the question of whether post-prandial hyperlipidemia, which rises concomitantly with post-prandial hyperglycemia, is the true risk factor.[79] Evidence suggests that post-prandial hypertriglyceridemia and hyperglycemia independently induce endothelial dysfunction through oxidative stress.[80] It is now well recognized that endothelial dysfunction is one of the first stages and earliest markers in the development of cardiovascular disease.[81] Recent studies demonstrate both independent and cumulative effects of post-prandial hypertriglyceridemia and hyperglycemia on endothelial function, with oxidative stress as the common mediator.[72,73] This lends credence to the idea of a direct atherogenic role for post-prandial hyperglycemia independent of the role of lipids.

CONCLUSIONS

The evidence described proves that hyperglycemia can acutely induce alterations of normal human homeostasis. It should be noted that acute increases of glucose serum level cause alterations in both healthy (normoglycemic) subjects and also in diabetic subjects who have basic hyperglycemia. On the basis of this evidence we can hypothesize that the acute effects of glucose serum levels can add to those produced by chronic hyperglycemia, thus contributing to a final picture of complicated diabetes. The precise relevance of this phenomenon is not exactly comprehensible and quantifiable at the moment, but based on the tendency to rapid variations of hyperglycemia in the lives of diabetic patients, primarily in the post-prandial phase, it is proper to think that it may exert an influence on the onset of complications. Epidemiological studies[3] and preliminary intervention studies seem to support this hypothesis.[27–30]

DCCT in type 1 diabetes[82] and UKPDS in type 2 diabetes[5] both attest to the importance of long-term glycemic control through HbA1c for the prevention of complications. However, the authors of the DCCT pointed out further that HbA1c only is not a sufficient parameter to explain the onset of such complications and suggested that post-prandial hyperglycemic excursions could reasonably favor the onset of diabetic complications.[82]

Evidence shows that post-prandial glucose serum level is the major determinant of HbA1c level after mean daily blood glucose[83–86] and that reducing post-prandial hyperglycemia significantly reduces HbA1c levels in type 2 diabetic patients.[87,88] On the basis of this evidence, it seems obvious that if post-prandial hyperglycemia is important to determine the level of HbA1c, which is fundamental to determining the extent of the risk of diabetic complications, it can be supposed that post-prandial glucose serum level will favor it as much.

Evidence also suggests that post-prandial excursions of blood glucose may be involved in the development of diabetic complications, particularly cardiovascular complications, but are not the only factors.[89,90] Many questions remain unanswered regarding the definition of post-prandial glucose and, perhaps most importantly, whether post-prandial hyperglycemia has a unique role in the pathogenesis of diabetic vascular complications and should be a specific target of therapy.

However, this alarmingly suggestive body of evidence for a harmful effect of post-prandial hyperglycemia on diabetic complications has been sufficient to influence guidelines from key professional bodies including the World Health Organization,[91] the American Diabetes Association,[92] the American College of Endocrinology,[93] the International Diabetes Federation,[94] the Canadian Diabetes

Association,[95] and more recently a large task force of European scientific societies focusing on cardiovascular disease.[96]

Therefore the real question, as recently underlined also by the American Diabetes Association,[97] seems to be that "because CVD is the major cause of morbidity and mortality in patients with diabetes, and in type 2 diabetes in particular, understanding the impact on CVD events of treatment directed at specifically lowering post-prandial glucose is crucial." Future studies must be designed specifically to evaluate this fundamental issue that may significantly change the therapeutic approach to diabetes.

REFERENCES

1. Kannel WB, McGee DL. Diabetes and cardiovascular diseases: the Framingham Study. *JAMA* 241, 2035, 1979.
2. Laakso M. Hyperglycemia and cardiovascular disease in type 2 diabetes. *Diabetes* 48, 937, 1999.
3. Bonora E and Muggeo M. Post-prandial blood glucose as a risk factor for cardiovascular disease in type II diabetes: epidemiological evidence. *Diabetologia* 44, 2107, 2001.
4. Stratton IM et al. Association of glycaemia with macrovascular and microvascular complications of type 2 diabetes (UKPDS 35): prospective observational study. *Br Med J* 321, 405, 2000.
5. United Kingdon Prospective Diabetes Study Group. Intensive blood-glucose control with sulphonylureas or insulin compared with conventional treatment and risk of complications in patients with type 2 diabetes (UKPDS 33). *Lancet* 352, 837, 1998.
6. United Kingdom Prospective Diabetes Study Group. Effect of intensive blood-glucose control with metformin on complications in overweight patients with type 2 diabetes (UKPDS 34). *Lancet* 352, 854, 1998.
7. Vinik AI and Jenkins DJ. Dietary fiber in management of diabetes. *Diabetes Care* 11, 160, 1988.
8. Wolever TMS et al. Variation of post-prandial plasma glucose, palatability, and symptoms associated with a standardized mixed test meal versus 75 g oral glucose. *Diabetes Care* 21, 336, 1998.
9. de Vegt F et al. Hyperglycaemia is associated with all-cause and cardiovascular mortality in the Hoorn population: the Hoorn study. *Diabetologia* 42, 926, 1999.
10. Donahue RP, Abbott RD, Reed DM, and Yano K. Post-challenge glucose concentration and coronary heart disease in men of Japanese ancestry: Honolulu Heart Program. *Diabetes* 36, 689, 1987.
11. Lowe LP, Liu K, Greenland P, Metzger BE, Dyer AR, and Stamler J. Diabetes, asymptomatic hyperglycemia, and 22-year mortality in black and white men: the Chicago Heart Association Detection Project in Industry Study. *Diabetes Care* 20, 163, 1997.
12. The DECODE study group on behalf of the European Diabetes Epidemiology Group. Glucose tolerance and mortality: comparison of WHO and American Diabetes Association diagnostic criteria. *Lancet* 354, 617, 1999.

13. Coutinho M, Gerstein HC, Wang Y, and Yusuf S. The relationship between glucose and incident cardiovascular events. a metaregression analysis of published data from 20 studies of 95,783 individuals followed for 12.4 years. *Diabetes Care* 22, 233, 1999.

14. Balkau B et al. High blood glucose concentration is a risk factor for mortality in middle-aged nondiabetic men: 20-year follow-up in the Whitehall Study, the Paris Prospective Study, and the Helsinki Policemen Study. *Diabetes Care* 21, 360, 1998.

15. Hanefeld M et al. The DIS Group. Risk factors for myocardial infarction and death in newly detected NIDDM: Diabetes Intervention Study, 11-year follow-up. *Diabetologia* 39, 1577, 1996.

16. Hanefeld M et al. Post-prandial plasma glucose is an independent risk factor for increased carotid intima-media thickness in non-diabetic individuals. *Atherosclerosis* 144, 229, 1999.

17. Temelkova-Kurktschiev TS et al. Post-challenge plasma glucose and glycemic spikes are more strongly associated with atherosclerosis than fasting glucose and glycosylated hemoglobin level. *Diabetes Care* 23, 1830, 2000.

18. Bellodi G et al. Hyperglycemia and prognosis of acute myocardial infarction in patients without diabetes mellitus. *Am J Cardiol* 64, 885, 1989.

19. O'Sullivan JJ, Conroy RM, Robinson K, Hickey N, and Mulcahy R. In-hospital prognosis of patients with fasting hyperglycemia after first myocardial infarction. *Diabetes Care* 14, 758, 1991.

20. Gray CS et al. The prognostic value of stress hyperglycaemia and previously unrecognized diabetes in acute stroke. *Diabet Med* 4, 237, 1987.

21. Gray CS, French JM, Bates D, Cartlidge NE, Venables GS, and James OF. Increasing age, diabetes mellitus and recovery from stroke. *Postgrad Med J* 65, 720, 1989.

22. Capes SE, Hunt D, Malmberg K, and Gerstein HC. Stress hyperglycaemia and increased risk of death after myocardial infarction in patients with and without diabetes: a systematic overview. *Lancet* 355, 773, 2000.

23. Malmberg K, Norhammar A, Wedel H, and Ryden L. Glycometabolic state at admission. important risk marker of mortality in conventionally treated patients with diabetes mellitus and acute myocardial infarction: long-term results from the diabetes and insulin-glucose infusion in acute myocardial infarction (DIGAMI) study. *Circulation* 99, 2626, 1999.

24. Marfella R, Nappo F, De Angelis L, Siniscalchi M, Rossi F, and Giugliano D. The effect of acute hyperglycaemia on QTc duration in healthy man. *Diabetologia* 43, 571, 2000.

25. Gokhroo R and Mittal SR. Electrocardiographic correlates of hyperglycemia in acute myocardial infarction. *Int J Cardiol* 22, 267, 1989.

26. Chiasson JL, Josse RG, Gomis R, Hanefeld M, Karasik A, and Laakso M. Acarbose for prevention of type 2 diabetes mellitus: STOP-NIDDM randomised trial. *Lancet* 359, 2072, 2002.

27. Chiasson JL, Josse RG, Gomis R, Hanefeld M, Karasik A, and Laakso M. Acarbose treatment and the risk of cardiovascular disease and hypertension in patients with impaired glucose tolerance: STOP-NIDDM trial. *JAMA* 290, 486, 2003.

28. Hanefeld M et al. Acarbose slows progression of intima-media thickness of the carotid arteries in subjects with impaired glucose tolerance. *Stroke* 35, 1073, 2004.

29. Hanefeld M, Cagatay M, Petrowitsch T, Neuser D, Petzinna D, and Rupp M. Acarbose reduces the risk for myocardial infarction in type 2 diabetic patients: meta-analysis of seven long-term studies. *Eur Heart J* 25, 10, 2004.

30. Esposito K, Giugliano D, Nappo F, and Marfella R. Regression of carotid atherosclerosis by control of post-prandial hyperglycemia in type 2 diabetes mellitus. *Circulation* 29, 2978, 2004.

31. Tsai EC, Hirsch IB, Brunzell JD, and Chait A. Reduced plasma peroxyl radical trapping capacity and increased susceptibility of LDL to oxidation in poorly controlled IDDM. *Diabetes* 43, 1010, 1994.

32. Jenkins AJ et al. LDL from patients with well-controlled IDDM is not more susceptible to *in vitro* oxidation. *Diabetes* 45, 762, 1996.

33. Diwadkar VA, Anderson JW, Bridges SR, Gowri MS, and Oelgten PR. Post-prandial low density lipoproteins in type 2 diabetes are oxidized more extensively than fasting diabetes and control samples. *Proc Soc Exp Biol Med* 222, 178, 1999.

34. Ceriello A et al. Meal-induced oxidative stress and low-density lipoprotein oxidation in diabetes: possible role of hyperglycemia. *Metabolism* 48, 1503, 1999.

35. Jorgensen RG, Russo L, Mattioli L, and Moore WV. Early detection of vascular dysfunction in type I diabetes. *Diabetes* 37, 292, 1988.

36. Marfella R et al. Glutathione reverses systemic hemodynamic changes by acute hyperglycemia in healthy subjects. *Am J Physiol* 268, E1167, 1995.

37. Kawano H et al. Hyperglycemia rapidly suppresses flow-mediated endothelium-dependent vasodilation of brachial artery. *J Am Coll Cardiol* 34, 146, 1999.

38. Giugliano D et al. Vascular effects of acute hyperglycemia in humans are reversed by L-arginine: evidence for reduced availability of nitric oxide during hyperglycemia. *Circulation* 95, 1783, 1997.

39. Shige H et al. Endothelium-dependent flow-mediated vasodilation in the post-prandial state in type 2 diabetes mellitus. *Am J Cardiol* 84, 1272, 1999.

40. Ceriello A. Coagulation activation in diabetes mellitus: the role of hyperglycaemia and therapeutic prospects. *Diabetologia* 36, 1119, 1993.

41. Jones RL and Peterson CM. Reduced fibrinogen survival in diabetes mellitus: a reversible phenomenon. *J Clin Invest* 63, 485, 1979.

42. Jones RL. Fibrinopeptide A in diabetes mellitus: relation to levels of blood glucose, fibrinogen disappearance, and hemodynamic changes. *Diabetes* 34, 836, 1985.

43. Ceriello A et al. Hyperglycemia may determine fibrinopeptide A plasma level increase in humans. *Metabolism* 38, 1162, 1989.

44. Ceriello A et al. Hyperglycemia-induced thrombin formation in diabetes: possible role of oxidative stress. *Diabetes* 44, 924, 1995.

45. Ceriello A, Giugliano D, Quatraro A, Dello Russo P, and Torella R. Blood glucose may condition factor VII levels in diabetic and normal subjects. *Diabetologia* 31, 889, 1988.

46. Sakamoto T et al. Rapid change of platelet aggregability in acute hyperglycemia: detection by a novel laser-light scattering method. *Thromb Haemost* 83, 475, 2000.

47. Ceriello A et al. Post-meal coagulation activation in diabetes mellitus: effect of acarbose. *Diabetologia* 39, 469, 1996.

48. Ruosladti E. Integrins. *J Clin Invest* 187, 1, 1991.

49. Lopes-Virella MF and Virella G. Immune mechanism of atherosclerosis in diabetes mellitus. *Diabetes* 41, Suppl. 2, 86, 1992.

50. Blann AD and McCollum CN. Circulating endothelial cell/leukocyte adhesion molecules in atherosclerosis. *Thromb Haemost* 72, 151, 1994.

51. Cominacini L et al. Elevated levels of soluble E-selectin in patients with IDDM and NIDDM: relation to metabolic control. *Diabetologia* 38, 1122, 1995.
52. Ceriello A et al. Increased circulating ICAM-1 levels in type 2 diabetic patients: possible role of metabolic control and oxidative stress. *Metabolism* 45, 498, 1996.
53. Ceriello A et al. Hyperglycemia-induced circulating ICAM-1 increase in diabetes mellitus: possible role of oxidative stress. *Horm Metab Res* 30, 146, 1998.
54. Marfella R et al. Circulating adhesion molecules in humans: role of hyperglycemia and hyperinsulinemia. *Circulation* 101, 2247, 2000.
55. Plutzky J. Inflammation in atherosclerosis and diabetes mellitus. *Rev Endocr Metab Disord* 5, 255, 2004.
56. Esposito K et al. Inflammatory cytokine concentrations are acutely increased by hyperglycemia in humans: role of oxidative stress. *Circulation* 106, 2067, 2002.
57. Nappo F et al. Post-prandial endothelial activation in healthy subjects and in type 2 diabetic patients: role of fat and carbohydrate meals. *J Am Coll Cardiol* 39,1145, 2002.
58. Brownlee M. Biochemistry and molecular cell biology of diabetic complications. *Nature* 414, 813, 2001.
59. Ceriello A. New insights on oxidative stress and diabetic complications may lead to a "causal" antioxidant therapy. *Diabetes Care* 26, 1589, 2003.
60. Title LM, Cummings PM, Giddens K, and Nassar BA. Oral glucose loading acutely attenuates endothelium-dependent vasodilation in healthy adults without diabetes: effect prevented by vitamins C and E. *J Am Coll Cardiol* 36, 2185, 2000.
61. Beckman JA, Goldfine AB, Gordon MB, and Creager MA. Ascorbate restores endothelium-dependent vasodilation impaired by acute hyperglycemia in humans. *Circulation* 103, 1618, 2001.
62. Ceriello A et al. Antioxidant defenses are reduced during oral glucose tolerance test in normal and non-insulin dependent diabetic subjects. *Eur J Clin Invest* 28, 329, 1998.
63. Tessier D, Khalil A, and Fulop T. Effects of an oral glucose challenge on free radical/antioxidant balance in an older population with type II diabetes. *J Gerontol* 54, 541, 1999.
64. Konukoglu D, Hatemi H, Ozer EM, Gonen S, and Akcay T. The erythrocyte glutathione levels during oral glucose tolerance test. *J Endocrinol Invest* 20, 471, 1997.
65. Ceriello A et al. Meal-generated oxidative stress in type 2 diabetic patients. *Diabetes Care* 21, 1529, 1998.
66. Ceriello A et al. The post-prandial state in type 2 diabetes and endothelial dysfunction: effects of insulin aspart. *Diabet Med* 21, 171, 2004.
67. Ceriello A et al. Role of hyperglycemia in nitrotyrosine post-prandial generation. *Diabetes Care* 25, 1439, 2002.
68. Beckman JS and Koppenol WH. Nitric oxide, superoxide, and peroxynitrite: the good, the bad, and ugly. *Am J Physiol* 271, C1424, 1996.
69. Shishehbor MH et al. Association of nitrotyrosine levels with cardiovascular disease and modulation by statin therapy. *JAMA* 289, 1675, 2003.
70. Pennathur S, Wagner JD, Leeuwenburgh C, Litwak KN, and Heinecke JW. A hydroxyl radical-like species oxidizes cynomolgus monkey artery wall proteins in early diabetic vascular disease. *J Clin Invest* 107, 853, 2001.

71. Marfella R, Quagliaro L, Nappo F, Ceriello A, and Giugliano D. Acute hypergly-cemia induces an oxidative stress in healthy subjects (letter). *J Clin Invest* 108, 635, 2001.

72. Ceriello A et al. Evidence for an independent and cumulative effect of post-pran-dial hypertriglyceridemia and hyperglycemia on endothelial dysfunction and oxi-dative stress generation: effects of short- and long-term simvastatin treatment. *Circulation* 106, 1211, 2002.

73. Ceriello A et al. Effect of post-prandial hypertriglyceridemia and hyperglycemia on circulating adhesion molecules and oxidative stress generation and the possible role of simvastatin treatment. *Diabetes* 53, 701, 2004.

74. Ceriello A et al. Acute hyperglycemia induces nitrotyrosine formation and apop-tosis in perfused heart from rat. *Diabetes* 51, 1076, 2002.

75. Mihm MJ, Jing L, and Bauer JA. Nitrotyrosine causes selective vascular endothe-lial dysfunction and DNA damage. *J Cardiovasc Pharmacol* 36, 182, 2000.

76. Taskinen MR, Lahdenpera S, and Syvanne M. New insights into lipid metabolism in non-insulin-dependent diabetes mellitus. *Ann Med* 28, 335, 1996.

77. Karpe F, de Faire U, Mercuri M, Bond MG, Hellenius ML, and Hamsten A. Magnitude of alimentary lipemia is related to intima-media thickness of the common carotid artery in middle-aged men. *Atherosclerosis* 141, 307, 1998.

78. Cavallero E, Dachet C, Neufcou D, Wirquin E, Mathe D, and Jacotot B. Post-pran-dial amplification of lipoprotein abnormalities in controlled type II diabetic sub-jects: relationship to post-prandial lipemia and C-peptide/glucagon levels. *Metabolism* 43, 270, 1994.

79. Heine RJ and Dekker JM. Beyond post-prandial hyperglycemia: metabolic factors associated with cardiovascular disease. *Diabetologia* 45, 461, 2002.

80. Ceriello A and Motz E. Is oxidative stress the pathogenic mechanism underlying insulin resistance, diabetes, and cardiovascular disease? The common soil hypoth-esis revisited. *Arterioscler Thromb Vasc Biol* 24, 816, 2004.

81. De Caterina R. Endothelial dysfunctions. common denominators in vascular dis-ease. *Curr Opin Lipidol* 11, 923, 2000.

82. The Diabetes Control and Complications Trial Research Group. The relationship of glycemic exposure (HbA1c) to the risk of development and progression of retinopathy in the Diabetes Control and Complications Trial. *Diabetes* 44, 968, 1995.

83. Avignon A, Radauceanu A, and Monnier L. Nonfasting plasma glucose is a better marker of diabetic control than fasting plasma glucose in type 2 diabetes. *Diabetes Care* 20, 1822, 1997.

84. Soonthornpun S, Rattarasarn C, Leelawattana R, and Setasuban W. Post-prandial plasma glucose: a good index of glycemic control in type 2 diabetic patients having near-normal fasting glucose levels. *Diabetes Res Clin Pract* 46, 23, 1999.

85. Monnier L, Lapinski H, and Colette C. Contributions of fasting and post-prandial plasma glucose increments to the overall diurnal hyperglycemia of type 2 diabetic patients: variations with increasing levels of HbA(1c). *Diabetes Care* 26, 881, 2003.

86. Rohlfing CL et al. Defining the relationship between plasma glucose and HbA(1c): analysis of glucose profiles and HbA(1c) in the Diabetes Control and Complica-tions Trial. *Diabetes Care* 25, 275, 2002.

87. Bastyr EJ 3rd et al. Therapy focused on lowering post-prandial glucose, not fasting glucose, may be superior for lowering HbA1c. *Diabetes Care* 23, 1236, 2000.

88. Home PD, Lindholm A, and Hylleberg B, Round P. Improved glycemic control with insulin aspart. a multicenter randomized double-blind crossover trial in type 1 diabetic patients. *Diabetes Care* 21, 1904, 1998.

89. Shichiri M, Kishikawa H, Ohkubo Y, and Wake N. Long-term results of Kumamoto Study on optimal diabetes control in type 2 diabetic patients. *Diabetes Care* 23, Suppl. 2, B21, 2000.

90. Singleton JR, Smith AG, Russell JW, and Feldman EL. Microvascular complications of impaired glucose tolerance. *Diabetes* 52, 2867, 2003.

91. World Health Organization. Definition, diagnosis and classification of diabetes mellitus and its complications. Part 1. Geneva, 1999.

92. American Diabetes Association. Standards of medical care in diabetes. *Diabetes Care* 27, Suppl.1, S15, 2004.

93. American College of Endocrinology. Consensus statement on guidelines for glycemic control. *Endocr Pract* 8, 5, 2002.

94. Alberti KG and Gries FA. Management of non-insulin-dependent diabetes mellitus in Europe: a consensus view. *Diabet Med* 5, 275, 1988.

95. Canadian Diabetes Association, Clinical Practice Guidelines Expert Committee. Clinical practice guidelines for the prevention and management of diabetes in Canada. *Can J Diabetes* Suppl. 2, 27, 2003.

96. De Backer G et al. Third Joint Task Force of European and Other Societies on Cardiovascular Disease Prevention in Clinical Practice. European guidelines on cardiovascular disease prevention in clinical practice. *Eur Heart J* 24, 1601, 2003.

97. American Diabetes Association. Postprandial blood glucose. *Diabetes Care* 24, 775, 2001.

9 Obesity and Inflammation: Implications for Atherosclerosis

John Alan Farmer

CONTENTS

INTRODUCTION

Age-adjusted mortality rates for cardiovascular disease have been significantly declining in the United States over the past 20 years. The encouraging reduction is due to a complex interaction of continued improvements in diagnostic imaging, medical therapy, and surgical techniques. The pathogenesis of atherosclerosis is complex and best regarded as a syndrome currently lacking a unifying hypothesis that explains all aspects of the initiation and progression of vascular disease. However, the recognition and modification of potential cardiovascular risk factors that have been statistically related to coronary disease (in particular factors whose modification has been demonstrated to reduce clinical event rates in prospective controlled clinical trials) have gained credence as a means to reduce the risk from atherosclerosis.

The improvements in cardiovascular morbidity and mortality rates are threatened by changing patterns in obesity and diabetes. The incidence and prevalence of obesity are significantly increasing in the United States and represent a major health hazard due to the independent risks directly related to increased body mass index *per se* in addition to the associated potential risks for the development of diabetes, dyslipidemia, and hypertension.[1]

Individuals may be classified as normal, overweight, obese, or morbidly obese on the basis of body mass index (weight in kilograms over body surface area in square meters). Normality determined by body mass index is considered below 25. A subject may be classified as overweight if the body mass index falls between 25 and 30. Obesity is defined as a body mass index in excess of 30 and morbid obesity requires a body mass index above 40.

Recent epidemiologic data concerning the prevalence of obesity in the United States utilized the National Health and Nutrition Education Survey (NHANES).[2] The prevalence of subjects who qualified as overweight or obese in the period between years 2003 and 2004 was compared to an earlier phase of the survey that analyzed subjects in 1999 and 2000. The prevalence of obesity increased from 27.5 to 31.1% in male subjects between the designated time periods. In comparison, the prevalence of obesity in women remained relatively constant at approximately 33%. The prevalence of overweight and obese children and adolescents was disturbingly high (17.1%). The overall prevalence of obesity in the general United States population was determined to be 32.2%.

The increasing prevalence of obesity and related conditions has led to significant health and economic concerns due to the dramatic impacts on both direct and indirect health care costs. The total economic costs of obesity in the United States are estimated to account for expenditures of approximately $60 billion.[3] Although the interpretation of the actuarial data became controversial after intense scrutiny, the Centers for Disease Control previously estimated that approximately 400,000 deaths per year can be directly attributed to obesity. Furthermore, the health risks associated with increased body weight may overtake the quantitative impact of tobacco as a contributor to overall cardiac risk assuming the present trends continue.

Obesity had been previously regarded simply as a passive deposition of excess energy stored in adipose tissue. However, an increasing body of clinical and experimental evidence has implicated adipose tissue as a highly active endocrine organ that is intimately involved in a variety of clinically important metabolic pathways that subsequently increase the risk for atherosclerosis. Despite a clear univariate statistical correlation between increased body mass index and total or cardiovascular mortality, the independent cause-and-effect relationship is difficult to establish with definite certainty due to multiple potential confounding factors including degree of physical activity, regional distribution of fat (truncal versus gluteal), socioeconomic status, and associated metabolic disorders such as hypertension, hyperlipidemia, and diabetes. Age may also play a role in the impact of obesity on mortality; the statistical relationship may attenuate over time and become less prominent in older age groups.[4] This review will focus on the role of the adipocyte in inflammation and the potential impact on cardiovascular risk.

OBESITY AND CARDIOVASCULAR RISK

Obesity has been statistically associated with profound reductions in lifespan in both men and women. For example, a 20-year-old Caucasian man with a body mass index in excess of 45 is estimated to lose 13 years of life due to medical problems directly related to obesity.[5] Cardiovascular disease (including cerebrovascular accident and sudden cardiac death) is a major component of the increased mortality associated with obesity.

However, significant clinical and statistical problems are encountered in the quantification of the impact of obesity on cardiovascular mortality due to the common association of a variety of intimately interrelated and frequently coexistent metabolic conditions such as diabetes, dyslipidemia, and hypertension. Additionally, individuals who are classified as overweight purely on the basis of body mass index may be relatively healthy if free of associated metabolic disorders.

The concept that overweight or obese individuals who are physically fit and do not have associated metabolic conditions may not demonstrate increased cardiovascular risk has been proposed although considerable controversy exists due to methodological problems and the lack of large-scale prospective controlled clinical trials.[6] However, it is clear that a lean physically fit individual has the lowest overall risk of developing cardiovascular disease. The presence and quantitative contributions of lifestyle, diet, exercise patterns, and degrees of physical fitness are difficult to determine in epidemiologic studies and may potentially confound the interpretation of observational data.

The duration and intensity of physical activity are also significant determinants of body weight, glucose tolerance, and blood pressure but the quantitation of physical fitness is frequently problematic in large scale observational studies. The relationship of cardiovascular fitness, mortality, and obesity is conflicting. Studies have demonstrated that the highest relative risk of all-cause mortality was evident in obese individuals who were physically unfit.[7] However, subjects who were of normal weight and physically unfit appeared to have higher relative risks of mortality when compared to obese individuals with optimal cardio-respiratory fitness. The concept of the metabolically healthy obese (MHO) individual has been proposed as a means to eliminate confounding findings but is difficult to define and has not traditionally been evaluated in epidemiologic studies.

Multiple observational trials performed in men have linked physical fitness and health benefits with resultant reductions in risk from cardiovascular disease.[8] The benefits from cardiovascular fitness have also been demonstrated in women. However, conflicting results relative to the contributions of physical fitness and obesity may be gender-related. The relationship of physical fitness, obesity, and cardiovascular events in roughly 1000 women was determined over a 4-year observational period.[9] Women who were demonstrated to have low levels of physical activity independent of obesity were significantly more likely to have major adverse cardiovascular events during the 48-month follow-up period. The results were interpreted to imply that physical fitness and metabolic health may be more important than obesity *per se* for the determination of cardiovascular risk in women. The presence of conflicting data has generated a significant debate concerning the relative role of "fitness versus fatness" in cardiovascular and total mortality and has not been completely resolved.

Diabetes mellitus has been termed a coronary equivalent due to the high risks of developing cardiac events in diabetic subjects. The role of physical fitness and alteration of body mass index as a means to reduce the risk of developing diabetes has been recently analyzed.[10] The relative contribution of body mass index as a means to predict the subsequent risk of development of type II diabetes appeared to be stronger than the degree of physical activity. Additionally, improved physical activity appeared to have little effect upon the relationship between changes in body mass index and development of diabetes. However, definite clinical endpoints were not evaluated and the method employed to report physical activity may not have been optimal. Thus the relative roles of physical activity, metabolic health, and obesity are difficult to separate in observational studies and represent significant problems for related risk factors such as dyslipidemia and hypertension.

Despite the methodological difficulties in the analysis of the quantitative independent contribution of obesity to global cardiovascular risk, it is clear that metabolic factors associated with obesity play a major role in the subsequent incidence of cardiac events. The term *metabolic syndrome* describes a constellation of cardiovascular risk factors (including truncal obesity) that appear to share common metabolic pathways as a means to identify and target high risk individuals for aggressive multifactorial therapeutic interventions. The National Cholesterol Education Program has established diagnostic criteria and requires any three of the five following risk factors[11] to be present in order to make the diagnosis of metabolic syndrome (Figure 9.1):

1. Abdominal obesity is defined as waist circumference measured at the umbilicus and is considered to be positive if in excess of 40" in men and 35" in women.
2. Fasting triglycerides must be in excess of 150 mg per deciliter.
3. Fasting glucose: blood sugar must be in excess of 110 mg per deciliter.

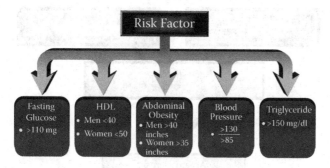

FIGURE 9.1　Clinical diagnosis of the metabolic syndrome.

4. HDL cholesterol must be less than 40 mg per deciliter in men and less than 50 mg per deciliter in women.
5. Blood pressure must be in excess of 130 mm of mercury systolic and 85 mm of mercury diastolic.

The clinical utility of the concept of the metabolic syndrome as a specific disease entity has been challenged by the American Diabetes Association and the European Association for the Study of Diabetes.[12] However, the use of the *syndrome* term does not necessarily imply a specific disease entity but refers to a grouping of clinical symptoms or conditions that appear to coexist in individuals out of proportion to the prevalence in the general community. The major cardiovascular risk factors share multiple metabolic pathways that have been linked to the process of atherosclerosis. The Common Soil Hypothesis of atherosclerosis is based on the premise that factors such as obesity, hypertension, diabetes, and dyslipidemia share pathways that may play a role both in the preclinical manifestations of atherosclerosis and the risk of progression to occlusive vascular disease.(Figure 9.2). The clinical factors required for the definition of the metabolic syndrome are characterized by evidence of inflammation and oxidative stress that is demonstrable across the spectrum of atherosclerosis and may represent preclinical targets for risk stratification or therapeutic interventions.[13]

CLINICAL MARKERS OF INFLAMMATION

The advent of clinical markers has established a central role for inflammation as a primary pathogenetic mechanism in the initiation and progression of atherosclerosis. The concept that atherosclerosis is an inflammatory disease has gained credence over the past decade. The recognition that obesity is also associated with a significant inflammatory component has been established from a large body of clinical evidence that emanated from experimental, epidemiologic, genetic, and pathologic studies.

Additionally, markers of inflammation have been utilized to link obesity with other cardiovascular risk factors such as diabetes, hypertension, and dyslipidemia.

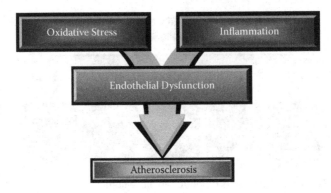

FIGURE 9.2   The common soil hypothesis.

Sensitive, reproducible, and specific markers of inflammation are necessary to establish accurate risk stratification and potentially to monitor response to therapy. While the bulk of clinical and experimental evidence has centered around the role of C-reactive protein in the inflammatory state associated with obesity and atherosclerosis, a number of other pro- and anti-inflammatory proteins have been employed for clinical stratification including leptin, tumor necrosis factor-α, adiponectin, interleukins-1, -6, and -10, resistin, fibrinogen, plasminogen activator inhibitor-1, and others.

C-Reactive Protein

C-reactive protein is one of the original inflammatory markers that was systematically analyzed for a potential role in risk stratification for coronary atherosclerosis. An extensive body of epidemiologic and clinical trial evidence has been accumulated concerning its role in both vascular disease and obesity. Additionally, the utility of C-reactive protein as a marker for cardiovascular risk stratification has continued to expand as improved methods of analysis have been developed. The original assays for C-reactive protein exhibited wide ranges of normality. Considerable intra-individual variability was problematic in analyzing the clinical relevance of relatively minor changes occurring over time.

The subsequent availability of high sensitivity assays has allowed the analysis of changes of C-reactive protein within the normal range and significantly improved the clinical utility of this marker in cardiac risk stratification. Multiple epidemiologic studies in primary prevention have demonstrated that a single measurement of C-reactive protein is a strong predictor of future vascular events and is independent of the traditional cardiac risk factors.[14] C-reactive protein levels are increased in several cardiac risk factors (including truncal obesity) required by the National Cholesterol Education Program for the diagnosis of the metabolic syndrome and has added prognostic information to traditional risk stratification.[15]

C-reactive protein has been demonstrated to be an independent predictor of future cardiac risk and adds prognostic information obtained from traditional lipid screening and the Framingham Risk Score in subjects classified as candidates for primary prevention. The clinical utility of quantifying C-reactive protein levels in risk stratification in secondary prevention is more controversial because these subjects require an aggressive multifaceted therapeutic approach that would include interventions that lower inflammatory markers. However, C-reactive protein has been demonstrated to be an independent predictor of future cardiovascular risk in patients with angiographically documented coronary artery disease and is also correlated with the number of angiographically determined complex coronary artery lesions in subjects with acute coronary syndromes.[16] The role that C-reactive protein plays in risk stratification in primary prevention has been clearly defined over the past decade. However, increasing evidence indicates that it is also a direct participant in the atherosclerotic process.[17]

C-reactive protein is a complex molecule composed of 523 kilodalton subunits and has been classified as a member of the pentraxin family. The liver is a major site of the synthesis of C-reactive protein which is at least partially under the modulation of interleukin-6 produced by adipose cells.[18] C-reactive protein was determined to be a major component of the immune response system and is directly involved in the primary migration of inflammatory cells into the subendothelial space.[19] The recognition that C-reactive protein is an active participant in the inflammatory process has raised the possibility of a direct role in the pathogenesis of atherosclerosis.

C-reactive protein has been demonstrated to mediate a variety of pro-inflammatory and atherosclerotic effects on human aortic endothelial cells following recognition and binding to FC gamma receptors (CD 32 and CD 64).[20] It has also been demonstrated to be localized in the vascular intima within regions that are prone to the development of atherosclerosis.[21] Additionally, C-reactive protein deposition precedes the appearance of monocytes in the earliest stages of atherosclerosis.

The monocytes that subsequently give rise to macrophage scavenger cells have been demonstrated to express a specific receptor for C-reactive protein. The administration of monoclonal antibodies directed at the receptor eliminates C-reactive protein-mediated monocyte chemotaxis and supports the major role it plays in the recruitment of inflammatory cells in coronary artery disease. The localization of C-reactive protein within the subendothelial space is also associated with the expression of a variety of cellular adhesion molecules, monocyte chemotactic protein (MCP-1), and plasminogen activator inhibitor (PAI-1). C-reactive protein levels thus appear to be intimately involved with the early phases of atherosclerosis and are associated with increased risk across the spectrum of vascular disease.

Additionally, C-reactive protein has been demonstrated to be elevated in subjects with increased body mass index and obesity which substantiates the role of inflammation in obesity. The epidemiologic association of elevated levels of C-reactive protein and cardiovascular risk is robust.[22,23] The presence of elevated C-reactive protein in obese and insulin-resistant subjects also appears to predict the subsequent risk of development of type 2 diabetes which also implicates inflammation as a potential mechanism in the pathogenesis of the metabolic syndrome.[24]

Hygienic measures such as weight loss in obesity have been demonstrated to lower circulating levels of inflammatory markers such as C-reactive protein by regulating genes involved with the inflammatory responses in white adipose tissue.[25] The elevations of C-reactive protein in obesity link the progressive accumulation of adipose tissue with inflammation and may provide correlation with cardiovascular risk in obese subjects.

LEPTIN

Leptin was one of the original hormones isolated from adipose tissue and the discovery subsequently initiated extensive investigations that elucidated its role in the complex metabolic pathways characterizing obesity. Leptin is directly synthesized within adipocytes and is a central physiologic factor in the regulation of both appetite and energy expenditure. It plays a physiologic role in the regulation of body mass index by increasing energy expenditure and reducing caloric intake to maintain metabolic balance.

Genetic models such as ob/ob mice that lack the gene encoding for the synthesis of leptin demonstrate significant increases in body mass index that are also associated with insulin resistance and high subsequent risks for development of type II diabetes.[26]

Additionally, the administration of leptin in these models has been demonstrated to result in reductions in both caloric intake and body mass index. The physiologic action of leptin in the regulation of caloric intake and body mass requires the presence of normal hormonal levels, receptor function, and signal transduction. Rare genetic conditions associated with morbid obesity have been demonstrated to be correlated with either reduced levels of circulating leptin or associated receptor abnormalities.[27] Leptin has also been utilized as a therapeutic agent and its administration to subjects with genetically mediated deficiency states resulted in reductions in body mass index plus improvements in both inflammatory and metabolic parameters.[28]

However, significant differences are present in the forms of obesity prevalent in the general population compared to subjects with genetically mediated congenital leptin deficiency. The increased body mass in obese individuals is associated with a significant capacity to synthesize leptin due to the presence of large numbers of adipocytes. Leptin resistance is common in subjects with coexistent insulin resistance, diabetes, and obesity. The precise mechanism that underlies the high circulating levels of leptin and resistance to the normal physiologic function of this hormone is complex and multifactorial. The presence of progressively increasing levels of leptin appears to result in down-regulation of receptor number or activity, with subsequent resistance to the normal physiologic activity of the hormone.

The chronic administration of a diet enriched in calories accounted for by saturated fat had been demonstrated to be associated with a subsequent increase in circulating leptin levels despite a progressive expansion of body mass. The induced increase in body mass index secondary to caloric excess has been correlated with a suppression of signaling for leptin due to a progressively increased expression of SOCS-3 (suppressor of cytokine signaling-3) that significantly alters the normal physiologic and regulatory effects within hypothalamic centers in the brain.[29]

Leptin also has been demonstrated to have significant metabolic effects in peripheral tissues that were initially identified in genetic conditions such as congenital lipodystrophy. Experimental models with this uncommon genetic condition are associated with minimal adipose tissue deposits in peripheral tissues, resistance to insulin, and progressive and persistent hyperinsulinemia. The administration of leptin in subjects with congenital lipodystrophy was demonstrated to restore the metabolic abnormalities to normal.[30] Leptin is now established as a major hormone produced by adipocytes and plays a central role in the regulation of energy homeostasis, insulin sensitivity, and body mass index.

ADIPONECTIN

Adiponectin is an adipocyte-derived protein shown to express beneficial physiologic effects in clinical conditions associated with the development of coronary artery disease including significant anti-inflammatory effects[31] (Figure 9.3). It has been demonstrated to be interrelated with obesity, insulin resistance, endothelial function and atherosclerosis. Adiponectin may reverse a variety of inflammatory and metabolic abnormalities which are associated with obesity by increasing the sensitivity of insulin in the peripheral tissues with resultant improvement in glucose homeostasis.[32]

Adiponectin has also been demonstrated to mediate an improvement in endothelial function that may be at least partially due to its potent inherent anti-inflammatory effect. Adiponectin has been shown to reduce the synthesis of the potent tumor necrosis factor-α pro-inflammatory cytokine associated with endothelial dysfunction. Management of endothelial function associated with

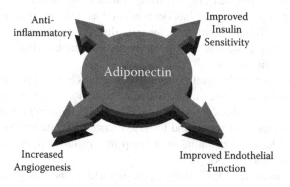

FIGURE 9.3   Adiponectin and atherosclerosis.

dyslipidemia or insulin resistance may be enhanced by therapy with drugs such as PPAR (peroxisome proliferator-activated receptor) agonists such as fenofibrate or insulin sensitizers such as thiazolidinediones that also increase adiponectin levels.[33,34] Circulating adiponectin levels have been evaluated in obese male and female subjects and proved to be inversely related to body mass index although the relationship may be more pronounced in women.[35]

Reduction in the circulating level of adiponectin is demonstrable in multiple stages of atherosclerosis ranging from endothelial dysfunction to advanced obstructive vascular disease. The earliest stage of atherosclerosis is characterized by inflammatory cell binding to the dysfunctional endothelium which is mediated by the production of cellular adhesion molecules. The production of adhesion molecules is mediated by a variety of pro-inflammatory cytokines such as tumor necrosis factor-α and has been demonstrated to be reduced by adiponectin. The inverse relationship between adiponectin and tumor necrosis factor-α links the protective effect of this adipocyte derived protein to the preclinical phases of coronary artery disease.[36]

Adiponectin may also play a protective role in more advanced phases of vascular obstruction by increasing angiogenesis in chronic ischemia. Experimental studies have been performed in an adiponectin knock-out mouse model that employed a chronic hind limb ischemia preparation. The depletion of adiponectin resulted in a significant impairment of vascular repair of the involved ischemic area. Adenovirus-mediated supplementation of adiponectin resulted in a significant improvement in angiogenic repair which implicates a potential role for this adipocyte-derived protein in the development of collateral flow in chronic ischemia.[37] Adiponectin thus has significant anti-inflammatory activity and is reduced in obesity and obstructive vascular disease.

TUMOR NECROSIS FACTOR-α

Tumor necrosis factor-α is a pro-inflammatory cytokine whose levels are increased in such seemingly disparate clinical conditions as collagen vascular diseases and congestive heart failure. Modification of the activity of tumor necrosis factor by receptor blockers has been utilized as a therapeutic target in rheumatoid arthritis. Tumor necrosis factor-α is directly synthesized in adipocytes and is felt to play a significant role in the maintenance of both the mass and metabolism of adipose tissue.[38]

The localization of the synthesis of tumor necrosis factor-α within the adipose tissue also allows for a potential autocrine or paracrine effect that may alter the numbers and sizes of adipocytes in addition to systemic pro-inflammatory effects. Tumor necrosis factor-α was found to be associated with enhanced apoptosis of stem cells that have the potential to develop into mature adipocytes in addition to fully differentiated adipocytes.[39]

The mechanism by which tumor necrosis factor-α induces programmed cell death is complex and multifactorial. The pro-inflammatory activity of tumor necrosis factor-α is associated with a variety of proteolytic enzymes that may be involved in progressive and irreversible cellular degradation.[40] Additionally, up-regulation of the capase gene may also be related to increased levels of tumor necrosis factor-α and has been demonstrated to be inhibited by the administration of potent anti-inflammatory agents such as glucocorticoids (e.g., dexamethasone).[41]

Tumor necrosis factor-α also plays a significant role in lipid metabolism and the resultant metabolic fate of circulating lipoproteins. The degradation of tri-glyceride-rich lipoproteins such as very low density lipoprotein is modulated by activation of lipoprotein lipase and results in the generation of free fatty acids. The local availability of free fatty acids within adipose tissue is a significant determinant of the triglyceride contents of individual adipocytes. Free fatty acids may be accumulated within adipocytes and subsequently synthesized into triglycerides under the enzymatic regulation of Acyl-CoA synthetase. Tumor necrosis factor-α may exert significant effects on fatty acid metabolism and lipid contents of adipocytes by direct effects on the activity of lipoprotein lipase which alters substrate availability for triglyceride synthesis.

The intrinsic enzymatic activity of lipoprotein lipase is significantly reduced by tumor necrosis factor-α with a resultant decrease in the metabolism of tri-glyceride-rich lipoproteins. The impaired degradation of triglyceride-rich lipo-proteins decreases the potential for enhanced flux of free fatty acids into adipo-cytes and results in the alteration of intracellular lipid concentration.[43] Additionally, tumor necrosis factor-α is associated with direct down-regulation of acyl Co-A synthetase which may also contribute to the degree of triglyceride synthesis and lipid accumulation within adipocytes.[43]

The role of tumor necrosis factor-α in the manifestations of inflammation in obesity in human subjects is complex. The local synthesis of tumor necrosis factor-α in the adipocytes of obese subjects is increased and sustained weight loss has been demonstrated to reduce circulating levels.[44] However, conflicting results have been reported concerning the relationship of circulating tumor necro-sis factor-α and body mass index that may be secondary to methodologic prob-lems. Additionally, genetic models for obesity have demonstrated that increased circulating levels of tumor necrosis factor-α are associated only with extreme increases in body weight.[45]

Obesity and increased body mass index are frequently associated with the development of insulin resistance and adult onset diabetes. Inflammation has been postulated to play a significant role in the pathogenesis of insulin resistance in obese subjects. The role of tumor necrosis factor-α in the pathogenesis of insulin resistance has been evaluated in experimental studies. Obese subjects with asso-ciated insulin resistance showed increased circulating levels of tumor necrosis factor-α that were compatible with the possibility of a cause-and-effect relation-ship. Tumor necrosis factor-α may induce insulin resistance by several mecha-nisms including down-regulation of the insulin receptor, decreased GLUT 4 synthesis, and altered lipolytic activity.[46] The interrelationship between obesity and tumor necrosis factor is complex and a hypothesis that explains all aspects remains elusive due to multiple genetic, gender, and metabolic issues. However, it is clear that tumor necrosis factor-α plays a central role in adipocyte function and mass and may have implications for insulin resistance and diabetes.

HEMOSTATIC PARAMETERS

Obesity has been correlated with a variety of abnormalities of the hemostatic system including alterations in the levels of fibrinogen, von Willebrand factor, and plasminogen activator inhibitor. Fibrinogen, which is primarily synthesized in the liver, is intimately related to the coagulation cascade and additionally is an acute phase reactant. Fibrinogen is thus closely linked to inflammation and is demonstrated to be increased following a variety of inflammatory stimuli.

Obesity has been demonstrated to be associated with increased circulating levels of fibrinogen.[47] However, the relationship between fibrinogen levels and body mass index or localization (truncal versus peripheral) of adipose tissue is controversial. The metabolic syndrome (for which truncal obesity is one of the diagnostic criteria) has been correlated with abnormalities in multiple coagulation parameters. Fibrinogen levels are increased in individuals fulfilling criteria for a diagnosis of metabolic syndrome which may contribute to the increased risk of hypercoagulability associated with the classical risk factors such as hypertension, diabetes, and dyslipidemia.[48]

Endothelial dysfunction is associated with a variety of abnormalities in the coagulation and fibrinolytic systems. The von Willebrand factor is a large glyco-protein that is synthesized at least partially in endothelial cells and may represent a clinical marker for endothelial dysfunction. However, the relationship between von Willebrand factor and obesity is complex and difficult to elucidate because the glycoprotein is not directly synthesized by adipocytes. However, the levels of von Willebrand factor antigen have been significantly correlated with the degree of visceral fat in subjects who are either overweight or definitely obese.[49]

Obesity is frequently associated with alterations of the balance between thrombotic and fibrinolytic parameters. The primary inhibitor of tissue plasmi-nogen activation is mediated by plasminogen activator inhibitor-1 (PAI-1) which decreases the rate of breakdown of fibrin clots (Figure 9.4). Risk factors such as obesity, hypertension, hypertriglyceridemia, and diabetes are frequently charac-terized by elevations of circulating levels of plasminogen activator inhibitor. The presence of a relative excess of circulating plasminogen activator inhibitor is associated with an increased risk of occlusive intravascular thrombus due to impaired fibrinolysis. Plasminogen activator inhibitor is synthesized in response to inflammation by endothelial cells, adipocytes, hepatocytes, and platelets.[50] Additionally, the levels of plasminogen activator inhibitor have been correlated with both obesity and increased body mass index which establishes an association of obesity, hypercoagulability, and inflammation.[51]

Abnormalities of plasminogen activator inhibitor levels are further correlated with the degree of visceral fat, which emphasizes the role of the distribution of adipose tissue in cardiovascular risk.[52] The relationship between visceral fat accumulation and abnormalities in the fibrinolytic pathway is emphasized by the

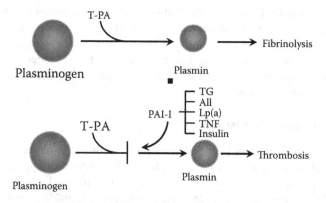

FIGURE 9.4 Endothelial dysfunction and fibrinolysis.

fact that reductions in plasminogen activator inhibitor are best correlated with a decrease in the amount of visceral adipose rather than body weight per se.[53]

The current definitions of the metabolic syndrome do not include components of the hemostatic or fibrinolytic system. However, the risk factors (e.g., diabetes, hypertension, obesity, and dyslipidemia) included in the criteria for diagnosis have inflammatory bases and are also frequently associated with either hypercoagulability or impaired fibrinolysis.[54] Markers of inflammation or fibrinolysis may be included in the criteria for metabolic syndrome as further studies evolve and more sensitive and specific indicators of cardiac risk are evaluated.

RESISTIN

Resistin is a recently described pro-inflammatory cytokine that is produced within adipose tissue and has been linked to both obesity and the risk for development of diabetes mellitus.[55] However, the role of resistin in the characteristic obesity of humans is controversial despite the compelling data from animal models. Resistin is a cysteine-rich polypeptide that is directly synthesized by adipocytes.[56]

Circulating levels of resistin were shown to be increased in experimental models that demonstrated impaired insulin sensitivity and diabetes. Additionally, circulating resistin levels were reduced by insulin sensitizing agents such as thiazolidinediones.[57] The increased levels of resistin in obesity have been demonstrated to be generated by direct synthesis from activated macrophages that are localized within adipose tissue and regulated by PPAR gamma activators.[58]

The glucose intolerance associated with obesity may also be linked to increased levels of resistin via complex metabolic and inflammatory pathways. Experimental studies utilizing transgenic mice with high circulating concentrations of resistin in the presence of normal weight showed increased glucose production in the setting of a hyperinsulinemic–euglycemic clamp.[59] The fact that elevated levels of resistin are associated with impaired glucose homeostasis is at least partially explained by the associated increased expression of the phosphoenolpyruvate carboxykinase hepatic enzyme that serves as a key enzyme in glucose metabolism. The relationship between resistin and inflammation has been demonstrated by an association with inflammatory mediators such as tumor necrosis factor-α and C-reactive protein.[60,61] Resistin has been hypothesized to play multiple roles in obesity and diabetes secondary to the induction of inflammation and a secondary alteration of the metabolism of glucose due to the associated insulin resistance.

OTHER INFLAMMATORY MARKERS

A variety of other markers have been proposed as clinically relevant parameters that have utility in the identification of inflammation and potential cardiovascular risk stratification. Interleukin-6, interleukin-10, interleukin-1, adhesion molecules, and visfatin have all been identified as potentially valuable clinical markers. However, the bulk of clinical and experimental data related to obesity, diabetes, and inflammation were derived from the identification of C-reactive protein and improved high sensitivity measurements that have allowed accumulation of a large body of clinical and experimental evidence that substantiates the role of inflammation in obesity

ATHEROSCLEROSIS AND INFLAMMATION

The concept of atherosclerosis as an inflammatory disorder is supported by an ever-increasing body of pathologic, experimental, epidemiologic, and clinical evidence.[62] The delineation of the role that inflammation plays as a primary factor in the initiation and progression of vascular disease has led to enhanced understanding of the basic biologic properties of atherosclerotic plaques, risk stratification, and therapeutic strategies.

The cellular elements involved in inflammation can be demonstrated to be present to variable degrees in multiple stages in the atherosclerotic process. Additionally, inflammation also can be related to the classic cardiovascular risk factors including obesity. The process by which inflammatory cells are localized into the subendothelial space and contribute to atherosclerosis is complex and requires cellular recognition, binding, migration, and transformation. The normal endothelium is not associated with significant adherence of inflammatory cells. However, the dysfunctional endothelium expresses a variety of adhesion molecules such as vascular cell adhesion molecule-1 (VCAM-1) and members of the selectin (E and P) family.[63]

The progressive expression of adhesion molecules on the dysfunctional endothelium provides the means for the recognition, binding, tethering, and transmigration of circulating inflammatory cells. The accumulation of inflammatory cells is an initial phase of atherosclerosis. Additionally, cytokines and chemokines are small proteins that are intimately involved in the transmigration and concentration of inflammatory cells into vascular regions prone to develop atherosclerosis.[64] Cytokines are involved in enhanced migration of several inflammatory cell lines of which the monocyte is the prototype example.

Monocyte chemoattractant protein (MCP-1) is a primary regulator of the accumulation and concentration of monocytes into the subendothelial spaces in the initial stages of atherosclerosis. Monocytes play a significant role in the cellular inflammatory response and also can transform into scavenger cells that are involved in the recognition and binding of oxidized low density lipoprotein with subsequent generation of lipid-laden foam cells. The capacity of the monocyte to transform into a macrophage scavenger cell line is under the regulation of a variety of colony stimulating factors.[65] The monocyte scavenger receptor recognizes both native and modified low-density lipoprotein. However, the binding and uptake of non-modified low-density lipoprotein is quantitatively minimal.

In contrast, low-density lipoproteins modified by processes such as oxidation or glycation are progressively internalized at a relatively rapid rate by macrophages and are associated with progressive lipid accumulation and the subsequent development of atherosclerosis. In contrast to the classic LDL receptor, the number or activity of the scavenger receptor is not down-regulated by progressive accumulation of intracellular lipid.

The macrophage that originates from the inflammatory cell line also plays a significant role in the progression and ultimate fate of the atherosclerotic plaque. The monocyte–macrophage cellular elements are gradually depleted due either to apoptosis or the cytotoxic effects of a variety of noxious stimuli. The degradation of cellular constituents results in the release of lipids and proteinaceous debris into the plaque and thus has implications for vulnerability and rupture. Macrophages also play a significant role in the local degree of oxidative stress demonstrated to be secondary to the capacity to synthesize cytotoxic reactive oxygen species.[66] The concentration and rate of production of reactive oxidative species are major contributors to the initiation and progression of atherosclerosis effects on lipid oxidation and cellular damage.

The degree of inflammation also plays a significant role in the vulnerability of the atherosclerotic plaque and subsequent risk for rupture. High-grade inflammation is associated with a progressive reduction in plaque stability and resultant increased vulnerability. Macrophages have the capacity to elaborate a variety of proteolytic enzymes that may be involved in the net balance between intraplaque matrix synthesis and degradation. Collagenase and gelatinase are matrix metalloproteinases that demonstrate the capacity to progressively degrade the stabilizing matrix protein constituents and result in a progressively increased risk for rupture due to reduction in the tensile strength of the plaque.[67]

Increased plaque vulnerability is characterized by an enhanced risk of erosion or frank rupture of the fibrous cap. Destruction of the structural integrity of the fibrous cap results in exposure of platelets and coagulation factors to the highly thrombogenic lipid-rich core localized within the atherosclerotic plaque. Procoagulants such as tissue factor are produced within the lipid-laden foam cell and are associated with increased risk for vascular occlusion due to enhanced synthesis of activated clotting factors VII and X.[68] Macrophages are also involved in the synthesis of inhibitors to plasminogen activators (e.g., PAI-I), resulting in a decreased capacity to lyse a potentially occlusive intravascular thrombus.[69]

Inflammation is thus a major contributor to all phases of atherosclerosis from endothelial dysfunction to acute myocardial infarction. The major cardiac risk factors such as diabetes, hypertension, and dyslipidemia are all associated with a degree of inflammation and oxidative stress. Recent evidence has centered around the role of adipocytes in inflammation and a large body of work relates obesity to the inflammatory response.

OBESITY AND INFLAMMATION

The major cardiac risk factors frequently coexist and share multiple common metabolic pathways. Diabetes, hypertension, the use of tobacco products, and dyslipidemia have all been demonstrated to be associated with inflammation. However, the concept that obesity is characterized by active inflammation is a relatively new concept that initially was not widely accepted. The previous perception of obesity was that adipose tissue simply represented a passive storage depot for excess energy resulting from increased caloric intake relative to energy expenditure.

Adipose tissue was subsequently demonstrated to be a metabolically active endocrine organ and the resultant increase in body mass index was implicated in the pathogenesis of diabetes, hypertension, and dyslipidemia. Adipose tissue has been shown to be intimately involved in a variety of inflammatory processes and the mechanisms have been partially clarified. Adipose tissue is the source of a number of cytokines that regulate the production of inflammatory mediators including C-reactive protein, tumor necrosis factor-α, interleukin-6, etc.

Insulin resistance is also thought to play a significant role in the pathogenesis of several cardiovascular risk factors in obese subjects. Individuals with elevated circulating levels of insulin due to impaired peripheral resistance have increased free fatty acid release from adipocytes and enhanced delivery to the liver. The increased bioavailability of free fatty acids as a synthetic substrate results in enhanced hepatic production of very low-density lipoprotein. Additionally, the enzymatic activity of lipoprotein lipase is decreased in obese subjects with insulin resistance. The reduced catabolic rates in combination with hepatic overproduction of triglyceride-rich particles results in an atherogenic dyslipidemic phenotype characterized as the lipid triad (hypertriglyceridemia, low HDL, and small dense LDL) (Figure 9.5 and 9.6).

Additionally, hyperinsulinemia is associated with vascular smooth hypertrophy, enhanced sympathetic activity, and increased renal sodium reabsorption with the resultant development of systemic hypertension.[70] Adipocytes have been implicated as the initial sites of inflammation in obesity and predate the involvement of other organ systems. C-reactive protein has been demonstrated to modulate the movement of inflammatory cells into adipose tissue via the production of a variety of chemokines. Monocyte migration and subsequent transformation into macrophages within adipose tissue are key factors in the self-perpetuating low-grade inflammation associated with obesity.

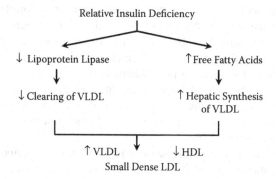

FIGURE 9.5 Insulin deficiency and dyslipidemia.

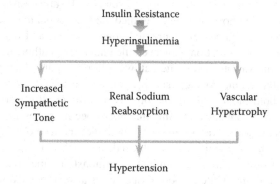

FIGURE 9.6 Insulin levels and hypertension.

The demonstration that adipocytes and monocytes share pathways in the immune response system emphasizes the role that adipose tissue plays in the interaction between inflammation and atherosclerosis. Additionally, the cellular precursors to mature adipocytes were demonstrated to have the ability, similar to the monocyte–macrophage system, to act as phagocytes. The adipocytes localized within fat depots may thus participate in a primary manner in the inflammatory response, systemic hypertension, and atherosclerosis.

The inflammatory response within adipose tissue may at least be partially expressed under genetic control. Experimental models that utilized animals with genetically transmitted obesity allowed the examination of factors that control and modulate inflammatory states associated with increases in body mass. Multiple genes have been implicated in the inflammatory response associated with obesity. Genes that regulate the monocyte–macrophage system have been demonstrated to be present in white adipose tissue. The experimental studies in the obese mouse models have demonstrated the expression of a large number of transcripts that can be correlated with body mass.

Biochemical markers for macrophages were significantly correlated with both body mass and the size of the adipocytes. Analysis of the genes that had the highest correlations with obesity demonstrated that approximately 30% were involved in encoding for proteins that were characteristic of macrophages.[71]

Tumor necrosis factor-α is a pro-inflammatory cytokine demonstrated to be produced by macrophages and is localized within adipose tissue. The potential thus exists for a self-perpetuating cycle of increased macrophage infiltration coupled with enhanced synthesis of cytokines within adipose tissue and a resultant persistent inflammatory milieu.

The establishment of an inflammatory environment within adipose tissue may also play a significant role in the pathogenesis of insulin resistance. However, significant differences may exist among genetically mediated obesity, insulin resistance, and an increased body mass that is purely secondary to excessive dietary intake of calories. Chronic inflammation has been linked to the pathogenesis of both type 2 diabetes and insulin resistance.[72] Insulin resistance is due to a failure of circulating hormonal levels to mediate the normal metabolic handling of glucose despite progressively increased concentrations and has been related to the presence of inflammation. However, the underlying pathogenesis and interrelationship of the various metabolic pathways is not totally clear.

Inflammation and macrophage-specific genes have been demonstrated to play direct roles in both models of genetic obesity and the obesity associated with a high fat diet.[73] Macrophages and inflammatory-specific genes are significantly up-regulated in the white adipose tissues of both diet-induced obesity and genetic mouse models. The up-regulated genes precede a significant increase in the level of circulating insulin. Histologic examination demonstrates that macrophage infiltration is present in the white adipose tissue and is also associated with evidence of cellular destruction of adipocytes.

Therapy with rosiglitazone, an insulin sensitizing agent, was associated with a significant down-regulation of the macrophage-originated genes. Thus, inflammation and macrophage infiltration may play significant roles in both diet-induced obesity and genetically induced obesity and result in progressive insulin resistance due to a chronic inflammatory and cytotoxic response initiated in adipose tissue.

MODIFICATION OF INFLAMMATION IN OBESITY

The identification of obesity as a condition associated with a definite inflammatory component has significant clinical implications relative to targets of therapy (i.e., modification of body mass index in and of itself, improving the function of adipose tissue, and normalizing the levels of inflammatory markers). Therapeutic interventions that alter the inflammatory response may also be associated with restoration of the normal physiologic function of adipose tissue.

The initial and primary foci of therapy in obesity involve lifestyle modification with restriction of calories and saturated fat coupled with increased energy expenditure through physical activity. Increased physical activity has been demonstrated to decrease body mass index, improve dysfunctional adipose tissue, and reduce levels of inflammatory markers. However, subjects who are obese but metabolically healthy may require alternative approaches to improve long-term cardiovascular risk. As risk factor stratification proceeds beyond tabulation of the classical risk factors, modification of inflammatory markers may possibly become a primary target for therapy with pharmacologic or lifestyle interventions. The role of modification of inflammatory markers as targets for therapy has been addressed in a variety of clinical trials utilizing lifestyle or pharmacologic interventions.

C-Reactive Protein

C-reactive protein is a major inflammatory marker and a number of lifestyle or pharmacologic interventions have been demonstrated to reduce circulating levels of this marker. The effect of weight loss on C-reactive protein has been prospectively determined in controlled clinical trials utilizing a variety of dietary interventions. The role of implementation of a Mediterranean style diet low in calories but associated with an increase in monounsaturated fat coupled with an increase in physical activity and designed to decrease body weight by 10% has been analyzed.

Body mass index, interleukin-6, and C-reactive protein were all significantly reduced by the combination of diet and lifestyle intervention. Additionally, adiponectin was increased, thus demonstrating that diet-induced weight loss and enhanced physical activity improved the balance of pro- and anti-inflammatory mediators.[74] Risk factors that compromise the components of the metabolic syndrome have inflammation as a common denominator.

The impact of the institution of a Mediterranean style diet on endothelial function and markers of inflammation in subjects who fulfill the criteria for the diagnosis of the metabolic syndrome has been investigated. The dietary intervention consisted of an increase in the consumption of vegetables, whole grains, fruits, olive oil, and nuts. The Mediterranean diet resulted in significant reductions of C-reactive protein and improvements in insulin resistance. Endothelial function score as analyzed by blood pressure and platelet aggregation responses to l-arginine was also improved by dietary therapy. The components of the metabolic syndrome were normalized to a greater degree in subjects who were randomized to receive intensive dietary interventions.[75]

The effect of statin therapy on inflammatory markers had been presumed to be secondary to lipid lowering. However, statins have been clearly demonstrated to possess a variety of pleiotropic or non-lipid effects. Statins may play a significant role in leukocyte trafficking and T cell activation by binding to a novel allosteric site within the β-2 integrin leukocyte function associated antigen-1.[76] Statin therapy has also been shown to reduce C-reactive protein levels and has been correlated with clinical outcomes in prospective clinical studies (albeit in post-hoc analyses).

The role of statin therapy in lipid lowering and alteration of inflammatory markers was initially analyzed in the Cholesterol and Recurrent Events (CARE) trial that prospectively compared pravastatin to placebo in subjects with relatively normal cholesterol levels who had suffered acute myocardial infarctions.[77] The highest cardiovascular risks occurred in subjects who had significant evidence of inflammation and were randomized to receive placebo. Pravastatin therapy reduced markers of inflammation (C-reactive protein and serum amyloid A) in addition to beneficial effects on lipid profiles.

The utilization of C-reactive protein in predicting response to statin therapy in the primary prevention of atherosclerosis was the subject of the Air Force/Texas Coronary Atherosclerosis Prevention Study (AFCAPS/TEXCAPS) analyzing the effect of lovastatin in 5742 relatively low risk subjects over 5 years.[78] The effect of lovastatin was analyzed after stratification of subjects by the ratio of plasma cholesterol:HDL and levels of C-reactive protein. Lovastatin reduced coronary events irrespective of C-reactive protein levels in subjects with elevated lipid ratios. However, it was also effective in subjects whose lipid ratios were below the median but associated with elevated C-reactive protein levels, inferring benefits of statin therapy in subjects with evidence of inflammation despite the presence of normal lipid levels.

C-reactive protein is associated with an increased density of the AT-1 angiotensin receptor, and modulators of the renin angiotensin system may be associated with an anti-inflammatory effect. Losartan, a renin angiotensin receptor blocker, has been associated with reductions in levels of C-reactive protein although this may not be a class effect.[79,80]

Diabetes and insulin resistance have been correlated with a variety of markers of inflammation including C-reactive protein that additionally may play a role in pathogenesis.[81] The treatment of insulin resistance and diabetes with insulin sensitizing agents such as rosiglitazone may demonstrate beneficial affects on both circulating glucose levels and markers of inflammation such as C-reactive protein.[82] Statins have been demonstrated to reduce the morbidity and mortality in diabetes in multiple large scale trials and are also effective even if baseline lipids are relatively normal.[83,84] Additionally, HMG CoA reductase inhibitor therapy has been demonstrated to reduce the incidence of diabetes in subjects in primary prevention, implicating a possible non-lipid effect of statin therapy.[85] The multivariate predictors of the development of diabetes in the West of Scotland Coronary Prevention Study were glucose, body mass index, and triglyceride levels. Interestingly, randomization to pravastatin therapy resulted in a 30% reduction in the incidence of diabetes. The precise mechanism involved in the reduction of the incidence of diabetes could not be delineated but a potential benefit from an anti-inflammatory effect was postulated.

Additionally, angiotensin-converting enzyme inhibition has demonstrated clinical benefits across the spectrum of atherosclerosis in multiple clinical trials. Angiotensin II has been shown to demonstrate a significant inflammatory effect and the use of ramipril was demonstrated to decrease the incidence of diabetes in the Heart Outcomes Prevention Evaluation (HOPE) study.[86] Additionally, in an extension of the original HOPE trial that continued 30 months after the termination of the initial trial, ramipril use was shown to be associated with sustained reductions in the incidence of new onset diabetes.

Tumor Necrosis Factor-α

Tumor necrosis factor-α is well established as a pro-inflammatory cytokine although the effects of interventions to modify circulating levels in obesity or cardiovascular disease lack the broad data base associated with C-reactive protein. However, tumor necrosis factor-α is directly synthesized in adipose tissue and may be a target for therapy in obesity.

Clinical studies focused on the role of weight reduction in obese subjects as a means to reduce insulin resistance, utilizing hypocaloric diets and increased physical activity, have demonstrated decreases in circulating levels of adipokines including tumor necrosis factor-α.[87] However, the individual studies were small and the reduction in tumor necrosis factor approached but did not reach statistical significance.

Hypolipidemic therapy with statins has been demonstrated to reduce tumor necrosis factor-α levels. Interestingly, the reduction in levels also correlated with a significant reduction in matrix metalloproteinase levels, which provides a link among statin therapy, inflammation, and plaque stability.[88] The increase in tumor necrosis factor-α levels associated with obesity and insulin resistance may be successfully managed by treatment with PPAR gamma agonists.

The anti-inflammatory effect of rosiglitazone has been evaluated in both diabetic and non-diabetic obese subjects. Tumor necrosis factor-α levels were reduced in obese subjects compatible with significant anti-inflammatory effects of the PPAR gamma agonists. Tumor necrosis factor has been established as a valuable marker of inflammation and may be reduced by a variety of lifestyle and pharmacologic interventions that improve lipids, insulin sensitivity, and blood pressure in addition to potential anti-inflammatory effects. However, further clinical trials are necessary to demonstrate the clinical effects of modifying circulating levels of tumor necrosis factor-α.

Hemostatic Factors

Hemostatic factors such as fibrinogen and plasminogen activator inhibitor are increased in obesity and are also linked to inflammation. However, lifestyle and pharmacologic interventions produced variable effects on these parameters and a consensus view of beneficial interventions has not been established due to conflicting results. However, in some studies, statin therapy was demonstrated to improve hemostatic parameters.

Atorvastatin has been demonstrated to improve global fibrinolytic activity and reduce plasminogen activator inhibitor coupled with an increase in tissue plasminogen activator levels in hypercholesterolemic patients.[89] The effect of simvastatin on hemostatic and fibrinolytic factors has been studied in diabetic patients. In addition to the beneficial effects on total and low-density lipoprotein cholesterol levels, simvastatin was demonstrated to reduce circulating plasminogen activator inhibitor concentration and prothrombinase activity.[90]

Additionally, lifestyle interventions have been demonstrated to beneficially alter fibrinolytic activity in some studies of obese patients. Reduction in weight and increased physical activity in obese subjects with insulin resistance have been demonstrated to result in improvement in levels of plasminogen activator inhibitor.[91] Hemostatic factors may be related to an imbalance in fibrinolytic activity and impaired capacity to lyse intravascular thrombi. While hemostatic factors represent attractive targets in obesity due to their relationship with inflammation and may be modified by pharmacologic interventions or lifestyle changes, a consensus view regarding their role as a modifiable risk factor remains controversial and will require larger prospective studies.

SUMMARY

The incidence and prevalence of obesity are markedly increasing in the industrialized world and have reached epidemic proportions. Obesity had been regarded as a passive depot for excess energy. Recent studies have demonstrated that adipose tissue is a dynamic and metabolically active organ system with multiple endocrine functions. Obesity is significantly interrelated with a variety of classical cardiac risk factors including diabetes, hypertension, dyslipidemia, and hemostatic abnormalities. Adipose tissue has been demonstrated to produce a number of pro-inflammatory cytokines that establish a self perpetuating low-grade inflammatory state. Lifestyle modifications such as caloric restriction and increased physical activity have been demonstrated to reduce both body weight and a variety of inflammatory markers.

Pharmacologic interventions such as statin therapy, modulators of the renin angiotensin system, and insulin sensitizers have been demonstrated to exhibit not only effects on blood pressure, lipids and glucose, but they also significantly reduce inflammatory mediators. Thus obesity should be viewed as a highly active metabolic state with strong components of inflammation and oxidative stress that predispose a patient to coronary artery disease and should represent an active target for interventions designed to reduce risks of coronary artery disease.

REFERENCES

1. Mokdad, A.H. et al. The continuing increase of diabetes in the U.S. *Diabetes Care* 24, 412 (2001).
2. Ogden, C.L. et al. Prevalence of overweight and obesity in the United States, 1999–2004. *JAMA* 295, 1549 (2006).
3. National Task Force on the Prevention and Treatment of Obesity. Long-term pharmacotherapy in the management of obesity. *JAMA* 276, 1907 (1996).
4. Stevens, J. et al. The effect of age on the association between body-mass index and mortality. *New Engl J Med* 338, 1 (1998).

5. Fontaine, K.R., Redden, D.T., Wang, C., Westfall, A.O., and Allison, D.B. Years of life lost due to obesity. *JAMA* 289, 187 (2003).

6. Hu, F.B. et al. Adiposity as compared with physical activity in predicting mortality among women. *New Engl J Med* 351, 2694 (2004).

7. Wei, M. et al. Relationship between low cardiorespiratory fitness and mortality in normal-weight, overweight, and obese men. *JAMA* 282, 1547 (1999).

8. Lee, C.D., Blair, S.N., and Jackson, A.S. Cardiorespiratory fitness, body composition, and all-cause and cardiovascular disease mortality in men. *Am J Clin Nutr* 69, 373 (1999).

9. Wessel, T.R. et al. Relationship of physical fitness vs body mass index with coronary artery disease and cardiovascular events in women. *JAMA* 292, 1179 (2004).

10. Weinstein, A.R. et al. Relationship of physical activity vs body mass index with type 2 diabetes in women. *JAMA* 292, 1188 (2004).

11. Grundy, S.M. et al. Diagnosis and management of the metabolic syndrome: an American Heart Association/National Heart, Lung, and Blood Institute Scientific statement. *Circulation* 112, 2735 (2005).

12. Kahn, R., Buse, J., Ferrannini, E., and Stern, M. The metabolic syndrome: time for a critical appraisal: joint statement from the American Diabetes Association and the European Association for the Study of Diabetes. *Diabetes Care* 28, 2289 (2005).

13. Stern, M.P. Diabetes and cardiovascular disease: the "common soil" hypothesis. *Diabetes* 44, 369 (1995).

14. Ridker, P.M. Clinical application of C-reactive protein for cardiovascular disease detection and prevention. *Circulation* 107, 363 (2003).

15. Ridker, P.M., Buring, J.E., Cook, N.R., and Rifai, N. C-reactive protein, the metabolic syndrome, and risk of incident cardiovascular events: an 8-year follow-up of 14,719 initially healthy American women. *Circulation* 107, 391 (2003).

16. Arroyo-Espliguero, R. et al. C-reactive protein elevation and disease activity in patients with coronary artery disease. *Eur Heart J* 25, 401 (2004).

17. Jialal, I., Devaraj, S., and Venugopal, S.K. C-reactive protein: risk marker or mediator in atherothrombosis? *Hypertension* 44, 6 (2004).

18. Moshage, H.J. et al. The effect of interleukin-1, interleukin-6 and its interrelationship on the synthesis of serum amyloid A and C-reactive protein in primary cultures of adult human hepatocytes. *Biochem Biophys Res Commun* 155, 112 (1988).

19. Han, K.H. et al. C-reactive protein promotes monocyte chemoattractant protein-1-mediated chemotaxis through upregulating CC chemokine receptor 2 expression in human monocytes. *Circulation* 109, 2566 (2004).

20. Devaraj, S., Du Clos, T.W., and Jialal, I. Binding and internalization of C-reactive protein by Fc gamma receptors on human aortic endothelial cells mediates biological effects. *Arterioscler Thromb Vasc Biol* 25, 1359 (2005).

21. Torzewski, M. et al. C-reactive protein in the arterial intima: role of C-reactive protein receptor-dependent monocyte recruitment in atherogenesis. *Arterioscler Thromb Vasc Biol* 20, 2094 (2000).

22. Jialal, I. and Devaraj, S. Inflammation and atherosclerosis: the value of the high-sensitivity C-reactive protein assay as a risk marker. *Am J Clin Pathol* 116, Suppl S108 (2001).

23. Rifai, N. and Ridker, P.M. High-sensitivity C-reactive protein: a novel and promising marker of coronary heart disease. *Clin Chem* 47, 403 (2001).

24. Muntner, P., He, J., Chen, J., Fonseca, V., and Whelton, P.K. Prevalence of nontraditional cardiovascular disease risk factors among persons with impaired fasting glucose, impaired glucose tolerance, diabetes, and the metabolic syndrome: analysis of Third National Health and Nutrition Examination Survey (NHANES III). *Ann Epidemiol* 14, 686 (2004).

25. Clement, K. et al. Weight loss regulates inflammation-related genes in white adipose tissue of obese subjects. *FASEB J* 18, 1657 (2004).

26. Halaas, J.L. et al. Weight-reducing effects of the plasma protein encoded by the obese gene. *Science* 269, 543 (1995).

27. Montague, C.T. et al. Congenital leptin deficiency is associated with severe early-onset obesity in humans. *Nature* 387, 903 (1997).

28. Farooqi, I.S. et al. Beneficial effects of leptin on obesity, T cell hypo-responsiveness, and neuroendocrine/metabolic dysfunction of human congenital leptin deficiency. *J Clin Invest* 110, 1093 (2002).

29. Munzberg, H., Flier, J.S., and Bjorbaek, C. Region-specific leptin resistance within the hypothalamus of diet-induced obese mice. *Endocrinology* 145, 4880 (2004).

30. Shimomura, I., Hammer, R.E., Ikemoto, S., Brown, M.S., and Goldstein, J.L. Leptin reverses insulin resistance and diabetes mellitus in mice with congenital lipodystrophy. *Nature* 401, 73 (1999).

31. Dunajska, K. et al. Plasma adiponectin concentration in relation to severity of coronary atherosclerosis and cardiovascular risk factors in middle-aged men. *Endocrine* 25, 215 (2004).

32. Yamauchi, T. et al. The fat-derived hormone adiponectin reverses insulin resistance associated with both lipoatrophy and obesity. *Nat Med* 7, 941 (2001).

33. Tilg, H. and Wolf, A.M. Adiponectin: a key fat-derived molecule regulating inflammation. *Expert Opin Ther Targets* 9, 245 (2005).

34. Koh, K.K et al. Beneficial effects of fenofibrate to improve endothelial dysfunction and raise adiponectin levels in patients with primary hypertriglyceridemia. *Diabetes Care* 28, 1419 (2005).

35. Arita, Y. et al. Paradoxical decrease of an adipose-specific protein, adiponectin, in obesity. *Biochem Biophys Res Commun* 257, 79 (1999).

36. Ouchi, N. et al. Novel modulator for endothelial adhesion molecules: adipocyte-derived plasma protein adiponectin. *Circulation* 100, 2473 (1999).

37. Shibata, R. et al. Adiponectin stimulates angiogenesis in response to tissue ischemia through stimulation of amp-activated protein kinase signaling. *J Biol Chem* 279, 28670 (2004).

38. Kern, P.A., Ranganathan, S., Li, C., Wood, L., and Ranganathan, G. Adipose tissue tumor necrosis factor and interleukin-6 expression in human obesity and insulin resistance. *Am J Physiol Endocrinol Metab* 280, E745 (2001).

39. Prins, J.B. et al. Tumor necrosis factor-alpha induces apoptosis of human adipose cells. *Diabetes* 46, 1939 (1997).

40. Qian, H. et al. TNFalpha induces and insulin inhibits caspase 3-dependent adipocyte apoptosis. *Biochem Biophys Res Commun* 284, 1176 (2001).

41. Zhang, H.H., Kumar, S., Barnett, A.H., and Eggo, M.C. Dexamethasone inhibits tumor necrosis factor-alpha-induced apoptosis and interleukin-1 beta release in human subcutaneous adipocytes and preadipocytes. *J Clin Endocrinol Metab* 86, 2817 (2001).

42. Kern, P.A. et al. The expression of tumor necrosis factor in human adipose tissue. Regulation by obesity, weight loss, and relationship to lipoprotein lipase. *J Clin Invest* 95, 2111 (1995).

43. Memon, R.A., Feingold, K.R., Moser, A.H., Fuller, J., and Grunfeld, C. Regulation of fatty acid transport protein and fatty acid translocase mRNA levels by endotoxin and cytokines. *Am J Physiol* 274, E210 (1998).

44. Hotamisligil, G.S., Arner, P., Caro, J.F., Atkinson, R.L., and Spiegelman, B.M. Increased adipose tissue expression of tumor necrosis factor-alpha in human obesity and insulin resistance. *J Clin Invest* 95, 2409 (1995).

45. Koistinen, H.A. et al. Subcutaneous adipose tissue expression of tumour necrosis factor-alpha is not associated with whole body insulin resistance in obese nondiabetic or in type-2 diabetic subjects. *Eur J Clin Invest* 30, 302 (2000).

46. Moller, D.E. Potential role of TNF-alpha in the pathogenesis of insulin resistance and type 2 diabetes. *Trends Endocrinol Metab* 11, 212 (2000).

47. Avellone, G. et al. Evaluation of cardiovascular risk factors in overweight and obese subjects. *Int Angiol* 13, 25 (1994).

48. Anand, S.S. et al. Relationship of metabolic syndrome and fibrinolytic dysfunction to cardiovascular disease. *Circulation* 108, 420 (2003).

49. Mertens, I. and Van Gaal, L.F. Visceral fat as a determinant of fibrinolysis and hemostasis. *Semin Vasc Med* 5, 48 (2005).

50. Lyon, C.J. and Hsueh, W.A. Effect of plasminogen activator inhibitor-1 in diabetes mellitus and cardiovascular disease. *Am J Med* 115, Suppl 8A, 62S (2003).

51. De Pergola, G. et al. Increase in both pro-thrombotic and anti-thrombotic factors in obese premenopausal women: relationship with body fat distribution. *Int J Obes Relat Metab Disord* 21, 527 (1997).

52. Mertens, I. et al. Visceral fat is a determinant of PAI-1 activity in diabetic and non-diabetic overweight and obese women. *Horm Metab Res* 33, 602 (2001).

53. Janand-Delenne, B. et al. Visceral fat as a main determinant of plasminogen activator inhibitor 1 level in women. *Int J Obes Relat Metab Disord* 22, 312 (1998).

54. Juhan-Vague, I., Thompson, S.G., and Jespersen, J. Involvement of the hemostatic system in the insulin resistance syndrome: study of 1500 patients with angina pectoris. *Arterioscler Thromb* 13, 1865 (1993).

55. McTernan, P.G., Kusminski, C.M., and Kumar, S. Resistin. *Curr Opin Lipidol* 17, 170 (2006).

56. Steppan, C.M. et al. A family of tissue-specific resistin-like molecules. *Proc Natl Acad Sci USA* 98, 502 (2001).

57. Rajala, M.W., Obici, S., Scherer, P.E., and Rossetti, L. Adipose-derived resistin and gut-derived resistin-like molecule-beta selectively impair insulin action on glucose production. *J Clin Invest* 111, 225 (2003).

58. Patel, L. et al. Resistin is expressed in human macrophages and directly regulated by PPAR gamma activators. *Biochem Biophys Res Commun* 300, 472 (2003).

59. Rangwala, S.M. et al. Abnormal glucose homeostasis due to chronic hyper-resistinemia. *Diabetes* 53, 1937 (2004).

60. Kaser, S. et al. Resistin messenger-RNA expression is increased by pro-inflammatory cytokines *in vitro*. *Biochem Biophys Res Commun* 309, 286 (2003).

61. Shetty, G.K., Economides, P.A., Horton, E.S., Mantzoros, C.S., and Veves, A. Circulating adiponectin and resistin levels in relation to metabolic factors, inflammatory markers, and vascular reactivity in diabetic patients and subjects at risk for diabetes. *Diabetes Care* 27, 2450 (2004).

62. Ross, R. Atherosclerosis is an inflammatory disease. *Am Heart J* 138, S419 (1999).

63. Iiyama, K. et al. Patterns of vascular cell adhesion molecule-1 and intercellular adhesion molecule-1 expression in rabbit and mouse atherosclerotic lesions and at sites predisposed to lesion formation. *Circ Res* 85, 199 (1999).

64. van Furth, R. Human monocytes and cytokines. *Res Immunol* 149, 719 (1998).

65. Fogelman, A.M., Haberland, M.E., Seager, J., Hokom, M., and Edwards, P.A. Factors regulating the activities of the low density lipoprotein receptor and the scavenger receptor on human monocyte-macrophages. *J Lipid Res* 22, 1131 (1981).

66. Heinecke, J.W. Mechanisms of oxidative damage by myeloperoxidase in atherosclerosis and other inflammatory disorders. *J Lab Clin Med* 133, 321 (1999).

67. Galis, Z.S., Sukhova, G.K., Lark, M.W., and Libby, P. Increased expression of matrix metalloproteinases and matrix degrading activity in vulnerable regions of human atherosclerotic plaques. *J Clin Invest* 94, 2493 (1994).

68. Osterud, B. Tissue factor expression by monocytes: regulation and pathophysiological roles. *Blood Coagul Fibrinolysis* 9, Suppl 1, S9 (1998).

69. Toschi, V. et al. Tissue factor modulates the thrombogenicity of human atherosclerotic plaques. *Circulation* 95, 594 (1997).

70. Semplicini, A. et al. Interactions between insulin and sodium homeostasis in essential hypertension. *Am J Med Sci* 307, Suppl 1, S43 (1994).

71. Weisberg, S.P. et al. Obesity is associated with macrophage accumulation in adipose tissue. *J Clin Invest* 112, 1796 (2003).

72. Stolar, M.W. Insulin resistance, diabetes, and the adipocyte. *Am J Health Syst Pharm* 59, Suppl 9, S3 (2002).

73. Xu, H. et al. Chronic inflammation in fat plays a crucial role in the development of obesity-related insulin resistance. *J Clin Invest* 112, 1821 (2003).

74. Esposito, K. et al. Effect of weight loss and lifestyle changes on vascular inflammatory markers in obese women: a randomized trial. *JAMA* 289, 1799 (2003).

75. Esposito, K. et al. Effect of a mediterranean-style diet on endothelial dysfunction and markers of vascular inflammation in the metabolic syndrome: a randomized trial. *JAMA* 292, 1440 (2004).

76. Liao, J.K. Clinical implications for statin pleiotropy. *Curr Opin Lipidol* 16, 624 (2005).

77. Ridker, P.M. et al. Inflammation, pravastatin, and the risk of coronary events after myocardial infarction in patients with average cholesterol levels: Cholesterol and Recurrent Events (CARE) Investigators. *Circulation* 98, 839 (1998).

78. Ridker, P.M. et al. Measurement of C-reactive protein for the targeting of statin therapy in the primary prevention of acute coronary events. *New Engl J Med* 344, 1959 (2001).

79. Koh, K.K. et al. Comparison of effects of losartan, irbesartan, and candesartan on flow-mediated brachial artery dilation and on inflammatory and thrombolytic markers in patients with systemic hypertension. *Am J Cardiol* 93, 1432, A10 (2004).

80. Wang, C.H. et al. C-reactive protein up-regulates angiotensin type 1 receptors in vascular smooth muscle. *Circulation* 107, 1783 (2003).

81. Haffner, S.M. The metabolic syndrome: inflammation, diabetes mellitus, and cardiovascular disease. *Am J Cardiol* 97, 3A (2006).

82. Haffner, S.M. et al. Effect of rosiglitazone treatment on nontraditional markers of cardiovascular disease in patients with type 2 diabetes mellitus. *Circulation* 106, 679 (2002).

83. MRC/BHF Heart Protection Study of cholesterol lowering with simvastatin in 20,536 high-risk individuals: a randomised placebo-controlled trial. *Lancet* 360, 7 (2002).

84. Colhoun, H.M. et al. Primary prevention of cardiovascular disease with atorvastatin in type 2 diabetes in the Collaborative Atorvastatin Diabetes Study (CARDS) multicentre randomised placebo-controlled trial. *Lancet* 364, 685 (2004).

85. Freeman, D.J. et al. Pravastatin and the development of diabetes mellitus: evidence for a protective treatment effect in the West of Scotland Coronary Prevention Study. *Circulation* 103, 357 (2001).

86. Yusuf, S. et al. Effects of an angiotensin-converting-enzyme inhibitor, ramipril, on cardiovascular events in high-risk patients: Heart Outcomes Prevention Evaluation (HOPE) Study Investigators. *New Engl J Med* 342, 145 (2000).

87. Monzillo, L.U. et al. Effect of lifestyle modification on adipokine levels in obese subjects with insulin resistance. *Obes Res* 11, 1048 (2003).

88. Koh, K.K. et al. Comparative effects of diet and statin on NO bioactivity and matrix metalloproteinases in hypercholesterolemic patients with coronary artery disease. *Arterioscler Thromb Vasc Biol* 22, e19 (2002).

89. Orem, C. et al. The effects of atorvastatin treatment on the fibrinolytic system in dyslipidemic patients. *Jpn Heart J* 45, 977 (2004).

90. Ludwig, S. et al. Impact of simvastatin on hemostatic and fibrinolytic regulators in type 2 diabetes mellitus. *Diabetes Res Clin Pract* 70, 110 (2005).

91. Hoekstra, T., Geleijnse, J.M., Schouten, E.G., and Kluft, C. Plasminogen activator inhibitor-type 1: its plasma determinants and relation with cardiovascular risk. *Thromb Haemost* 91, 861 (2004).

10 Oligomeric Composition of Adiponectin and Obesity

T. Bobbert and Joachim Spranger

CONTENTS

INTRODUCTION

Obesity is among the most frequently encountered metabolic diseases worldwide. Moreover, its incidence and prevalence are rising rapidly.[1,2] More than half the world population is considered overweight.[3] Being overweight constitutes a health risk because it is associated with several co-morbidities including dyslipidemia, hypertension, type 2 diabetes, and atherosclerotic cardiovascular disease.[4,5] Adipose tissue was initially believed to be only a fat storage organ, but it is now acknowledged to be an active participant in energy homeostasis and other physiological functions. Adipose tissue is known to express and secrete a variety of novel adipocytokines that have been implicated in the development of insulin resistance and atherosclerosis.[6,7] Dysregulation of adipocytokine production is directly involved in the pathophysiology of metabolic syndrome, and normalization of plasma concentrations of adipocytokines reverses the phenotype of metabolic syndrome.[8,9]

ADIPOCYTES AND ADIPONECTIN

Adipocytes secrete a variety of polypeptides such as leptin, resistin, and adiponectin. Adiponectin, the gene product of the adipose tissue's most abundant gene transcript,[10] may be a link between obesity and the development of insulin resistance. cDNAs for adiponectin were first identified independently in different mouse cells[11,12] and by large-scale random sequencing of a 3′-directed human adipose tissue cDNA library.[10] Subsequently, adiponectin was purified from human plasma,[13] where it circulates in considerable concentrations. Especially in the regulation of the glucose and lipid metabolism, adiponectin has been shown to play an important role.[11–13]

Unlike other adipose-derived proteins, plasma levels of adiponectin have been found to be decreased in a number of deranged metabolic states including obesity,[14] dyslipidemia,[15] type 2 diabetes, and insulin resistance.[12,16,17] Many studies have demonstrated that reduced circulating adiponectin levels can be partially elevated after induction of weight loss in obese and insulin-resistant subjects.[17,18] Apart from a connection to obesity or diabetes, further parameters like age, sex hormones, and glucocorticoids may play parts in the regulation of adiponectin levels.[19,20]

Adiponectin is composed of a carboxyl-terminal globular domain and an amino-terminal collagenous domain.[21,22] It belongs to the soluble collagen superfamily and has structural homology with collagen VIII, X, complement factor C1q,[16] and the tumor necrosis (TNF) family.[10,21] This kind of structure is known to form characteristic multimers.[23,24]

Gel filtration and velocity gradient sedimentation studies revealed adiponectin circulating in serum to form several different molecular weight species; the largest species was more than several hundred kilodaltons in size.[11,13,14] Scherer et al. noted that adiponectin from 3T3-L1 adipocytes forms trimers, hexamers, and larger multimers.[11] Tsao et al. and Arita et al. analyzed multimer formation of adiponectin in serum by gel filtration chromatography and showed adiponectin to be separable into three species.[14,25]

Kadowaki et al. showed a method of evaluating the multimer formation of adiponectin.[26] Using SDS-PAGE under non-reducing and non-heat-denaturing conditions, they separated multimers of adiponectin from various sources into three species; LMW trimers, MMW multimers, and HMW multimers. This classification is not used in all publications and depends partially on the determination method used. Also the MMW fractions or hexamers are added to the LMW oligomers and predictions or correlations were calculated by the ratio of HMW to MMW plus LMW species.

Oligomer formation of adiponectin depends on the disulfide bond formation mediated by Cys-39.[27] Adiponectin was reported to be an α-2,8-linked disialic acid-containing glycoprotein, although the biological functions of the disialic acid epitope of adiponectin remain to be elucidated.[28] The regulation of adiponectin multimerization and secretion occurs also via changes in post-translational modifications (PTMs). PTMs identified in murine and bovine adiponectin include hydroxylation of multiple conserved proline and lysine residues and glycosylation of hydroxylysines.

BIOLOGICAL ACTIVITIES OF ADIPONECTIN MULTIMERS

Discussion of the biological activities of the different adiponectin multimers has proven controversial. HMW multimer levels appear to be higher in women compared to men[26] and adiponectin exerts multiple metabolic actions at multiple tissue sites. The isolated globular domain of adiponectin stimulates fatty acid oxidation in skeletal muscle, whereas full length adiponectin synergizes with insulin to inhibit hepatic glucose production.[16,29,30] In mice, disruption of the adiponectin locus leading to its ablation resulted in impaired fatty acid clearance, increased tumor necrosis factor-α levels, and aggravated insulin resistance in animals fed high fat diets.[31,32]

The HMW and hexameric adiponectin can activate transcription factor NF-κB in undifferentiated or differentiated C2C12 cells, but trimeric adiponectin and the isolated globular domain of adiponectin cannot. The isolated globular domain of adiponectin, but not full-length adiponectin hexamer, enhances muscle fatty acid oxidation by inactivating acetyl-CoA carboxylase following stimulation of AMP-activated protein kinase (AMPK).[33,34] Waki et al. reported specifically that the HMW isoform promotes AMP-activated protein kinase in hepatocytes.[26] In contrast, Tsao et al. recently reported that only trimers activate AMPK in muscle, whereas hexamers and HMW form activated NF-κB.[35]

Differences in the tissue-specific expression patterns of two adiponectin receptors may contribute to these divergent activities.[36] Bobbert et al. investigated the effects of moderate weight reduction by a lifestyle intervention on adiponectin oligomer composition and its relation to glucose and fat metabolism.[37] While HMW and MMW adiponectins increased, a decreased amount of LMW adiponectin was found after weight reduction. Total adiponectin and especially HMW adiponectin correlated strongly with HDL cholesterol while correlations with other markers of glucose or fat metabolism were weak.

Many studies have demonstrated a correlation between total adiponectin and HDL cholesterol. Most studies suggested that hypoadiponectinemia is more closely related to adiposity and dyslipidemia rather than insulin sensitivity.[38] Indeed, Bobbert et al. found a strong correlation between HMW adiponectin and HDL cholesterol, which suggests that the relationship between total adiponectin and HDL cholesterol is primarily driven by HMW adiponectin rather than total adiponectin (Figure 10.1).

HDL cholesterol is basically generated from lipid-free apolipoprotein A-I or lipid-poor pre-ß1-HDL as a precursor. These precursors are partially produced by the liver and it is well known that adiponectin oligomers specifically affect liver metabolism.[25] The importance of considering adiponectin oligomers is supported by a multivariate analysis revealing that HMW adiponectin explained about 30% of the variability of HDL cholesterol (Figure 10.2). These results confirm those of Baratta and co-workers, who demonstrated that adiponectin is correlated

a)

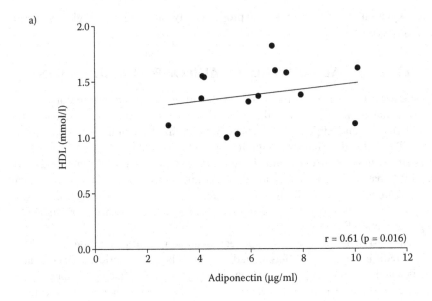

b)

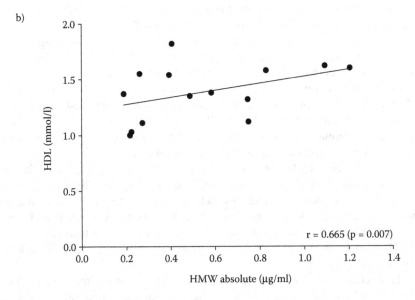

FIGURE 10.1 Correlation between (a) total adiponectin and HDL and (b) absolute concentrations of adiponectin oligomers and HDL in participants of a weight reduction program after moderate weight loss.

with serum lipid improvement independently of insulin sensitivity changes after weight loss.[39]

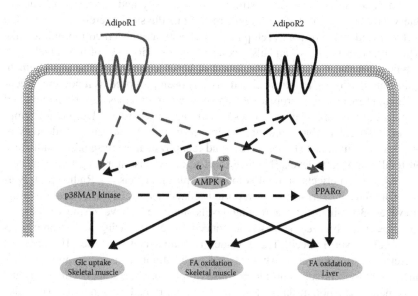

FIGURE 10.2 Adiponectin-dependent intracellular pathways.

Different adiponectin oligomers and the varying appearances of the AdipoR1 and AdipoR2 adiponectin receptors may be responsible for the different actions of adiponectin oligomers on fat or glucose metabolism. However, the precise molecular mechanism for the oligomer-specific effect is still unclear. The mechanism that regulates adiponectin oligomer composition has not been identified. Data from intervention studies with thiazolidinediones suggest that the process is PPAR-γ-dependent.[40] Because PPAR-γ is activated by negative energy balance and weight reduction,[41] the effect on adiponectin oligomers may also depend on PPAR-γ-related pathways. Tsuchida et al. showed the influence of PPAR-α and γon the expression of adiponectin and AdipoR1 and AdipoR2. They showed that activation of PPAR-γ or food restriction increased the ratio of HMW to total adiponectin and that activation of PPARα did not affect the ratio.

RESULTS OF WEIGHT LOSS

The results of Bobbert et al. are in agreement with other studies showing that moderate weight loss results in relatively small changes of circulating adiponectin levels.[42] One study of six patients and a 3-month follow-up investigated the effects of weight reduction on adiponectin oligomers. This study primarily aimed to investigate the effects of HMW adiponectin on endothelial cell apoptosis. Due to the small number of participants, the precise relationship of adiponectin oligomers to metabolic changes after weight loss was not further evaluated.[43]

Physical activity correlates strongly with obesity and is apart from dietary interventions a major part of weight reduction. Physical exercise is associated with reduced risks for the development of obesity-associated co-morbidities like type 2 diabetes mellitus[44] and also reduces the mortality risks of individuals with impaired glucose tolerance to the levels of healthy persons.[45] The improvement of insulin sensitivity by physical activity has been proposed as a possible mechanism of these effects.[46] From a mechanistic view, adipokines have been identified as potential mediators between obesity and insulin sensitivity. Despite the temptation to speculate that circulating adiponectin may be affected by degrees of physical activity, most studies have found that physical exercise has no influence on total adiponectin levels. Neither short term exercise nor long term physical training exerted effects on total adiponectin plasma levels.[47,48] Rather controversially, a study by Jurimae et al. reported reductions directly after acute rowing. However, 30 min after the rowing levels increased above resting values,[49,50] Kriketos et al. observed increased adiponectin levels following two or three bouts of exercise over 1 week. These values remained elevated after 10 weeks of exercise.[49,50] Bobbert et al. showed that total adiponectin and oligomers were unchanged by acute and chronic exercise (in press). Total adiponectin levels and oligomers were not different for trained or untrained persons and were also unaltered by different training intensities. It is therefore unlikely that changes of adiponectin oligomers are responsible for the beneficial effects of sustained physical exercise on lipid and glucose metabolism. However, total adiponectin and again specifically HMW oligomers correlated with HDL cholesterol in this study, which supported the hypothesis that HMW adiponectin specifically mediates the positive effects of adiponectin.

CENTRAL NERVOUS SYSTEM EFFECTS

In addition to the peripheral influence of adiponectin on human metabolism, central nervous effects also play an important role in the regulation of human energy balance and metabolism. A growing body of evidence suggests that adiponectin directly affects energy balance by increasing thermogenesis.[51] Adiponectin has recently been reported to generate a negative energy balance by increasing energy expenditure.[52]

Spranger et al. showed that neither radiolabeled non-glycosylated nor glycosylated globular adiponectin crossed the blood–brain barrier (BBB) in mice.[53] In addition, adiponectin and adiponectin oligomers were not detectable in human cerebrospinal fluid using various established methods. Using murine cerebral microvessels, they demonstrated expression of adiponectin receptors that were up-regulated during fasting in brain endothelium. Interestingly, treatment with adiponectin reduced secretion of centrally active interleukin-6 from brain endothelial cells — a phenomenon paralleled by similar trends of other pro-inflammatory cytokines. These data suggest that direct effects of endogenous adiponectin on central nervous system pathways are unlikely to exist. However, the identification of adiponectin receptors on brain endothelial cells and the finding of a modified secretion pattern of centrally active substances from BBB cells provide an alternate explanation as to how adiponectin may evoke effects on energy metabolism.

SUMMARY

Adiponectin plays an important role in human metabolism, although the exact function of adiponectin is still unclear. The influences of adiponectin and adiponectin oligomers on peripheral glucose and lipid metabolism serve as foci of current clinical research. Some *in vitro* analyses showed first hints of adiponectin oligomer-dependent intracellular signalling cascades and initial human data demonstrated the different peripheral and central nervous system actions of total adiponectin and the different adiponectin oligomers.

REFERENCES

1. Caterson ID and Gill TP. Obesity: epidemiology and possible prevention. *Best Pract Res Clin Endocrinol Metab* 2002; 16: 595.
2. Formiguera X and Canton A. Obesity: epidemiology and clinical aspects. *Best Pract Res Clin Gastroenterol* 2004; 18: 1125.
3. Field AE et al. Impact of overweight on the risk of developing common chronic diseases during a 10-year period. *Arch Intern Med* 2001; 161: 1581.
4. Linton MF and Fazio S. A practical approach to risk assessment to prevent coronary artery disease and its complications. *Am J Cardiol* 2003; 92: 19i.
5. Scott CL. Diagnosis, prevention, and intervention for the metabolic syndrome. *Am J Cardiol* 2003; 92: 35i.
6. Matsuzawa Y, Funahashi T, and Nakamura T. Molecular mechanism of metabolic syndrome X: contribution of adipocytokines adipocyte-derived bioactive substances. *Ann NY Acad Sci* 1999; 892: 146.
7. Funahashi T et al. Role of adipocytokines on the pathogenesis of atherosclerosis in visceral obesity. *Intern Med* 1999; 38: 202.
8. Hotamisligil GS, Shargill NS, and Spiegelman BM. Adipose expression of tumor necrosis factor-alpha: direct role in obesity-linked insulin resistance. *Science* 1993; 259: 87.
9. Shimomura I et al. Leptin reverses insulin resistance and diabetes mellitus in mice with congenital lipodystrophy. *Nature* 1999; 401: 73.
10. Maeda K, Okubo K, Shimomura I, Funahashi T, Matsuzawa Y, and Matsubara K. cDNA cloning and expression of a novel adipose specific collagen-like factor, apM1 (adipose most abundant gene transcript 1). *Biochem Biophys Res Commun* 1996; 221: 286.

11. Scherer PE, Williams S, Fogliano M, Baldini G, and Lodish HF. A novel serum protein similar to C1q produced exclusively in adipocytes. *J Biol Chem* 1995; 270: 26746.

12. Hu E, Liang P, and Spiegelman BM. AdipoQ is a novel adipose-specific gene dysregulated in obesity. *J Biol Chem* 1996; 271: 10697.

13. Nakano Y, Tobe T, Choi-Miura NH, Mazda T, and Tomita M. Isolation and characterization of GBP28, a novel gelatin-binding protein purified from human plasma. *J Biochem (Tokyo)* 1996; 120: 803.

14. Arita Y et al. Paradoxical decrease of an adipose-specific protein, adiponectin, in obesity. *Biochem Biophys Res Commun* 1999; 257: 79.

15. Matsubara M, Maruoka S, and Katayose S. Decreased plasma adiponectin concentrations in women with dyslipidemia. *J Clin Endocrinol Metab* 2002; 87: 2764.

16. Yamauchi T et al. The fat-derived hormone adiponectin reverses insulin resistance associated with both lipoatrophy and obesity. *Nat Med* 2001; 7: 941.

17. Hotta K et al. Plasma concentrations of a novel, adipose-specific protein, adiponectin, in type 2 diabetic patients. *Arterioscler Thromb Vasc Biol* 2000; 20: 1595.

18. Yang WS et al. Weight reduction increases plasma levels of an adipose-derived anti-inflammatory protein, adiponectin. *J Clin Endocrinol Metab* 2001; 86: 3815.

19. Reinehr T, Roth C, Menke T, and Andler W. Adiponectin before and after weight loss in obese children. *J Clin Endocrinol Metab* 2004; 89: 3790.

20. Diez JJ and Iglesias P. The role of the novel adipocyte-derived hormone adiponectin in human disease. *Eur J Endocrinol* 2003; 148: 293.

21. Shapiro L and Scherer PE. The crystal structure of a complement-1q family protein suggests an evolutionary link to tumor necrosis factor. *Curr Biol* 1998; 8: 335.

22. Yokota T et al. Adiponectin, a new member of the family of soluble defense collagens, negatively regulates the growth of myelomonocytic progenitors and the functions of macrophages. *Blood* 2000; 96: 1723.

23. Crouch E, Persson A, Chang D, and Heuser J. Molecular structure of pulmonary surfactant protein D (SP-D). *J Biol Chem* 1994; 269: 17311.

24. McCormack FX et al. The Cys6 intermolecular disulfide bond and the collagen-like region of rat SP-A play critical roles in interactions with alveolar type II cells and surfactant lipids. *J Biol Chem* 1997; 272: 27971.

25. Tsao TS, Murrey HE, Hug C, Lee DH, and Lodish HF. Oligomerization state-dependent activation of NF-kappa B signaling pathway by adipocyte complement-related protein of 30 kDa (Acrp30). *J Biol Chem* 2002; 277: 29359.

26. Waki H et al. Impaired multimerization of human adiponectin mutants associated with diabetes. Molecular structure and multimer formation of adiponectin. *J Biol Chem* 2003; 278: 40352.

27. Pajvani UB et al. Structure-function studies of the adipocyte-secreted hormone Acrp30/adiponectin: implications for metabolic regulation and bioactivity. *J Biol Chem* 2003; 278: 9073.

28. Sato C, Yasukawa Z, Honda N, Matsuda T, and Kitajima K. Identification and adipocyte differentiation-dependent expression of the unique disialic acid residue in an adipose tissue-specific glycoprotein, adipo Q. *J Biol Chem* 2001; 276: 28849.

29. Berg AH, Combs TP, Du X, Brownlee M, and Scherer PE. The adipocyte-secreted protein Acrp30 enhances hepatic insulin action. *Nat Med* 2001; 7: 947.

30. Fruebis J et al. Proteolytic cleavage product of 30-kDa adipocyte complement-related protein increases fatty acid oxidation in muscle and causes weight loss in mice. *Proc Natl Acad Sci USA* 2001; 98: 2005.

31. Maeda N et al. Diet-induced insulin resistance in mice lacking adiponectin/ACRP30. *Nat Med* 2002; 8: 731.

32. Kubota N et al. Disruption of adiponectin causes insulin resistance and neointimal formation. *J Biol Chem* 2002; 277: 25863.

33. Tomas E et al. Enhanced muscle fat oxidation and glucose transport by ACRP30 globular domain: acetyl-CoA carboxylase inhibition and AMP-activated protein kinase activation. *Proc Natl Acad Sci USA* 2002; 99: 16309.

34. Yamauchi T et al. Adiponectin stimulates glucose utilization and fatty-acid oxidation by activating AMP-activated protein kinase. *Nat Med* 2002; 8: 1288.

35. Tsao TS et al. Role of disulfide bonds in Acrp30/adiponectin structure and signaling specificity: different oligomers activate different signal transduction pathways. *J Biol Chem* 2003; 278: 50810.

36. Yamauchi T et al. Cloning of adiponectin receptors that mediate antidiabetic metabolic effects. *Nature* 2003; 423: 762.

37. Bobbert T et al. Changes of adiponectin oligomer composition by moderate weight reduction. *Diabetes* 2005; 54: 2712.

38. Kazumi T, Kawaguchi A, Hirano T, and Yoshino G. Serum adiponectin is associated with high-density lipoprotein cholesterol, triglycerides, and low-density lipoprotein particle size in young healthy men. *Metabolism* 2004; 53: 589.

39. Baratta R et al. Adiponectin relationship with lipid metabolism is independent of body fat mass: evidence from both cross-sectional and intervention studies. *J Clin Endocrinol* Metab 2004; 89: 2665.

40. Pajvani UB et al. Complex distribution, not absolute amount of adiponectin, correlates with thiazolidinedione-mediated improvement in insulin sensitivity. *J Biol Chem* 2004; 279: 12152.

41. Verreth W et al. Weight loss-associated induction of peroxisome proliferator-activated receptor-alpha and peroxisome proliferator-activated receptor-gamma correlate with reduced atherosclerosis and improved cardiovascular function in obese insulin-resistant mice. *Circulation* 2004; 110: 3259.

42. Abbasi F et al. Plasma adiponectin concentrations do not increase in association with moderate weight loss in insulin-resistant, obese women. *Metabolism* 2004; 53: 280.

43. Kobayashi H et al. Selective suppression of endothelial cell apoptosis by the high molecular weight form of adiponectin. *Circ Res* 2004; 94: e27.

44. Helmrich SP, Ragland DR, Leung RW, and Paffenbarger RS, Jr. Physical activity and reduced occurrence of non-insulin-dependent diabetes mellitus. *New Engl J Med* 1991; 325: 147.

45. Eriksson KF and Lindgarde F. No excess 12-year mortality in men with impaired glucose tolerance who participated in the Malmo Preventive Trial with diet and exercise. *Diabetologia* 1998; 41: 1010.

46. Kirwan JP and del Aguila LF. Insulin signalling, exercise and cellular integrity. *Biochem Soc Trans* 2003; 31, Pt 6: 1281.

47. Hulver MW et al. Adiponectin is not altered with exercise training despite enhanced insulin action. *Am J Physiol Endocrinol Metab* 2002; 283: E861.

48. Kraemer RR et al. Adiponectin responses to continuous and progressively intense intermittent exercise. *Med Sci Sports Exerc* 2003; 35: 1320.

49. Kriketos AD et al. Exercise increases adiponectin levels and insulin sensitivity in humans. *Diabetes Care* 2004; 27: 629.

50. Jurimae J, Purge P, and Jurimae T. Adiponectin is altered after maximal exercise in highly trained male rowers. *Eur J Appl Physiol* 2005; 93: 502.

51. Seeley RJ, D'Alessio DA, and Woods SC. Fat hormones pull their weight in the CNS. *Nat Med* 2004; 10: 454.

52. Qi Y et al. Adiponectin acts in the brain to decrease body weight. *Nat Med* 2004; 10: 524.

53. Spranger J et al. Adiponectin does not cross the blood-brain barrier but modifies cytokine expression of brain endothelial cells. *Diabetes* 2006; 55: 141.

11 Insulin-Stimulated Reactive Oxygen Species and Insulin Signal Transduction

Barry J. Goldstein, Kalyankar Mahadev, and Xiangdong Wu

CONTENTS

INTRODUCTION

Cellular reactive oxygen species (ROS; superoxide and H_2O_2), especially when chronically raised to high levels and associated with hyperglycemia, are widely recognized to play an important pathophysiological role in the chronic complications of diabetes as well as in the development of the disease.[1–3] In contrast, the transient generation of smaller amounts of ROS is triggered in cells in response to stimulation with a variety of growth factors, cytokines, and hormones including insulin, and facilitates their respective signaling cascades. The involvement of an oxidation step in the action of insulin has been suggested for decades, but only recently have potential molecular mechanisms been identified for these effects.

Among the signaling enzymes potentially susceptible to inhibition by biochemical oxidation are those that contain reduced cysteine thiol side chains essential for their catalytic activities, including the family of protein–tyrosine phosphatases (PTPs) and other important signal regulators. Recently, we identified a role for the NADPH oxidase homolog known as Nox4 in the rapid generation of ROS in insulin-stimulated cells.[4–6] A full understanding of this signaling network may potentially provide a novel means of facilitating insulin action in states of insulin resistance and differentially regulating some of the pleiotropic cellular actions of insulin.

REGULATION OF REVERSIBLE TYROSINE PHOSPHORYLATION IN INSULIN SIGNALING BY TYROSINE PHOSPHATASES

Insulin is a critical regulator of pleiotropic cellular responses and resistance to the action of insulin in peripheral tissues is a fundamental defect in type 2 diabetes.[7,8] Insulin action is initiated by binding to a specific plasma membrane receptor that encodes a tyrosine-specific protein kinase that autophosphorylates the receptor and its substrate proteins in cells.[9,10] These phosphorylated tyrosine motifs act as docking scaffolds for the binding and activation of a variety of signaling and adaptor proteins that are linked to downstream insulin responses.

Specific PTPs regulate the steady-state balance of reversible protein–tyrosine phosphorylation in the insulin signaling cascade, in concert with the insulin receptor kinase.[5] In addition to serving as steady-state regulators, PTPs appear to be *required* for receptor deactivation since purified insulin receptors retain their autophosphorylation state after insulin is removed from the ligand binding site *in vitro,* and *in vivo,* dissociation of insulin from the receptor is followed by its rapid dephosphorylation and deactivation of its kinase activity.[11]

In particular, PTP1B has become an important target for therapeutic intervention in disease states associated with clinical insulin resistance.[12–14] This single-domain intracellular PTP has been convincingly shown in two knock-out mouse models to negatively regulate the insulin action cascade *in vivo.*[15,16] The cellular basis of these *in vivo* findings is well documented.[12] PTP1B is active *in vitro* against the autophosphorylated insulin receptor[17,18] and it also has a relatively high specific activity toward IRS-1 compared to other candidate PTPs.[19,20] Several studies have characterized the unique molecular interactions underlying the close interaction between the insulin receptor and PTP1B that is facilitated by the presence of a second phosphotyrosine binding site in the PTP1B catalytic region that interacts with the multiple phosphotyrosine residues of the receptor kinase.[21–23]

The PTPs comprise an extensive group of homologous proteins involved in a variety of signal transduction pathways.[24] Receptor and non-receptor forms of the classical tyrosine-specific PTPs have in common a ~230 amino acid phosphatase domain that contains the tightly conserved signature catalytic motif VHCSxGxGR[T/S]G.[25] The non-classical, dual-specificity (tyrosine and serine/threonine) phosphatases (e.g., MAP kinase phosphatase and others)[26] and PTEN (phosphatase and tensin homolog deleted on chromosome 10), which dephosphorylates the 3-phosphate of inositol phospholipids generated by PI 3-kinase and exhibits dual phosphatase activity *in vitro* are structurally related, sharing a less tightly conserved catalytic motif that retains the essential $C(x)_5R$ core structure.[27] This sequence contains the reduced cysteine residue required for catalysis that is involved in the formation of a cysteinyl-phosphate intermediate.[28,29] Modification of this catalytic cysteine by oxidation or disulfide conjugation is a critically important mode of reversible and irreversible PTP regulation *in vivo*, including in the insulin action cascade.[30–32]

ROS AS SECOND MESSENGERS FOR CELLULAR TYROSINE KINASE SIGNALING

Superoxide and H_2O_2 are now well recognized to play an integral role in several growth factor and cytokine signal transduction pathways (recently reviewed in Rhee).[33] Superoxide $[O_2{}^{\bullet-}]$, hydroxyl $[{}^\bullet OH]$ ions, and H_2O_2 generated by cellular redox reactions have complex physiologies and are ultimately converted to H_2O + O_2 by cellular catalase, thioredoxin, glutathione peroxidase, and/or peroxiredoxins.

Relatively low levels of H_2O_2 generated in response to growth factor stimulation occur in a concerted fashion with specific signaling targets in cells, suggesting a role as a second messenger. H_2O_2 activates tyrosine phosphorylation cascades in cultured cells in a manner that mimics ligand-mediated signaling by PDGF and EGF. However, intracellular H_2O_2 generated transiently during stimulation of cells with growth factors has been convincingly demonstrated in seminal experiments by the groups of Finkel et al.[34] and Rhee et al.,[35] showing that autophosphorylation of PDGF and EGF receptors, respectively, and their distal signaling effects are dependent on post-receptor intracellular H_2O_2 production, not simply the addition of exogenous H_2O_2. As described below, numerous studies support the hypothesis that cellular oxidant signaling is mediated by discrete, localized redox circuitry, distinct from the notion of a generalized "oxidative stress" effect.[36]

NOVEL REGULATORY PARADIGM: PTPS ARE THIOL-DEPENDENT ENZYMES REGULATED BY CELLULAR ROS

In parallel with developments in the cellular physiology of H_2O_2 generation, studies have also characterized the biochemical inhibition of PTPs by progressive oxidation of their catalytic cysteine thiol moieties by cellular ROS to more inert forms *in vivo* (Figure 11.1).[30,32,37–39] The activity of PTP1B is dependent on the oxidation state of its cys-215 residue, which is required for catalytic activity *via* the formation of a phospho-enzyme intermediate.[29,37,40,41]

The catalytic cysteine residue in the PTP active site is particularly sensitive to oxidation because of hydrogen bonding of neighboring side chains, which lowers the thiol p_{Ka} to ~5.5, more than 3 units below that of a typical –SH group, rendering it in an ionized state at physiological pH. Compared to other typical protein sulfhydryl side chains, the catalytic PTP thiol can be readily oxidized by

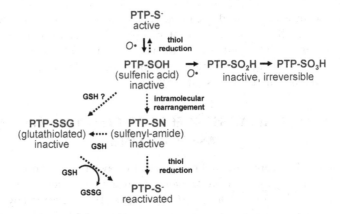

FIGURE 11.1 Regulation of PTP catalytic activity by oxidation, reduction, and conjugation. The catalytic cysteine residue of PTPs is especially reactive because of the low p_{Ka} of the sulfhydryl that favors a relatively ionized state of the cysteinyl hydrogen.[42] When subjected to ROS, including those elicited by cellular insulin stimulation, the cysteine side chain undergoes step-wise oxidation to increasingly inert forms.[32,38,106] The inactive sulfenic derivative may be reduced to regenerate the active thiol form of the protein. Alternatively, it may be directly conjugated with glutathione in the cell, producing a catalytically inert PTP derivative that can be reactivated by biochemical reduction or through the action of glutathione reductases.[46] Recently, the sulfenic derivative of PTP1B has been shown to undergo an intramolecular rearrangement, forming a novel sulfenyl–amide derivative that also sequesters the PTP in an inactive state.[47,48] The sulfenyl–amide form may actually be an obligate intermediate in this reaction scheme because its altered protein conformation opens a groove adjacent to the catalytic center that may render it particularly susceptible to reduction with cytosolic glutathione compared with the sulfenic derivative. (Reprinted from Goldstein, B.J. et al. *Diabetes* 54, 311, 2005. With permission from the American Diabetes Association.)

locally generated H_2O_2 even in the presence of high cytosolic concentrations (millimolar) of the cysteine-containing tripeptide glutathione (GSH).[42]

The catalytic cysteine thiol is initially oxidized to the sulfenic (–SOH) form that can be reversed by cellular enzymatic mechanisms or with reducing agents *in vitro* (Figure 11.1).[40,43] Sequential steps of progressive oxidation to the sulfinic (–SO$_2$H) and sulfonic (–SO$_3$H) forms can lead to irreversible PTP inactivation.[41,44,45] However, the partially oxidized sulfenic acid intermediate of PTP1B can also be rapidly converted to other forms that may stabilize the molecule and protect it from further irreversible oxidation.

One potentially stabilizing modification is conjugation with glutathione which may be enzymatically reactivated in cells by glutaredoxin.[46] The catalytic cysteine of PTP1B has recently been shown to be reversibly converted to a previously unknown intramolecular sulfenyl–amide species, in which it becomes linked to the main chain nitrogen of an adjacent residue, rendering the enzyme inactive and inducing large conformational changes that inhibit substrate binding.[47,48] This novel protein modification not only protects the enzyme catalytic site from irreversible oxidation to sulfonic acid but also permits redox regulation of the enzyme by promoting its reversible reduction by thiols. The conformation of the catalytic cleft assumes a more open structure in the sulfenyl–amide derivative of PTP1B and renders it particularly amenable to reduction by GSH.[47,48] This suggests that this unique protein derivative may, in fact, be an obligatory intermediate in the generation of the glutathionylated form of the oxidized enzyme.

GENERATION OF H_2O_2 BY CELLULAR INSULIN STIMULATION

The potential involvement of oxidant species in insulin signaling was initially explored more than 30 years ago, with the observation by Czech et al. that certain metal cations interacting with albumin could transfer electrons to a cellular target and enhance glucose utilization by adipocytes.[49–51] Livingston et al. also showed that polyamines and related insulin mimickers acted *via* the generation of H_2O_2[52] and that H_2O_2 stimulates lipid synthesis in adipocytes.[53] In complementary studies, insulin was also shown to stimulate the generation of H_2O_2 in adipocytes.[54]

An early characterization of the enzymology of this process revealed that insulin activated a plasma membrane enzyme system with the properties of an NADPH oxidase, resulting in the downstream production of H_2O_2.[55,56] Further biochemical studies of this activity showed that it accounted for insulin-stimulated ROS production in rat adipocyte plasma membranes[57,58] and was also present in 3T3-L1 adipocytes.[59] NADPH oxidase catalyzes the reduction of oxygen to superoxide radical: $2 O_2 + NADPH \rightarrow 2 O_2^{\cdot-} + NADP^+ + H^+$. While superoxide anions can react with thiols, they are rapidly converted spontaneously or by superoxide dismutase in the cell to generate H_2O_2.[42]

INSULIN-STIMULATED H_2O_2 GENERATION NEGATIVELY REGULATES PTP1B

One hypothesis drawn from the diverse research data described above suggests that plasma membrane oxidase activity stimulated by insulin generates cellular ROS which, in turn, facilitates the insulin signaling cascade via the oxidative inhibition of cellular PTP activity, in particular involving PTP1B.[5,6] We reported that in the 3T3-L1 adipocyte cell model, insulin stimulated the generation of cellular ROS in minutes within the high physiologic range of insulin concentrations.[30,31]

Blocking insulin-stimulated ROS with diphenyleneiodonium (DPI), an inhibitor of cellular NADPH oxidase activity, or catalase, reduced the insulin-stimulated autophosphorylation of the insulin receptor and the IRS proteins consistent with the notion that the oxidant signal inhibited cellular PTPs that serve as negative regulators of the insulin signaling cascade. That the enhancement of insulin signaling by the oxidant signal was associated with PTP inhibition was also shown by a novel approach that includes sample handling and analysis under anaerobic conditions to preserve the endogenous activity of PTPs isolated from cultured cells and avoiding cysteine oxidation that occurs on exposure to air.[60] In HepG2 hepatoma cells, stimulation with 100 nM insulin for 5 min reduced overall PTP activity in the cell homogenate, and biochemical reduction of the enzyme samples with dithiothreitol prior to PTP assay fully restored the reduced PTP activity of the insulin-treated samples, indicating that they had been reversibly oxidized and inactivated by insulin exposure.[30]

Similar effects were observed in 3T3-L1 adipocytes. Insulin-stimulated generation of H_2O_2 also affected the specific activity of endogenous PTP1B isolated from intact cells. In 3T3-L1 adipocytes, insulin treatment also potently reduced the activity of immunoprecipitated PTP1B that was also reversible toward control levels by preincubation with dithiothreitol prior to assay.[30] In the continued presence of insulin, this effect was sustained for at least 10 min. Catalase pretreatment of the cells abolished the insulin-induced inhibition of PTP1B. Oxidative inactivation of PTP1B is thus associated with enhanced insulin signal transduction via the insulin-stimulated H_2O_2 signal.

Further support of the notion of ROS serving as an enhancer of insulin action was provided by McClung and colleagues[61] who generated a line of mice overexpressing cellular glutathione peroxidase and showed that they developed hyperinsulinemia and hyperglycemia with insulin resistance and obesity. Insulin signaling was diminished in the tissues with reduced insulin-stimulated insulin receptor β-subunit phosphorylation and activation of Akt. Since excess glutathione peroxidase quenches intracellular ROS generation, it was hypothesized to interfere with insulin action by blocking PTP inactivation. Another group has also reported that an inability to generate ROS in response to insulin stimulation accounts for insulin insensitivity of ERK activation in SK-N-BE(2) neuroblastoma cells.[62]

DYNAMICS OF RECEPTOR-INDUCED ROS, PTP INHIBITION, AND SIGNAL TRANSDUCTION

ROS production following ligand stimulation, including insulin, generates only a fraction of the ROS concentration observed in phagocytic cells and follows a brief time course, on the order of minutes.[30] These features apparently account for the signaling role of insulin-induced ROS compared to the chronic exposure to ROS in patients with hyperglycemia that is associated with organ dysfunction and chronic complications of diabetes mellitus.[63]

The low levels of ROS in insulin signaling imply specific cellular protein targets that are particularly susceptible to oxidative modification. This biochemical evidence, therefore, also supports the notion of a discrete network of "redox circuitry"[42,64,65] with temporal and spatial influences that are likely to correspond to other regulatory aspects of the insulin action pathway.[66] Work by Stone and colleagues[36] clearly describes the gradient of cellular responses to oxidative stimulation, with low levels eliciting signaling responses and cell proliferation without cellular oxidation of most free thiols such as glutathione. Increasing exposure to ROS is followed by growth arrest, cellular damage, thiol oxidation, and induced cell death.

Insight into the disposition of growth factor-induced ROS has recently been gleaned from elegant studies by Reynolds et al. who modeled EGF receptor activation and signal propagation with PTP inhibition by ROS.[67] ROS generated in response to EGF stimulation in MCF7 cells were spatially constrained to a layer below the plasma membrane.

Using reaction constants gleaned from published experimental work, including PTP inhibition by H_2O_2 and related effects, a model was developed that was most consistent with a bistable activation state for PTP and receptor tyrosine kinase activity. The formation of a reaction "wavefront" is postulated to involve local cycles of EGFR activation, H_2O_2 production, and PTP inhibition, which propagates along the plasma membrane. The model proposes that signal initiation involving oxidative inhibition of PTPs adds a feedback control loop to a reaction network that responds in an amplified and switch-like manner, especially at low levels of ligand stimulus.

Consistent with this model, they also showed experimentally that blocking ligand-dependent H_2O_2 generation with the NADPH oxidase inhibitor DPI abolished the propagation of receptor phosphorylation and the amplification of receptor activation at low concentrations of EGF, converting the system to a "stable" steady state by generating a more linear phosphorylation response to ligand stimulation.[67] Diminishing PTP inactivation with DPI also suppresses insulin receptor activation and several aspects of the downstream insulin signaling cascade.[31,68] Thus, it would be of interest to employ these types of novel imaging techniques to evaluate the role of similar regulatory networks in insulin-sensitive cells.

IDENTIFICATION OF THE NADPH OXIDASE HOMOLOG NOX4 AS A POTENTIAL MEDIATOR OF INSULIN-STIMULATED ROS

Using the prototypic NADPH oxidase (Nox) catalytic gp91*phox* subunit, Cheng and Lambeth cloned a small family of five homologous Nox catalytic subunits (reviewed in Lambeth[69]). Although Nox4 is prominently expressed in kidney,[70–72] we also showed that it was expressed among insulin-sensitive cell types including liver, skeletal muscles, and adipocytes.[4] Evidence for an integral role of Nox4 in the insulin-induced oxidant signal was obtained by adenovirus-mediated expression of Nox4 deletion constructs lacking either the NADPH binding domain or the combined FAD/NADPH domains that acted in a dominant-negative fashion and attenuated insulin-stimulated generation of H_2O_2.

Functionally, expression of the deletion constructs led to an inhibition of insulin receptor and IRS-1 tyrosine phosphorylation, activation of downstream serine kinases, and glucose uptake. Similar results were obtained with transfection of siRNA oligonucleotides that reduced Nox4 protein abundance.[4] Altogether, these results suggest that Nox4 overexpression potentiates, and reduced Nox4 mass diminishes, insulin signal transduction in 3T3-L1 cells.

ROS generation by Nox4 in adipocytes was also associated with oxidative inhibition of cellular PTP1B activity.[4] Overexpression of recombinant PTP1B inhibited insulin-stimulated tyrosine phosphorylation of the insulin receptor, which was significantly reversed by co-overexpression of active Nox4. The effect of overexpression of Nox4 on receptor autophosphorylation was closely associated with inhibition of PTP1B catalytic activity, measured in enzyme immunoprecipitates.

REGULATION OF NOX4 SIGNALING IN INSULIN ACTION CASCADE

Clarification of the regulation of Nox4 by growth factors has been elusive.[73] All the Nox catalytic subunits analogous to gp91*phox* have been shown to be physically associated with p22*phox*, an essential component of the flavocytochrome complex[74]; thus, regulation of p22 abundance may be a means of regulating Nox4 activity. The phagocyte NADPH oxidase has been extensively characterized; the regulatory proteins p47*phox* and p67*phox* have been shown to play a key role in protein complex formation with activation of Nox2.[69,75]

Since these proteins are not expressed in non-phagocytic cells, several groups have recently reported the cloning of related interacting proteins termed NOXO1 (Nox organizer 1 or p41) and NOXA1 (Nox activator 1 or p51) that are homologs of the phagocyte p47*phox* and p67*phox* proteins.[76–79] Superoxide production by Nox1 and Nox3 is enhanced by interactions with these regulatory subunits.[80–83] Nox5 is regulated by intracellular calcium levels.[84] NOXO1 and NOXA1 do not affect Nox4 activity and currently the mechanisms underlying the regulation of Nox4 activity are not known. NIH 3T3 fibroblasts transfected with Nox4 exhibit constitutively increased superoxide generation,[70] suggesting that Nox4 may not depend on activation by regulatory subunits. However, the mechanism of the rapid activation of Nox4 in insulin-stimulated ROS generation remains unexplained.

POTENTIAL ROLE OF RAC IN INSULIN-STIMULATED H_2O_2 AND PTP REGULATION

Rac is a key component of the NADPH oxidase complex in a variety of cell types.[73,85] A chimeric Rac1-p67*phox* protein increases Nox2 activity, expression of a dominant-negative Rac inhibits the rise in ROS seen after stimulation by growth factors or cytokines, and a constitutively active Rac stimulates ROS formation in NIH 3T3 cells and in renal mesangial cells.[86,87] Since superoxide generation in the renal cell system is likely to involve Nox4, these data also implicated Rac as a potential regulator of this Nox homolog.[87]

In insulin-sensitive cells, Rac is involved in distal insulin signaling to glucose transport.[88] However, to date, our experiments using overexpression of wild-type, dominant-negative, and constitutively active Rac in differentiated 3T3-L1 adipocytes have not demonstrated a role for Rac in insulin-stimulated receptor auto-phosphorylation or substrate (IRS-1) tyrosine phosphorylation.

Gαi2

Data from a variety of sources over the years have also linked the small G-protein Gαi2 with insulin action and potentially with insulin-stimulated NADPH oxidase activity (reviewed by Waters et al. and Malbon[89,90]). Insulin-stimulated plasma membrane NADPH oxidase was shown to be coupled to Gαi2[91] and more recently insulin stimulation led to protein association between Gαi2 and the insulin receptor.[92] Insulin receptor autophosphorylation was stimulated by activated Gαi2, and blocked by pretreatment with pertussis toxin, consistent with an earlier paper on Fao hepatoma cells.[93]

A recent paper also linked the attenuation of platelet activation by insulin with tyrosine phosphorylation of Gαi2 and complex formation between IRS-1 and Gαi2, but not other Gα subunits.[94] Other approaches including a series of studies by Malbon and colleagues in transgenic mice have supported a permissive role of Gαi2 in insulin signaling *in vivo*.[95–98] Interestingly, mice lacking Gαi2 also exhibited increased tissue PTP activity,[95] implicating a potential loss of insulin-stimulated NADPH oxidase activity in Gαi2-deficient animals with reduced oxidative inhibition of PTPs that regulate the insulin action pathway. Overall, these studies are consistent with the hypothesis that the regulation of tyrosine phosphorylation in the insulin signal cascade is propagated by a wave of H_2O_2, possibly generated by a link between the insulin receptor and Gαi2, coupled to cellular NADPH oxidase activity, which transiently inhibits PTP activities.

NOVEL TARGETS OF INSULIN-STIMULATED ROS THAT MAY POTENTIALLY INFLUENCE INSULIN ACTION

In addition to PTPs that have been implicated in the regulation of insulin signaling (e.g., PTP1B and LAR, as discussed above[11]), a number of additional cellular enzymes are potential targets of oxidative inhibition by insulin-induced ROS. The serine–threonine phosphatase PP2A implicated in the negative regulation of Akt by dephosphorylation of ser-473 has a redox-sensitive cysteine residue that is potentially susceptible to inhibition by H_2O_2.[99]

The dual-specificity (ser/thr and tyr) phosphatase MKP-1, which attenuates insulin-stimulated MAP kinase activity,[100] is also dependent on a reduced thiol for activity. The lipid phosphatase PTEN can modulate downstream insulin signaling,[101] and is also inactivated by oxidation of essential cysteine residues in its active site, which can be reactivated by thioredoxin in cells.[102,103] Further research is needed to determine the effects of the oxidative inhibition of these important signaling regulators on proximal and distal events in the insulin action cascade.

Rhee and his group have shown that even in a cellular milieu containing millimolar concentrations of slowly reactive thiols like GSH, only a limited set of proteins are rapidly oxidized by growth factor-stimulated ROS, including PTP1B and a few other proteins with reactive cysteines including protein disulfide isomerases, thioredoxin reductase, and creatine kinase.[104,105] We have confirmed similar findings following cellular insulin stimulation of insulin target cell types (Wu et al., unpublished data), and are currently employing novel fluorescently tagged thiol reagents (iodoacetamide and maleimide derivatives) and adapting proteomic methods for the differential labeling of protein thiol before and after cellular insulin stimulation.

Further work will help define the regulatory components and mechanism of NADPH oxidase activation by insulin in various insulin-sensitive cell types and the effects of insulin-stimulated ROS on the insulin action cascade with the identification of specific cellular targets susceptible to oxidative modification by insulin-stimulated ROS. Elucidation of these processes will determine how they are involved in the normal physiology of insulin signaling, whether they contribute to insulin-resistant disease states, and whether elements of this system may emerge as novel targets for pharmaceutical intervention.

SUMMARY

Reversible tyrosine phosphorylation plays an essential role in the regulation of transmission of the insulin signal at receptor and post-receptor sites in the insulin action pathway. PTPs, in particular PTP1B, and other thiol-sensitive signaling proteins are integral to the negative regulation of insulin signaling. A growing body of data over the past three decades has led to the appreciation that cellular stimulation with insulin generates ROS that can inhibit these negative regulators by oxidative biochemical alterations which, in turn, can facilitate the insulin signaling cascade. With the recognition that a small family of NADPH oxidase

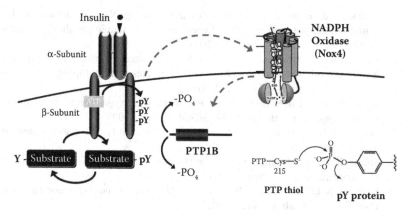

FIGURE 11.2 Insulin-induced ROS production and PTP regulation via Nox4 in insulin-sensitive cells. This figure illustrates the action of insulin to stimulate receptor tyrosine autophosphorylation which activates the receptor toward its cellular substrate (IRS) proteins. The receptor and IRS tyrosine phosphorylation require cellular PTP activity to return to a basal state. The insulin receptor is coupled to the Nox4 homolog of NADPH oxidase and stimulates the cellular generation of reactive oxygen species which, in turn, leads to oxidative inhibition of the thiol-dependent PTPs, including PTP1B, the major phosphatase for the insulin signaling cascade. The lower right portion illustrates the PTP reaction mechanism in which the reduced thiol side chain in the enzyme catalytic domain forms a phosphocysteine intermediate with the phosphotyrosine substrate. If the catalytic PTP thiol is oxidized, this reaction intermediate cannot form and the enzymatic reaction is blocked. See text for discussion and references.

homologs catalyzes the generation of ROS at the plasma membrane, we have recently provided evidence in the 3T3-L1 adipocyte system that Nox4, which is expressed in insulin-sensitive cell types, is a novel molecular target that may mediate this process (Figure 11.2).

ACKNOWLEDGMENT

This work was supported by National Institutes of Health grant RO1 DK43396 to Dr. Goldstein.

REFERENCES

1. Evans, J.L. et al. Oxidative stress and stress-activated signaling pathways: a unifying hypothesis of type 2 diabetes. *Endocr. Rev.* 23, 599, 2002.
2. Ceriello, A. and Motz, E. Is oxidative stress the pathogenic mechanism underlying insulin resistance, diabetes, and cardiovascular disease? The common soil hypothesis revisited. *Arterioscler. Thromb. Vasc. Biol.* 24, 816, 2004.
3. Houstis, N., Rosen, E.D., and Lander, E.S. Reactive oxygen species have a causal role in multiple forms of insulin resistance. *Nature* 440, 944, 2006.

4. Mahadev, K. et al. The NAD(P)H oxidase homolog Nox4 modulates insulin-stimulated generation of H2O2 and plays an integral role in insulin signal transduction. *Mol. Cell. Biol.* 24, 1844, 2004.

5. Goldstein, B.J., Mahadev, K., and Wu, X. Redox paradox: insulin action is facilitated by insulin-stimulated reactive oxygen species with multiple potential signaling targets. *Diabetes* 54, 311, 2005.

6. Goldstein, B.J. et al. Role of insulin-induced reactive oxygen species in the insulin signaling pathway. *Antioxid. Redox Signal.* 7, 1021, 2005.

7. Goldstein, B.J. Insulin resistance as the core defect in type 2 diabetes mellitus. *Am. J. Cardiol.* 90, 3, 2002.

8. Stumvoll, M., Goldstein, B.J., and van Haeften, T.W. Type 2 diabetes: principles of pathogenesis and therapy. *Lancet* 365, 1333, 2005.

9. Kido, Y., Nakae, J., and Accili, D. The insulin receptor and its cellular targets. J. Clin. Endocr. Metab. 86, 972, 2001.

10. White, M.F. IRS proteins and the common path to diabetes. *Am. J. Physiol. Endocrinol. Metab.* 283, E413, 2002.

11. Goldstein, B.J. Protein-tyrosine phosphatases and the regulation of insulin action, in *Diabetes Mellitus: A Fundamental and Clinical Text,* 3rd ed., LeRoith, D. et al., Eds., Lippincott, Philadelphia, 2003, p. 255.

12. Goldstein, B.J. Protein-tyrosine phosphatase 1B (PTP1B): a novel therapeutic target for type 2 diabetes mellitus, obesity and related states of insulin resistance. *Curr. Drug Targets Immune. Endocr. Metabol. Disord.* 1, 265, 2001.

13. Asante-Appiah, E. and Kennedy, B.P. Protein tyrosine phosphatases: the quest for negative regulators of insulin action. *Am. J. Physiol. Endocrinol. Metab.* 284, E663, 2003.

14. Tonks, N.K. PTP1B: from the sidelines to the front lines! *FEBS Lett.* 546, 140, 2003.

15. Elchebly, M., Cheng, A., and Tremblay, M.L. Modulation of insulin signaling by protein tyrosine phosphatases. *J. Molec. Med.* 78, 473, 2000.

16. Klaman, L.D. et al. Increased energy expenditure, decreased adiposity, and tissue-specific insulin sensitivity in protein-tyrosine phosphatase 1B-deficient mice. *Mol. Cell. Biol.* 20, 5479, 2000.

17. Hashimoto, N., Zhang, W.R., and Goldstein, B.J. Insulin receptor and epidermal growth factor receptor dephosphorylation by three major rat liver protein-tyrosine phosphatases expressed in a recombinant bacterial system. *Biochem. J.* 284, 569, 1992.

18. Walchli, S. et al. Identification of tyrosine phosphatases that dephosphorylate the insulin receptor: a brute force approach based on "substrate-trapping" mutants. *J. Biol. Chem.* 275, 9792, 2000.

19. Goldstein, B.J. et al. Tyrosine dephosphorylation and deactivation of insulin receptor substrate-1 by protein-tyrosine phosphatase 1B: possible facilitation by the formation of a ternary complex with the Grb2 adaptor protein. *J. Biol. Chem.* 275, 4283, 2000.

20. Calera, M.R., Vallega, G., and Pilch, P.F. Dynamics of protein–tyrosine phosphatases in rat adipocytes. *J. Biol. Chem.* 275, 6308, 2000.

21. Puius, Y.A. et al. Identification of a second aryl phosphate-binding site in protein-tyrosine phosphatase 1b: a paradigm for inhibitor design. *Proc. Natl. Acad. Sci. USA* 94, 13420, 1997.

22. Salmeen, A. et al. Molecular basis for the dephosphorylation of the activation segment of the insulin receptor by protein tyrosine phosphatase 1B. *Mol. Cell* 6, 1401, 2000.

23. Dadke, S. and Chernoff, J. Interaction of protein tyrosine phosphatase (PTP) 1B with its substrates is influenced by two distinct binding domains. *Biochem. J.* 364, 377, 2002.

24. Tonks, N.K. and Neel, B.G. Combinatorial control of the specificity of protein tyrosine phosphatases. *Curr. Opin. Cell Biol.* 13, 182, 2001.

25. Andersen, J.N. et al. Structural and evolutionary relationships among protein tyrosine phosphatase domains. *Mol. Cell Biol.* 21, 7117, 2001.

26. Keyse, S.M. Protein phosphatases and the regulation of mitogen-activated protein kinase signalling. *Curr. Opin. Cell Biol.* 12, 186, 2000.

27. Zhang, Z.Y. Protein tyrosine phosphatases: structure and function, substrate specificity, and inhibitor development. *Annu. Rev. Pharmacol. Toxicol.* 42, 209, 2002.

28. Denu, J.M. and Dixon, J.E. Protein tyrosine phosphatases: mechanisms of catalysis and regulation. *Curr. Opin. Chem. Biol.* 2, 633, 1998.

29. Zhang, Z.Y. Mechanistic studies on protein tyrosine phosphatases. *Progr. Nucl. Acid Res. Mol. Biol.* 73, 171, 2003.

30. Mahadev, K. et al. Insulin-stimulated hydrogen peroxide reversibly inhibits protein-tyrosine phosphatase 1B *in vivo* and enhances the early insulin action cascade. *J. Biol. Chem.* 276, 21938, 2001.

31. Mahadev, K. et al. Hydrogen peroxide generated during cellular insulin stimulation is integral to activation of the distal insulin signaling cascade in 3T3-L1 adipocytes. *J. Biol. Chem.* 276, 48662, 2001.

32. Meng, T.C., Fukada, T., and Tonks, N.K. Reversible oxidation and inactivation of protein tyrosine phosphatases *in vivo*. *Mol. Cell* 9, 387, 2002.

33. Rhee, S.G. et al. Intracellular messenger function of hydrogen peroxide and its regulation by peroxiredoxins. *Curr. Opin. Cell Biol.* 17, 183, 2005.

34. Sundaresan, M. et al. Requirement for generation of H_2O_2 for platelet-derived growth factor signal transduction. *Science* 270, 296, 1995.

35. Bae, Y.S. et al. Epidermal growth factor (EGF)-induced generation of hydrogen peroxide: role in EGF receptor-mediated tyrosine phosphorylation. *J. Biol. Chem.* 272, 217, 1997.

36. Stone, J.R. and Yang, S. Hydrogen peroxide: a signaling messenger. *Antioxid. Redox Signal.* 8, 243, 2006.

37. Lee, S.R. et al. Reversible inactivation of protein-tyrosine phosphatase 1b in a431 cells stimulated with epidermal growth factor. *J. Biol. Chem.* 273, 15366, 1998.

38. DeGnore, J.P. et al. Identification of the oxidation states of the active site cysteine in a recombinant protein tyrosine phosphatase by electrospray mass spectrometry using on-line desalting. *Rapid Commun. Mass Spectrom.* 12, 1457, 1998.

39. Meng, T.C. et al. Regulation of insulin signaling through reversible oxidation of the protein–tyrosine phosphatases TC45 and PTP1B. *J. Biol. Chem.* 279, 37716, 2004.

40. Denu, J.M. and Tanner, K.G. Specific and reversible inactivation of protein tyrosine phosphatases by hydrogen peroxide: evidence for a sulfenic acid intermediate and implications for redox regulation. *Biochemistry* 37, 5633, 1998.

41. Barrett, W.C. et al. Roles of superoxide radical anion in signal transduction mediated by reversible regulation of protein-tyrosine phosphatase 1B. *J. Biol. Chem.* 274, 34543, 1999.

42. Rhee, S.G. et al. Cellular regulation by hydrogen peroxide. *J. Am. Soc. Nephrol.* 14, S211, 2003.
43. Claiborne, A. et al. Protein sulfenic acids: diverse roles for an unlikely player in enzyme catalysis and redox regulation. *Biochemistry* 38, 15407, 1999.
44. Claiborne, A. et al. Protein sulfenic acid stabilization and function in enzyme catalysis and gene regulation. *FASEB J.* 7, 1483, 1993.
45. Skorey, K. et al. How does alendronate inhibit protein-tyrosine phosphatases. *J. Biol. Chem.* 272, 22472, 1997.
46. Barrett, W.C. et al. Regulation of PTP1B via glutathionylation of the active site cysteine 215. *Biochemistry* 38, 6699, 1999.
47. van Montfort, R.L. et al. Oxidation state of the active-site cysteine in protein tyrosine phosphatase 1B. *Nature* 423, 773, 2003.
48. Salmeen, A. et al. Redox regulation of protein tyrosine phosphatase 1B involves a sulphenyl-amide intermediate. *Nature* 423, 769, 2003.
49. Czech, M.P. and Fain, J.N. Cu++-dependent thiol stimulation of glucose metabolism in white fat cells. *J. Biol. Chem.* 247, 6218, 1972.
50. Czech, M.P., Lawrence, J.C., Jr., and Lynn, W.S. Evidence for electron transfer reactions involved in the Cu^{2+}-dependent thiol activation of fat cell glucose utilization. *J. Biol. Chem.* 249, 1001, 1974.
51. May, J.M. The insulin-like effects of low molecular weight thiols: role of trace metal contamination of commercial thiols. *Horm. Metab Res.* 12, 587, 1980.
52. Livingston, J.N., Gurny, P.A., and Lockwood, D.H. Insulin-like effects of polyamines in fat cells: mediation by H_2O_2 formation. *J. Biol. Chem.* 252, 560, 1977.
53. May, J.M. and de Haen, C. The insulin-like effect of hydrogen peroxide on pathways of lipid synthesis in rat adipocytes. *J. Biol. Chem.* 254, 9017, 1979.
54. May, J.M. and de Haen, C. Insulin-stimulated intracellular hydrogen peroxide production in rat epididymal fat cells. *J. Biol. Chem.* 254, 2214, 1979.
55. Mukherjee, S.P. and Lynn, W.S. Reduced nicotinamide adenine dinucleotide phosphate oxidase in adipocyte plasma membrane and its activation by insulin: possible role in the hormone's effects on adenylate cyclase and the hexose monophosphate shunt. *Arch. Biochem. Biophys.* 184, 69, 1977.
56. Mukherjee, S.P., Lane, R.H., and Lynn, W.S. Endogenous hydrogen peroxide and peroxidative metabolism in adipocytes in response to insulin and sulfhydryl reagents. *Biochem. Pharmacol.* 27, 2589, 1978.
57. Krieger-Brauer, H.I. and Kather, H. Human fat cells possess a plasma membrane-bound H_2O_2-generating system that is activated by insulin via a mechanism bypassing the receptor kinase. *J. Clin Invest.* 89, 1006, 1992.
58. Krieger-Brauer, H.I. and Kather, H. The stimulus-sensitive H_2O_2-generating system present in human fat-cell plasma membranes is multireceptor-linked and under antagonistic control by hormones and cytokines. *Biochem. J.* 307, 543, 1995.
59. Krieger-Brauer, H.I. and Kather, H. Antagonistic effects of different members of the fibroblast and platelet-derived growth factor families on adipose conversion and NADPH-dependent H_2O_2 generation in 3T3 L1-cells. *Biochem. J.* 307, 549, 1995.
60. Zhu, L. et al. Use of an anaerobic environment to preserve the endogenous activity of protein-tyrosine phosphatases isolated from intact cells. *FASEB J.* 15, 1637, 2001.

61. McClung, J.P. et al. Development of insulin resistance and obesity in mice over-expressing cellular glutathione peroxidase. *Proc. Natl. Acad. Sci. USA* 101, 8852, 2004.

62. Hwang, J.J. and Hur, K.C. Insulin cannot activate extracellular-signal-related kinase due to inability to generate reactive oxygen species in SK-N-BE(2) human neuroblastoma cells. *Mol. Cell* 20, 280, 2005.

63. Brownlee, M. Biochemistry and molecular cell biology of diabetic complications. *Nature* 414, 813, 2001.

64. Go, Y.M. et al. H_2O_2-dependent activation of GCLC-ARE4 reporter occurs by mitogen-activated protein kinase pathways without oxidation of cellular glutathione or thioredoxin-1. *J. Biol. Chem.* 279, 5837, 2004.

65. Dooley, C.T. et al. Imaging dynamic redox changes in mammalian cells with green fluorescent protein indicators. *J. Biol. Chem.* 279, 22284, 2004.

66. Saltiel, A.R. and Pessin, J.E. Insulin signaling in microdomains of the plasma membrane. *Traffic* 4, 711, 2003.

67. Reynolds, A.R. et al. EGFR activation coupled to inhibition of tyrosine phosphatases causes lateral signal propagation. *Nat. Cell. Biol.* 5, 447, 2003.

68. Mahadev, K. et al. Integration of multiple downstream signals determines the net effect of insulin on MAP kinase vs. PI 3-kinase activation: potential role of insulin-stimulated H_2O_2. *Cell Signal.* 16, 323, 2004.

69. Lambeth, J.D. NOX enzymes and the biology of reactive oxygen. *Nat. Rev. Immunol.* 4, 181, 2004.

70. Geiszt, M. et al. Identification of renox, an NAD(P)H oxidase in kidney. *Proc. Natl. Acad. Sci. USA* 97, 8010, 2000.

71. Cheng, G. et al. Homologs of gp91phox: cloning and tissue expression of Nox3, Nox4, and Nox5. *Gene* 269, 131, 2001.

72. Shiose, A. et al. A novel superoxide-producing NAD(P)H oxidase in kidney. *J. Biol. Chem.* 276, 1417, 2001.

73. Hordijk, P.L. Regulation of NADPH oxidases: role of Rac proteins. *Circ. Res.* 98, 453, 2006.

74. Ambasta, R.K., et al., Direct interaction of the novel Nox proteins with p22phox is required for the formation of a functionally active NADPH oxidase. *J. Biol. Chem.* 279, 45935, 2004.

75. Babior, B.M., Lambeth, J.D., and Nauseef, W. The neutrophil NADPH oxidase. *Arch. Biochem. Biophys.* 397, 342, 2002.

76. Banfi, B. et al. Two novel proteins activate superoxide generation by the NADPH oxidase NOX1. *J. Biol. Chem.* 278, 3510, 2003.

77. Takeya, R. et al. Novel human homologues of p47phox and p67phox participate in activation of superoxide-producing NADPH oxidases. *J. Biol. Chem.* 278, 25234, 2003.

78. Geiszt, M. et al. Proteins homologous to p47phox and p67phox support superoxide production by NAD(P)H oxidase 1 in colon epithelial cells. *J. Biol. Chem.* 278, 20006, 2003.

79. Cheng, G. and Lambeth, J.D. NOXO1, regulation of lipid binding, localization, and activation of Nox1 by the Phox homology (PX) domain. *J. Biol. Chem.* 279, 4737, 2004.

80. Cheng, G., Ritsick, D., and Lambeth, J.D. Nox3 regulation by NOXO1, p47phox, and p67phox. *J. Biol. Chem.* 279, 34250, 2004.

81. Banfi, B. et al. NOX3: a superoxide-generating NADPH oxidase of the inner ear. *J. Biol. Chem.* 279, 46065, 2004.
82. Cheng, G. and Lambeth, J.D. Alternative mRNA splice forms of NOXO1: differential tissue expression and regulation of Nox1 and Nox3. *Gene* 356, 118, 2005.
83. Cheng, G. et al. Nox1-dependent reactive oxygen generation is regulated by Rac1. *J. Biol. Chem.* epub M512751200, 2006.
84. Banfi, B. et al. Mechanism of Ca^{2+} activation of the NADPH oxidase 5 (NOX5). *J. Biol. Chem.* 279, 18583, 2004.
85. Lassegue, B. and Clempus, R.E. Vascular NAD(P)H oxidases: specific features, expression, and regulation. *Am. J. Physiol. Regul. Integr. Comp Physiol.* 285, R277, 2003.
86. Sundaresan, M. et al. Regulation of reactive-oxygen-species generation in fibroblasts by Rac1. *Biochem. J.* 318 (Pt. 2), 379, 1996.
87. Gorin, Y. et al. Nox4 mediates angiotensin II-induced activation of Akt/protein kinase B in mesangial cells. *Am. J. Physiol. Renal Physiol.* 285, F219, 2003.
88. JeBailey, L. et al. Skeletal muscle cells and adipocytes differ in their reliance on TC10 and Rac for insulin-induced actin remodeling. *Mol. Endocrinol.* 18, 359, 2004.
89. Waters, C., Pyne, S., and Pyne, N.J. The role of G-protein coupled receptors and associated proteins in receptor tyrosine kinase signal transduction. *Semin. Cell Dev. Biol.* 15, 309, 2004.
90. Malbon, C.C. Insulin signalling: putting the 'G-' in protein–protein interactions. *Biochem. J.* 380, e11, 2004.
91. Krieger-Brauer, H.I., Medda, P.K., and Kather, H. Insulin-induced activation of NADPH-dependent H_2O_2 generation in human adipocyte plasma membranes is mediated by Galphai2. *J. Biol. Chem.* 272, 10135, 1997.
92. Kreuzer, J., Nurnberg, B., and Krieger-Brauer, H.I. Ligand-dependent auto-phosphorylation of the insulin receptor is positively regulated by Gi-proteins. *Biochem. J.* 380, 831, 2004.
93. Muller-Wieland, D. et al. Pertussis toxin inhibits autophosphorylation and activation of the insulin receptor kinase. *Biochem. Biophys. Res. Commun.* 181, 1479, 1991.
94. Ferreira, I.A. et al. IRS-1 mediates inhibition of Ca^{2+} mobilization by insulin via the inhibitory G-protein Gi. *J. Biol. Chem.* 279, 3254, 2004.
95. Moxham, C.M. and Malbon, C.C. Insulin action impaired by deficiency of the G-protein subunit Gi-alpha-2. *Nature* 379, 840, 1996.
96. Chen, J.F. et al. Conditional, tissue-specific expression of Q205L G alpha i2 *in vivo* mimics insulin action. *J. Mol. Med.* 75, 283, 1997.
97. Zheng, X.L. et al. Expression of constitutively activated Gi-alpha-2 *in vivo* ameliorates streptozotocin-induced diabetes. *J. Biol. Chem.* 273, 23649, 1998.
98. Song, X. et al. G-alpha-i2 enhances *in vivo* activation of and insulin signaling to GLUT4. *J. Biol. Chem.* 276, 34651, 2001.
99. Guy, G.R. et al. Inactivation of a redox-sensitive protein phosphatase during the early events of tumor necrosis factor/interleukin-1 signal transduction. *J. Biol. Chem.* 268, 2141, 1993.
100. Kusari, A.B. et al. Insulin-induced mitogen-activated protein (MAP) kinase phosphatase-1 (MKP-1) attenuates insulin-stimulated MAP kinase activity: a mechanism for the feedback inhibition of insulin signaling. *Mol. Endocrinol.* 11, 1532, 1997.

101. Nakashima, N. et al. The tumor suppressor PTEN negatively regulates insulin signaling in 3T3-L1 adipocytes. *J. Biol. Chem.* 275, 12889, 2000.
102. Lee, S.R. et al. Reversible inactivation of the tumor suppressor PTEN by H2O2. *J. Biol. Chem.* 277, 20336, 2002.
103. Leslie, N.R. et al. Redox regulation of PI 3-kinase signalling via inactivation of PTEN. *EMBO J.* 22, 5501, 2003.
104. Wu, Y., Kwon, K.S., and Rhee, S.G. Probing cellular protein targets of H_2O_2 with fluorescein-conjugated iodoacetamide and antibodies to fluorescein. *FEBS Lett.* 440, 111, 1998.
105. Kim, J.R. et al. Identification of proteins containing cysteine residues that are sensitive to oxidation by hydrogen peroxide at neutral pH. *Anal. Biochem.* 283, 214, 2000.
106. Finkel, T. Oxidant signals and oxidative stress. *Curr. Opin. Cell Biol.* 15, 247, 2003.

12 Intracellular Signaling Pathways and Peroxisome Proliferator-Activated Receptors in Vascular Health in Hypertension and in Diabetes

Farhad Amiri, Karim Benkirane, and Ernesto L. Schiffrin

CONTENTS

INTRODUCTION

In both hypertension and diabetes mellitus significant changes that occur in the vasculature affect both large and small arteries and lead to cardiovascular events such as myocardial infarction, stroke, peripheral vascular disease, and compromise of renal function. In addition, diabetes also involves microvascular disease in different vascular beds including the retina and the kidney that contribute to the morbidity, and in the case of the kidney, the mortality associated with diabetes.

The degree and importance of vascular disease and its contribution to mortality in diabetes are such that diabetes has been called a vascular disease although its origin is undoubtedly metabolic. Moreover, hypertension and diabetes are often associated: approximately 20% of hypertensives may become diabetic and 80 to 90% of type 2 diabetic patients develop hypertension. Description of the nature of vascular disease in hypertension and diabetes, its mechanisms, and therapeutic targets, potential and already demonstrated, become accordingly extremely important. Some of these will be dealt with in this chapter, particularly in relation to signaling pathways and putative vascular protective effects of activation of peroxisome proliferator-activated receptors (PPARs).

Vascular injuries in large arteries in hypertension and diabetes differ mainly in intensity and are much more severe in diabetes. In large arteries, injuries of both the intima with development of atherosclerosis and of the media with arteriosclerosis occur. The description of the atherosclerotic process is beyond the scope of this chapter. It is sufficient to say that the severity of atherosclerosis in diabetes is such that it affects the peripheral circulation, leading to amputation of lower limbs, and also affects the coronary circulation where diabetes is considered a "coronary equivalent."[1]

In small (resistance) arteries that measure 150 to 300 μm in lumen diameter, hypertension is associated with remodeling changes that lead to increased peripheral resistance — the hallmark of essential hypertension. The remodeling of small arteries in hypertension is eutrophic — reduced outer and lumen diameters with increased media-to-lumen ratio and no significant increase in cross-sectional area of the media.[2-4] Associated with the structure changes are deposition in the media of extracellular matrix components such as collagen and fibronectin[5] and dysfunction of the endothelium.[6] Diabetes, on the other hand, involves hypertrophic remodeling in which the media-to-lumen ratio is also increased along with an increased media cross-section, achieving true media hypertrophy.[7,8] Deposition of collagen is less important. More frequently than in hypertension, endothelial dysfunction is the norm.[8,9]

In the United Kingdom Prospective Diabetes Study (UKPDS), tight blood pressure (BP) controls in hypertensive patients with type 2 diabetes reduced the risk of macrovascular disease, stroke, and deaths related to diabetes.[10] Most hypertension randomized clinical trials failed to show beneficial effects on cardiac ischemia expected from population studies. Thus, blood pressure lowering may not be enough to normalize remodeled arteries in hypertensive or diabetic subjects. In hypertensive patients, the extent and consequences of tissue ischemia (in the heart, kidney, or brain) are influenced by small vessel disease.[3] The intermediate coronary lesions that are frequent in hypertensive and diabetic subjects will produce changes in flow if small artery remodeling is present.

The renin–angiotensin–aldosterone system (RAAS) plays a critical role in the initiation and progression of cardiovascular disease.[11] The RAAS contributes to vascular remodeling of small arteries through different pathways. Angiotensin II (Ang II) can indirectly promote vascular remodeling through hemodynamic effects or directly activate myriad intracellular signaling pathways such as mitogen-activated protein kinase (MAPK), phosphoinositide-3 kinase (PI3K), Janus kinase/signal transducers and activators of transcription (JAK/STAT), and reactive oxygen species (ROS) through AT_1 receptors. Ang II may also transactivate growth factor receptors such as epidermal growth factor receptor (EGFR).[12] These pathways in turn initiate cascades in which different key proteins are stimulated, including nuclear factors that are responsible for cardiovascular gene transcription. The involvement of RAAS in hypertension and diabetes-related complications has been clarified through the use of a variety of RAAS inhibitors including but not limited to angiotensin converting enzyme (ACE) inhibitors, Ang receptor blockers (ARB), and aldosterone receptor antagonists.[3]

Insulin-sensitizing thiazolidinediones (TZDs) or glitazones may exert protective cardiovascular properties in part through the prevention of vascular remodeling.[13] For instance, TZDs such as pioglitazone and rosiglitazone reduced blood pressure in several hypertensive rodent models including Ang II-infused rats, stroke-prone spontaneously hypertensive rats (SHRSP), and deoxycorticosterone acetate (DOCA)-salt hypertensive rats.[14–16] Blood pressure lowering was accompanied by prevention of vascular remodeling and endothelial dysfunction and down-regulation of inflammatory mediators in these rodent models. Although the hypotensive effects of TZDs observed in rodents were not observed in humans, other beneficial effects such as prevention of vascular remodeling, improvement of endothelial function,[17,18] and down-regulation of inflammatory markers[19] have been reported.

TZDs mediate their effects through binding of PPAR-γ, a ligand-activated transcription factor belonging to the nuclear receptor superfamily. Forming a dimer with RXR and associated to co-activators and co-repressors, PPAR-γ binds to DNA PPAR response elements (PPREs) and regulates many genes implicated in carbohydrate metabolism, inflammation and thrombosis, endothelial function and cell growth.[20] Other PPAR isoforms include PPAR-α, which is activated by fatty acids and by lipid lowering fibrates and predominantly expressed in tissues exhibiting high fatty acid catabolism such as liver, heart, kidney, and skeletal muscle. Another isoform is PPAR-β/δ, which is expressed ubiquitously and is involved in fatty acid oxidation.[21]

The major beneficial effects of TZDs have been attributed to their actions on several protein kinases such has MAPK, PI3K, and ROS-generating enzymes, all of which have been implicated in vascular remodeling in hypertension.[14,22]

MAPK SIGNALING IN VASCULAR REMODELING

MAPK signaling has been largely studied for processes such as hyperplasia and hypertrophy associated with cell growth, and pro-inflammatory pathways, all of which are found in hypertensive and diabetic vascular remodeling. These serine/threonine kinases are subdivided into six subfamilies: extracellular signal-regulated kinases 1 and 2 (ERK1/2), p38, c-jun N-terminal protein kinase (JNK), ERK3, ERK5, and ERK6.[23] Due to elaborate cross-talk among these kinases, we will focus only on ERK1/2, p38 and JNK proteins because they are the most commonly studied (Figure 12.1).

Activation of MAPK is associated with cell growth, programmed cell death (apoptosis), cell transformation and differentiation, and cell contractility. ERK1/2 proteins can be activated by different growth factors whereas p38 and JNK proteins are generally activated by cytokines and cellular stress.[23] Ang II is a potent stimulator of MAPK pathways through various upstream second messenger systems.[12,24,25]

MAPK activation requires phosphorylation of threonine and tyrosine residues by other kinases such as mixed lineage kinases (MLKs), MAPK kinase, or mitogen extracellular-regulated kinase (MEK). MEKs are activated by serine/threonine MEK kinases (Raf-1, A-Raf and B-Raf), which are in turn activated by small protein G (Ras family) and other kinases.[26,27] Raf phosphorylation is influenced by different kinases such as c-Src, protein kinase C (PKC), protein kinase B (PKB), and p21 (rac/Cdc42)-activated protein kinase (PAK).[26]

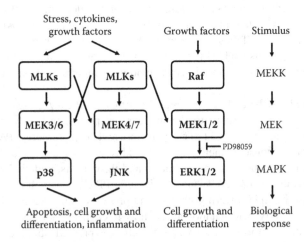

FIGURE 12.1 Mitogen-activated protein kinase (MAPK) signaling pathways and their responses. MLK = mixed lineage kinase (generic MEK). MEK = MAPK kinase. MEKK = MAPK kinase kinase.

ERK1 AND ERK2 SIGNALING, EFFECTS OF PPAR-γ, AND VASCULAR REMODELING

ERK1 and ERK2 are ubiquitous proteins highly expressed in different tissues of the cardiovascular system (blood vessels, kidneys, heart).[28] Additionally, differential regulation occurs in different cell types. ERK2 expression was significantly higher than ERK1 expression in immune cells,[26] but both enzymes are equally and highly expressed in endothelial cells and vascular smooth muscle cells (VSMCs).[29–31] Ang II through binding to AT_1 receptors activates signaling cascades such as the Src homology 2 domain (Shc), c-Src, growth factor receptor-bound protein 2 (Grb2), Son of sevenless (Sos), Ras-GTP, and MEK1, which then activate ERK1/2 through phosphorylation.[32]

Several studies using the MEK specific inhibitor PD98059 demonstrated an important role of ERK1/2 in the development and the maintenance of hypertension.[33,34] For instance, in hypertensive vascular remodeling, ERK1/2 activation led to activation of phospholipase A_2, cyclooxygenase (COX)-2, and p90 ribosomal S6 kinase (p90Rsk), along with translocation of nuclear receptors and reorganization of cytoskeletal proteins implicated in cell growth and migration.[26] More specifically, once ERK1/2 is phosphorylated, it will translocate to the nucleus and activate transcription of cell cycle genes.[23,26]

In VSMCs, ERK1/2 phosphorylation induced p90Rsk activation leading to ribosomal S6 protein phosphorylation and protein synthesis. In VSMCs derived from mesenteric arteries, we demonstrated the stimulatory effect of Ang II on cell hypertrophy, proliferation, and contractility.[29] Ang II-induced ERK1/2 activity was inhibited by PPAR-γ activation in conduit vessels whereas no effect was observed in mesenteric resistance vessels.[14] Unlike results of *in vivo* studies, *in vitro* acute stimulation of mesenteric VSMCs with Ang II-induced ERK1/2 activation was abrogated by PPAR-γ pre-stimulation with rosiglitazone (Figure 12.2).[14] Taken together, these results suggest that the extracellular environment and duration of stimulation (acute versus chronic) affect PPAR-γ-modulated changes in Ang II-induced ERK1/2 activation.

ERK1/2 may on the other hand inhibit PPAR-γ activity. *In vivo*, insulin, which constitutively activates MEKK1, induced in a ligand-independent manner PPAR-γ phosphorylation which increased TZD-dependent PPAR-γ trans-activation properties.[35] In contrast, both epidermal growth factor (EGF) and platelet derived growth factor (PDGF) reduced PPAR-γ transcriptional activity in a ligand-dependent manner in adipocytes contained in the vascular periadventitial fat that regulates vascular function in paracrine fashion and may play an important role in blood pressure regulation and vascular remodeling.[36]

Although mesenteric artery PPAR-γ activity was reduced in Ang II-infused rats, albeit in the absence of changes in PPAR-γ expression, treatment with rosiglitazone prevented these changes, suggesting a potent negative regulatory role of PPAR-γ activators.[14] Endogenous PPAR ligands such as linoleic and retinoic acid may stimulate ERK1/2 activity in adipocytes and aortic VSMCs, respectively,[37,38] whereas prostaglandin J2 (PGJ2) activated ERK1/2 through PI3K

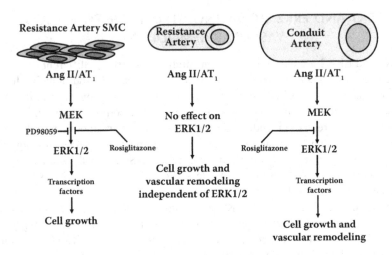

FIGURE 12.2 Angiotensin (Ang) II effects on extracellular-regulated kinase (ERK) 1 and 2 activation in resistance and conduit arteries.

signaling in aortic VSMCs. These data collectively demonstrate the countervailing effects of PPAR activators on MAPK signaling pathways implicated in cell growth and migration induced by vasoactive agents such as Ang II.

p38 SIGNALING IN VASCULAR REMODELING

p38 MAPK comprises six isoforms (p38α1 and α2, p38β1 and β2, p38δ, and p38γ): the α and β isoforms are ubiquitous; the δ isoform is expressed mainly in kidneys, lungs, and pancreas; and the γ isoform is highly expressed in skeletal muscle.[39] p38 MAPK can be activated by a variety of stimuli including cytokines, growth factors, osmotic shock, and different stressors. p38 MAPK has been implicated in several pathophysiological conditions such as atherosclerosis, hypertensive vascular remodeling, and ischemic cardiomyopathy.[26]

Ang II can stimulate vascular p38 MAPK, affecting inflammatory responses, apoptosis, cell growth inhibition, and contractility. Inhibition of p38 MAPK reduced these effects in aortic but not mesenteric VSMCs.[40–42] p38 MAPK cross-talks with other members of the MAPK family (including ERK1/2, MEK3, and MEK6; see Figure 12.1) and other signaling molecules (e.g., heat shock protein 27), which is important in cytoskeletal reorganization implicated in cell migration and vascular remodeling.[26,39] p38 MAPK is also regulated by PPAR-α and one of its co-activators, PGC-1, suggesting a regulatory role for p38 MAPK in PPAR-α-mediated lipid metabolism.[27] However, p38 MAPK does not seem to participate in PPAR-γ regulation, as inhibition of p38 MAPK failed to modulate PPAR-γ-induced AT$_1$ gene transcription.[43]

This further demonstrates differential effects of different MAPKs on the RAAS. Since p38 MAPK may also induce VSMC apoptosis, PPAR-α activators such as ω-3 fatty acids (e.g., docosahexaenoic acid) may stimulate VSMC apoptosis via a p38/PPAR-α-dependent manner.[44] As with ERK1/2, endogenous (retinoic acid) and synthetic PPAR ligands (various TZDs such as ciglitazone, troglitazone, and rosiglitazone) modulate p38 MAPK activity.[38,45,46] However, both ciglitazone and 15d-PGJ2 are potent stimulators of p38 MAPK activity in rat astrocytes but not in mesangial cells.[45,47] These data demonstrate that PPARs have the ability to directly and indirectly regulate MAPKs in a tissue-dependent manner.

ROLE OF JNK SIGNALING IN VASCULAR REMODELING

The third member of the MAPK family that we will address is c-Jun N-terminal kinase (JNK) which comprises JNK1, JNK2, and JNK3. JNKs can be stimulated by several cytokines, growth factors, stressors, and vasoactive agents (Ang II and endothelin-1).[26] Contrary to the pro-stimulatory effects of Ang II on cell growth via ERK1/2, JNK activation induced apoptosis and cell cycle inhibition.

Once activated, JNKs activate transcription factors such as c-Jun, c-Myc, PPAR-α, and PPAR-γ.[27,48] Ang II-induced JNK activation is dependent upon other factors such as Ca^{2+} mobilization and PKC-ξ activation, but is independent of c-Src activation.[49] JNK also seems to be activated in a cell-specific manner. In cardiac fibroblasts, Ang II stimulated JNK via PKC-independent proline-rich tyrosine kinase 2/Rac1, whereas in VSMCs this occurs through PAK, PKC, and Ca^{2+}.[50]

Contrary to the other MAPK proteins, the cellular effects of PPARs on the JNK pathway remain to be fully elucidated in VSMCs. PPAR-γ activation may induce JNK signaling involved in apoptosis. Taken together, all MAPK pathways are involved in molecular mechanisms that ultimately lead to development and maintenance of cardiovascular dysfunction such as the vascular remodeling observed in hypertension and diabetes.

PI3K SIGNALING PATHWAY AND VASCULAR REMODELING

The PI3K pathway is implicated in cardiovascular events such as cardiac and vascular remodeling, heart failure, and endothelial dysfunction.[51,52] The PI3K family is composed of 3′-OH inositol phosphorylating enzymes and includes members of three classes divided according to their molecular structures, activation mechanisms, and substrate specificities. Class I (A and B) is composed of heterodimeric regulatory and catalytic subunits while class II proteins are composed of single catalytic subunits (PI3K-C2α, PI3K-C2β, or PI3K-C2γ).[53] Class IA has three catalytic (p110α, β, and δ) and three regulatory subunits (p85α, p85β, and p55γ) expressed mostly in the heart and blood vessels. However, class IB contains only one protein kinase composed of a p110γ catalytic subunit and a p101 regulatory subunit, expressed predominantly in cardiomyocytes, fibroblasts, endothelial cells, and VSMCs.[53]

The importance of PI3K in Ang II-induced cardiac hypertrophy and VSMC proliferation has been established with the use of specific reversible (LY294002) and irreversible (wortmannin) inhibitors.[51] PI3K regulates cell growth and survival, cytoskeletal reorganization, ROS production, and glucose transport among other functions and plays a role in the progression of cardiac hypertrophy and dysfunction and in vascular remodeling.[52–54] PI3K is regulated by other kinases such as Rho kinase, which may contribute to endothelial dysfunction.[55] PI3K participates in endothelial nitric oxide synthase (eNOS) activation and thus in NO generation and vasodilation.[56]

Ang II-stimulated ERK1/2 activity was blunted by the PI3K inhibitor LY-294009 in mesenteric artery VSMCs from hypertensive rats but not normotensive rats, which may imply cross-talk between PI3K and ERK1/2 pathways.[29] PPAR-γ regulates PI3K activity. Rosiglitazone induced PI3K/p85α expression and activity in adipocytes and in conduit and resistance arteries, but not in skeletal muscle.[14,57,58] These data suggest that altered PI3K activity and expression may occur at the initiation and during the maintenance phase of hypertensive vascular disease.

ROLE OF Akt/PKB SIGNALING PATHWAY IN VASCULAR REMODELING

A key protein within the PI3K signaling pathway is Akt, also known as protein kinase B (PKB). In mammals, three Akt/PKB isoforms (Akt1/PKBα, Akt2/PKBβ, and Akt3/PKBγ) are highly expressed in cardiac, vascular, and endothelial cells.[53] They are regulated by PI3K, SH2-containing inositol phosphatase (SHIP2) and phosphoinositide-dependent kinase (PDK)-1 and -2.[59] Once activated, Akt/PKB modulates cell survival, growth, migration, and glucose metabolism.[60–63] Synthetic and endogenous PPAR ligands inhibit Akt/PKB in blood vessels and other tissues, possibly through the inhibition of eukaryotic initiation factor-4E-binding protein (4E-BP)-1, a protein implicated in cellular growth.[14,64]

ROS AND VASCULAR REMODELING

Reactive oxygen species (ROS) play important roles in physiological and pathophysiological conditions related to cell growth, differentiation, migration and signaling, extracellular matrix production and degradation, and inflammation in the cardiovascular system. ROS include superoxide anion ($\bullet O_2^-$), hydrogen peroxide (H_2O_2), hydroxyl radical ($\bullet OH^-$), and peroxynitrite ($ONOO^-$) as well as other free radicals such as NO.[22,65,66]

Ang II is a critical regulator of ROS generation in many tissues including VSMCs and cardiac, mesangial, and endothelial cells.[22] NAD(P)H oxidase is the most important vascular source of ROS. It is composed of five subunits: p22^{phox}, gp91^{phox} (membrane-bound), p47^{phox}, p67^{phox}, and a small G protein Rac1 or Rac2 (cytosolic). In addition to NAD(P)H oxidase, other sources of ROS include xanthine oxidase, uncoupled eNOS, and the mitochondrial respiratory chain.[22] ROS generation in response to pressor doses of Ang II plays a major role in the deleterious effects of this peptide. These effects are BP-independent effects of Ang II since similar BP elevation achieved by norepinephrine infusion did not increase vascular $\bullet O_2^-$ production.[67]

The pro-growth and hypertensive effects of ROS have also been shown to be mediated by $ONOO^-$ production via NO scavenging.[68] Many inhibitors of ROS production have been used to elucidate the detrimental role of ROS in hypertension and diabetes: protein nitration, lipid oxidation, and DNA degradation. The inhibitors include superoxide dismutase (SOD) mimetics (Tempol), $\bullet O_2^-$ scavengers (Tiron), specific inhibitors of NAD(P)H oxidase (apocynin, gp91ds-tat), SOD inhibitors (PEG-SOD), xanthine oxidase inhibitors (allopurinol) or mitochondrial chain inhibitors (thenotrifluoroacetone, carbonyl cyanide-m-chlorophenylhydrazone and rotenone).[69–71] Additionally, ROS interact with second messengers (e.g., intracellular Ca^{2+}), several kinases such as MAPK (ERK, p38, and JNK) and Akt/PKB, and transcription factors (NFκB and AP-1), all leading to the development of vascular inflammation and remodeling.[22,72,73]

Several studies have demonstrated beneficial vascular inhibitory effects of PPAR activators on ROS production. PPAR-α activators (docosahexaenoic acid, DHA, and fenofibrate) and PPAR-γ activators (rosiglitazone or pioglitazone) reduced ROS production in rats infused with Ang II and in endothelin-1-dependent hypertensive models such as DOCA-salt hypertensive rats.[16,74]

VASCULAR INFLAMMATION AND REMODELING

One of the major effects of activation of the signaling pathways described is the induction of vascular inflammation found in cardiovascular diseases such as hypertension and in diabetes.[75] Of the various systems involved, one of the major mechanisms is the pro-inflammatory effect of Ang II mediated via the AT_1 receptor.[75,76] These effects are mediated by increased expression of adhesion molecules [intercellular adhesion molecule (ICAM)-1, platelet endothelial cell adhesion molecule (PECAM), vascular cell adhesion molecule-1 (VCAM-1), and selectins] and transcription factors (NFκB, AP-1) on monocytes, endothelial cells, and VSMCs.[13,77,78]

In addition to increased adhesion molecules, Ang II stimulates monocyte chemotactic protein (MCP)-1 synthesis in monocytes/macrophages, endothelial cells, and VSMCs, causing accumulation of inflammatory cells and molecules in the vasculature that ultimately contributes to the development of endothelial dysfunction and vascular remodeling.[79] PPAR activators exert potent anti-inflammatory effects. For instance, PPAR-α activation with fenofibrate and gemfibrozil inhibited cytokine production (TNFα, interferon-γ, interleukin (IL)-6, IL-2, and IL-1β) and adhesion molecule synthesis, whereas eNOS and COX-2 expressions were increased in VSMCs.[80–82]

These effects were associated with inhibition of expression and activity of transcription factors NFκB and AP-1.[83,84] A PPAR-α activator prevented hypertension and vascular remodeling by a reduction in NAD(P)H oxidase activity, VCAM-1 and ICAM-1 expression, in Ang II infused rats.[74] PPAR-γ also has potent anti-inflammatory properties in addition to its insulin-sensitizing effects. Among beneficial effects reported are reductions in plasminogen activator inhibitor (PAI)-1 in microalbuminuria and in matrix metalloproteinase (MMP)-9 along with increased NO generation.[13,47,85] The anti-inflammatory vascular effects of PPAR-γ occur through inhibition of NFκB and AP-1, as well as down-regulation of NAD(P)H oxidase subunits.[13,86]

DUAL PPAR ACTIVATORS

Because of the beneficial cardiovascular effects of PPAR-α and PPAR-γ activators such as improvement of lipid metabolism, insulin sensitization, glucose metabolism, vascular remodeling and inflammation, combined activation of these nuclear receptors has been attempted. To date, at least eight dual PPARα/γ activators are being evaluated in studies ranging from preclinical to phase III. The activators have been shown to decrease circulating triglyceride concentrations in humans and transgenic human Apo A-I mice.[87,88]

Dual PPAR-α/γ activator effects on glucose metabolism and insulin sensitization are similar to those of PPAR-γ activators administered alone.[87,88] Two dual PPAR-α/γ activators, ragaglitazar and muraglitazar, demonstrated beneficial effects on vascular complications in hypertension and on metabolic abnormalities in diabetes, respectively.[89,90] However, muraglitazar administration in type 2 diabetic patients was associated with more death and major adverse cardiovascular events including myocardial infarction, stroke, and transient ischemic attacks, than use of a PPAR-γ activator.[91]

This result dampened interest in these dual PPAR activators, which may not really be superior to agents with single receptor activating capability. Because of these reported deleterious effects, we and others have used a different approach to stimulate both PPAR-α and PPAR-γ receptors. We administered concomitantly sub-therapeutic doses of PPAR-α and PPAR-γ activators that had beneficial effects when administered in full doses. At these lower doses, a combination of PPAR-α and PPAR-γ administration had beneficial effects in a rodent model of Ang II-induced hypertension on vascular function, including improvement of endothelial function, decreased ROS generation, and vascular inflammation. Seber et al. also found that concomitant administration of rosiglitazone and fenofibrate improved the atherogenic dyslipidemic profiles of type II diabetic patients with poor metabolic control.[92]

CONCLUSION

Hypertension is highly prevalent and one of the major causes of burden of disease, particularly cardiovascular morbidity and mortality. Diabetes is increasing worldwide, and often is associated with hypertension; it puts hypertensive subjects at the highest cardiovascular risk. Both conditions are associated with vascular injury that serves as the major mechanism for cardiovascular events that lead to myocardial infarction, stroke, amputations, and renal failure.

We have summarized here some of the pathways that participate in growth, inflammation, and oxidative stress that lead to atherosclerosis and vascular remodeling in hypertension and diabetes. PPAR-α and PPAR-γ appear to act as countervailing influences on one of the major activators of the pathways that trigger vascular disease, the RAAS. Preclinical and clinical data suggest that PPAR-α and PPAR-γ activators may exert important vascular protective effects. Although initial experience with agents endowed with combined PPAR-α and PPAR-γ stimulatory effects has been disappointing, mechanistic evidence suggests that agents with these properties that are able to affect the deleterious intracellular signaling pathways described in this chapter may be developed eventually and successfully contribute to reducing the burden of disease generated by hypertension and diabetes.

REFERENCES

1. Haffner SM, Lehto S, Ronnemaa T, Pyorala K, and Laakso M. Mortality from coronary heart disease in subjects with type 2 diabetes and in nondiabetic subjects with and without prior myocardial infarction. *New Engl J Med* 1998; 339: 229.

2. Schiffrin EL. Reactivity of small blood vessels in hypertension: relation with structural changes: state of the art lecture. *Hypertension* 1992; 19, Suppl:III.

3. Schiffrin EL. Remodeling of resistance arteries in essential hypertension and effects of antihypertensive treatment. *Am J Hypertens* 2004; 17: 1192.

4. Heagerty AM, Aalkjaer C, Bund SJ, Korsgaard N, and Mulvany MJ. Small artery structure in hypertension: dual processes of remodeling and growth. *Hypertension* 1993; 21: 391.

5. Intengan HD, Deng LY, Li JS, and Schiffrin EL. Mechanics and composition of human subcutaneous resistance arteries in essential hypertension. *Hypertension* 1999; 33, Pt 2: 569.

6. Deng LY, Li JS, and Schiffrin EL. Endothelium-dependent relaxation of small arteries from essential hypertensive patients: mechanisms and comparison with normotensive subjects and with responses of vessels from spontaneously hypertensive rats. *Clin Sci* 1995; 88: 611.

7. Rizzoni D et al. Structural alterations in subcutaneous small arteries of normotensive and hypertensive patients with non-insulin-dependent diabetes mellitus. *Circulation* 2001; 103: 1238.

8. Endemann DH et al. Persistent remodeling of resistance arteries in type 2 diabetic patients on antihypertensive treatment. *Hypertension* 2004; 43: 399.

9. Rizzoni D et al. Endothelial dysfunction in small resistance arteries of patients with non-insulin-dependent diabetes mellitus. *J Hypertens* 2001; 19: 913.

10. United Kingdom Prospective Diabetes Study Group. Tight blood pressure control and risk of macrovascular and microvascular complications in type 2 diabetes. *Br Med J* 1998; 317: 703.

11. Dzau VJ. Theodore Cooper Lecture: tissue angiotensin and pathobiology of vascular disease: a unifying hypothesis. *Hypertension* 2001; 37: 1047.

12. Touyz RM and Schiffrin EL. Signal transduction mechanisms mediating the physiological and pathophysiological actions of angiotensin II in vascular smooth muscle cells. *Pharmacol Rev* 2000; 52: 639.

13. Diep QN et al. Structure, endothelial function, cell growth, and inflammation in blood vessels of angiotensin II-infused rats: role of peroxisome proliferator-activated receptor-γ. *Circulation* 2002; 105: 2296.

14. Benkirane K, Viel EC, Amiri F, and Schiffrin EL. Peroxisome proliferator-activated receptor γ regulates angiotensin II-stimulated phosphatidylinositol 3-kinase and mitogen-activated protein kinase in blood vessels *in vivo*. *Hypertension* 2006; 47: 102.

15. Diep QN, Amiri F, Benkirane K, Paradis P, and Schiffrin EL. Long-term effects of the PPARgamma activator pioglitazone on cardiac inflammation in stroke-prone spontaneously hypertensive rats. *Can J Physiol Pharmacol* 2004; 82: 976.

16. Iglarz M, Touyz RM, Amiri F, Lavoie MF, Diep QN, and Schiffrin EL. Effect of peroxisome proliferator-activated receptor-α and -γ activators on vascular remodeling in endothelin-dependent hypertension. *Arterioscler Thromb Vasc Biol* 2003; 23: 45.

17. Sidhu JS, Kaposzta Z, Markus HS, and Kaski JC. Effect of rosiglitazone on common carotid intima-media thickness progression in coronary artery disease patients without diabetes mellitus. *Arterioscler Thromb Vasc Biol* 2004; 24: 930.

18. Wang TD, Chen WJ, Lin JW, Chen MF, and Lee YT. Effects of rosiglitazone on endothelial function, C-reactive protein, and components of the metabolic syndrome in nondiabetic patients with the metabolic syndrome. *Am J Cardiol* 2004; 93: 362.

19. Haffner SM, Greenberg AS, Weston WM, Chen H, Williams K, and Freed MI. Effect of rosiglitazone treatment on nontraditional markers of cardiovascular disease in patients with type 2 diabetes mellitus. *Circulation* 2002; 106: 679.

20. Schiffrin EL. Peroxisome proliferator-activated receptors and cardiovascular remodeling. *Am J Physiol Heart Circ Physiol* 2005; 288: H1037.

21. Berger J and Moller DE. The mechanisms of action of PPARs. *Annu Rev Med* 2002; 53: 409.

22. Touyz RM and Schiffrin EL. Reactive oxygen species in vascular biology: implications in hypertension. *Histochem Cell Biol* 2004; 122: 339.

23. Robinson MJ and Cobb MH. Mitogen-activated protein kinase pathways. *Curr Opin Cell Biol* 1997; 9: 180.

24. Touyz RM, He G, Deng LY, and Schiffrin EL. Role of extracellular signal-regulated kinases in angiotensin II-stimulated contraction of smooth muscle cells from human resistance arteries. *Circulation* 1999; 99: 392.

25. Touyz RM, He G, El Mabrouk M, Diep Q, Mardigyan V, and Schiffrin EL. Differential activation of extracellular signal-regulated protein kinase 1/2 and p38 mitogen-activated protein kinase by AT1 receptors in vascular smooth muscle cells from Wistar-Kyoto rats and spontaneously hypertensive rats. *J Hypertens* 2001; 19: 553.

26. Pearson G et al. Mitogen-activated protein (MAP) kinase pathways: regulation and physiological functions. *Endocr Rev* 2001; 22: 153.

27. Yang SH, Sharrocks AD, and Whitmarsh AJ. Transcriptional regulation by the MAP kinase signaling cascades. *Gene* 2003; 320: 3.

28. Fiebeler A and Haller H. Participation of the mineralocorticoid receptor in cardiac and vascular remodeling. *Nephron Physiol* 2003; 94: 47.

29. El Mabrouk M, Touyz RM, and Schiffrin EL. Differential ANG II-induced growth activation pathways in mesenteric artery smooth muscle cells from SHR. *Am J Physiol Heart Circ Physiol* 2001; 281: H30.

30. Kohlstedt K and Busse R, Fleming I. Signaling via the angiotensin-converting enzyme enhances the expression of cyclooxygenase-2 in endothelial cells. *Hypertension* 2005; 45: 126.

31. Kintscher U et al. Angiotensin II induces migration and Pyk2/paxillin phosphorylation of human monocytes. *Hypertension* 2001; 37: 587.

32. Touyz RM. Recent advances in intracellular signalling in hypertension. *Curr Opin Nephrol Hypertens* 2003; 12: 165.

33. Kim J et al. Mitogen-activated protein kinase contributes to elevated basal tone in aortic smooth muscle from hypertensive rats. *Eur J Pharmacol* 2005; 514: 209.

34. Rice KM, Kinnard RS, Harris R, Wright GL, and Blough ER. Effects of aging on pressure-induced MAPK activation in the rat aorta. *Pflugers Arch* 2005; 450: 192.

35. Zhang B et al. Insulin- and mitogen-activated protein kinase-mediated phosphorylation and activation of peroxisome proliferator-activated receptor gamma. *J Biol Chem* 1996; 271: 31771.

36. Verlohren S et al. Visceral periadventitial adipose tissue regulates arterial tone of mesenteric arteries. *Hypertension* 2004; 44: 271.

37. Rao GN, Alexander RW, and Runge MS. Linoleic acid and its metabolites, hydroperoxyoctadecadienoic acids, stimulate c-Fos, c-Jun, and c-Myc mRNA expression, mitogen-activated protein kinase activation, and growth in rat aortic smooth muscle cells. *J Clin Invest* 1995; 96: 842.

38. Teruel T, Hernandez R, Benito M, and Lorenzo M. Rosiglitazone and retinoic acid induce uncoupling protein-1 (UCP-1) in a p38 mitogen-activated protein kinase-dependent manner in fetal primary brown adipocytes. *J Biol Chem* 2003; 278: 263.

39. McMullen ME, Bryant PW, Glembotski CC, Vincent PA, and Pumiglia KM. Activation of p38 has opposing effects on the proliferation and migration of endothelial cells. *J Biol Chem* 2005; 280: 20995.

40. Ushio-Fukai M, Alexander RW, Akers M, and Griendling KK. p38 Mitogen-activated protein kinase is a critical component of the redox-sensitive signaling pathways activated by angiotensin II: role in vascular smooth muscle cell hypertrophy. *J Biol Chem* 1998; 273: 15022.

41. Meloche S, Landry J, Huot J, Houle F, Marceau F, and Giasson E. p38 MAP kinase pathway regulates angiotensin II-induced contraction of rat vascular smooth muscle. *Am J Physiol Heart Circ Physiol* 2000; 279: H741.

42. Touyz RM, He G, El Mabrouk M, and Schiffrin EL. p38 Map kinase regulates vascular smooth muscle cell collagen synthesis by angiotensin II in SHR but not in WKY. *Hypertension* 2001; 37: 574.

43. Sugawara A, Takeuchi K, Uruno A, Kudo M, Sato K, and Ito S. Effects of mitogen-activated protein kinase pathway and co-activator CREP-binding protein on peroxisome proliferator-activated receptor-gamma-mediated transcription suppression of angiotensin II type 1 receptor gene. *Hypertens Res* 2003; 26: 623.

44. Diep QN, Touyz RM, and Schiffrin EL. Docosahexaenoic acid, a peroxisome proliferator-activated receptor-alpha ligand, induces apoptosis in vascular smooth muscle cells by stimulation of p38 mitogen-activated protein kinase. *Hypertension* 2000; 36: 851.

45. Lennon AM, Ramauge M, Dessouroux A, and Pierre M. MAP kinase cascades are activated in astrocytes and preadipocytes by 15-deoxy-delta(12-14)-prostaglandin J(2) and the thiazolidinedione ciglitazone through peroxisome proliferator activator receptor gamma-independent mechanisms involving reactive oxygenated species. *J Biol Chem* 2002; 277: 29681.

46. Gardner OS, Shiau CW, Chen CS, and Graves LM. Peroxisome proliferator-activated receptor gamma-independent activation of p38 MAPK by thiazolidinediones involves calcium/calmodulin-dependent protein kinase II and protein kinase R: correlation with endoplasmic reticulum stress. *J Biol Chem* 2005; 280: 10109.

47. Hsueh WA and Law RE. PPAR- and atherosclerosis: effects on cell growth and movement. *Arterioscler Thromb Vasc Biol* 2001; 21: 1891.

48. Diradourian C, Girard J, and Pegorier JP. Phosphorylation of PPARs: from molecular characterization to physiological relevance. *Biochimie* 2005; 87: 33.

49. Liao DF, Monia B, Dean N, and Berk BC. Protein kinase C-zeta mediates angiotensin II activation of ERK1/2 in vascular smooth muscle cells. *J Biol Chem* 1997; 272: 6146.

50. Murasawa S et al. Angiotensin II initiates tyrosine kinase Pyk2-dependent signalings leading to activation of Rac1-mediated c-Jun NH2-terminal kinase. *J Biol Chem* 2000; 275: 26856.

51. Rocic P, Govindarajan G, Sabri A, and Lucchesi PA. A role for PYK2 in regulation of ERK1/2 MAP kinases and PI 3-kinase by ANG II in vascular smooth muscle. *Am J Physiol Cell Physiol* 2001; 280: C90.

52. Anderson KE and Jackson SP. Class I phosphoinositide 3-kinases. *Int J Biochem Cell Biol* 2003; 35: 1028.

53. Oudit GY et al. The role of phosphoinositide-3 kinase and PTEN in cardiovascular physiology and disease. *J Mol Cell Cardiol* 2004; 37: 449.

54. Hafizi S, Wang X, Chester AH, Yacoub MH, and Proud CG. ANG II activates effectors of mTOR via PI3-K signaling in human coronary smooth muscle cells. *Am J Physiol Heart Circ Physiol* 2004; 287: H1232.

55. Budzyn K, Marley PD, and Sobey CG. Opposing roles of endothelial and smooth muscle phosphatidylinositol 3-kinase in vasoconstriction: effects of rho-kinase and hypertension. *J Pharmacol Exp Ther* 2005; 313: 1248.

56. Andreozzi F, Laratta E, Sciacqua A, Perticone F, and Sesti G. Angiotensin II impairs the insulin signaling pathway promoting production of nitric oxide by inducing phosphorylation of insulin receptor substrate-1 on Ser312 and Ser616 in human umbilical vein endothelial cells. *Circ Res* 2004; 94: 1211.

57. Rieusset J et al. The expression of the p85alpha subunit of phosphatidylinositol 3-kinase is induced by activation of the peroxisome proliferator-activated receptor gamma in human adipocytes. *Diabetologia* 2001; 44: 544.

58. Rieusset J, Roques M, Bouzakri K, Chevillotte E, and Vidal H. Regulation of p85-alpha phosphatidylinositol-3-kinase expression by peroxisome proliferator-activated receptors (PPARs) in human muscle cells. *FEBS Lett* 2001; 502: 98.

59. Dong LQ and Liu F. PDK2: the missing piece in the receptor tyrosine kinase signaling pathway puzzle. *Am J Physiol Endocrinol Metab* 2005; 289: E187.

60. Shiojima I and Walsh K. Role of Akt signaling in vascular homeostasis and angiogenesis. *Circ Res* 2002; 90: 1243.

61. Shioi T et al. Akt/protein kinase B promotes organ growth in transgenic mice. *Mol Cell Biol* 2002; 22: 2799.

62. Li F, Zhang C, Schaefer S, Estes A, and Malik KU. ANG II-induced neointimal growth is mediated via c. *Am J Physiol Heart Circ Physiol* 2005; 289: H2592.

63. Ohashi H et al. Phosphatidylinositol 3-kinase/Akt regulates angiotensin II-induced inhibition of apoptosis in microvascular endothelial cells by governing survivin expression and suppression of caspase-3 activity. *Circ Res* 2004; 94: 785.

64. Benkirane K, Amiri F, Diep QN, El Mabrouk M, and Schiffrin EL. PPAR-γ inhibits ANG II-induced cell growth via SHIP2 and 4E-BP1. *Am J Physiol Heart Circ Physiol* 2006; 290: H390.

65. Griendling KK, Sorescu D, Lassegue B, and Ushio-Fukai M. Modulation of protein kinase activity and gene expression by reactive oxygen species and their role in vascular physiology and pathophysiology. *Arterioscler Thromb Vasc Biol* 2000; 20: 2175.

66. Paravicini TM, and Touyz RM. Redox signaling in hypertension. *Cardiovasc Res* 2006; 71: 247.

67. Rajagopalan S et al. Angiotensin II-mediated hypertension in the rat increases vascular superoxide production via membrane NADH/NADPH oxidase activation: contribution to alterations of vasomotor tone. *J Clin Invest* 1996; 97: 1916.

68. Muijsers RB, Folkerts G, Henricks PA, Sadeghi-Hashjin G, and Nijkamp FP. Peroxynitrite: a two-faced metabolite of nitric oxide. *Life Sci* 1997; 60: 1833.

69. Park JB, Touyz RM, Chen X, and Schiffrin EL. Chronic treatment with a superoxide dismutase mimetic prevents vascular remodeling and progression of hypertension in salt-loaded stroke-prone spontaneously hypertensive rats. *Am J Hypertens* 2002; 15: 78.

70. Rey FE, Cifuentes ME, Kiarash A, Quinn MT, and Pagano PJ. Novel competitive inhibitor of NAD(P)H oxidase assembly attenuates vascular O_2^- and systolic blood pressure in mice. *Circ Res* 2001; 89: 408.

71. Touyz RM, Yao G, Viel E, Amiri F., and Schiffrin EL. Angiotensin II and endothelin-1 regulate MAP kinases through different redox-dependent mechanisms in human vascular smooth muscle cells. *J Hypertens* 2004; 22: 1141.

72. Torrecillas G et al. The role of hydrogen peroxide in the contractile response to angiotensin II. *Mol Pharmacol* 2001; 59: 104.

73. Browatzki M et al. Angiotensin II stimulates matrix metalloproteinase secretion in human vascular smooth muscle cells via nuclear factor-kappaB and activator protein 1 in a redox-sensitive manner. *J Vasc Res* 2005; 42: 415.

74. Diep QN, Amiri F, Touyz RM, Cohn JS, Endemann D, Neves MF, and Schiffrin EL. PPAR-α activator effects on Ang II-induced vascular oxidative stress and inflammation. *Hypertension* 2002; 40: 866.

75. Savoia C and Schiffrin EL. Inflammation in hypertension. *Curr Opin Nephrol Hypertens* 2006; 15(2):152-158.

76. Victorino GP, Newton CR, and Curran B. Effect of angiotensin II on microvascular permeability. *J Surg Res* 2002; 104: 77.

77. Grafe M et al. Angiotensin II-induced leukocyte adhesion on human coronary endothelial cells is mediated by E-selectin. *Circ Res* 1997; 81: 804.

78. Pueyo ME et al. Angiotensin II stimulates endothelial vascular cell adhesion molecule-1 via nuclear factor-κB activation induced by intracellular oxidative stress. *Arterioscler Thromb Vasc Biol* 2000; 20: 645.

79. Cheng ZJ, Vapaatalo H, and Mervaala E. Angiotensin II and vascular inflammation. *Med Sci Monit* 2005; 11: RA194.

80. Moraes LA, Piqueras L, and Bishop-Bailey D. Peroxisome proliferator-activated receptors and inflammation. *Pharmacol Ther* 2006; 110: 371.

81. Forman BM, Chen J, and Evans RM. Hypolipidemic drugs, polyunsaturated fatty acids, and eicosanoids are ligands for peroxisome proliferator-activated receptors α and δ. *Proc Natl Acad Sci USA* 1997; 94: 4312.

82. Hu ZW, Kerb R, Shi XY, Wei-Lavery T, and Hoffman BB. Angiotensin II increases expression of cyclooxygenase-2: implications for the function of vascular smooth muscle cells. *J Pharmacol Exp Ther* 2002; 303: 563.

83. Delerive P et al. Peroxisome proliferator-activated receptor α negatively regulates the vascular inflammatory gene response by negative cross-talk with transcription factors NF-κB and AP-1. *J Biol Chem* 1999; 274: 32048.

84. Staels B et al. Activation of human aortic smooth-muscle cells is inhibited by PPAR- but not by PPAR- activators. *Nature* 1998; 393: 790.

85. Martens FM et al. Metabolic and additional vascular effects of thiazolidinediones. *Drugs* 2002; 62: 1463.

86. Hwang J et al. Peroxisome proliferator-activated receptor-γ ligands regulate endothelial membrane superoxide production. *Am J Physiol Cell Physiol* 2005; 288: C899.

87. Lohray BB et al. (-)3-[4-[2-(Phenoxazin-10-yl)ethoxy]phenyl]-2-ethoxypropanoic acid [(-)DRF 2725]: a dual PPAR agonist with potent antihyperglycemic and lipid modulating activity. *J Med Chem* 2001; 44: 2675.

88. Etgen GJ et al. A tailored therapy for the metabolic syndrome: the dual peroxisome proliferator-activated receptor-alpha/gamma agonist LY465608 ameliorates insulin resistance and diabetic hyperglycemia while improving cardiovascular risk factors in preclinical models. *Diabetes* 2002; 51: 1083.

89. Mamnoor PK et al. Antihypertensive effect of ragaglitazar: a novel PPARα and γ dual activator. *Pharmacol Res* 2006; 54: 129.

90. Harrity T et al. Muraglitazar, a novel dual (α/γ) peroxisome proliferator-activated receptor, improves diabetes and other metabolic abnormalities and perserves beta-cell function in db/db mice. *Diabetes* 2006; 55: 240.

91. Nissen SE, Wolski K, and Topol EJ. Effect of muraglitazar on death and major adverse cardiovascular events in patients with type 2 diabetes mellitus. *JAMA* 2005; 294: 2581.

92. Seber S, Ucak S, Basat O, and Altuntas Y. The effect of dual PPAR α/γ stimulation with combination of rosiglitazone and fenofibrate on metabolic parameters in type 2 diabetic patients. *Diabetes Res Clin Pract* 2006; 71: 52.

13 Role of Uncoupling Protein 2 in Pancreatic β Cell Function: Secretion and Survival

Jingyu Diao, Catherine B. Chan, and Michael B. Wheeler

CONTENTS

ABSTRACT

Uncoupling protein (UCP) 2 has been considered a negative modulator of insulin secretion. In response to stress stimuli such as hyperlipidemia and inflammation, the pancreatic β cell up-regulates UCP2 expression, which results in decreased insulin secretion. Fatty acids and superoxide regulate UCP2 activity. In addition to influencing insulin secretion, UCP2 may play a role in β cell survival and proliferation.

INTRODUCTION

Regulation of cellular adenosine triphosphate (ATP) production and rapid adjustments in ATP levels are essential for most cells during exposure to growth stimuli or stressful conditions such as fuel deficiency, oxidative stimuli, and inflammation. ATP is created by ATP synthase in mitochondria through coupling of the proton motive force (PMF) or electrochemical H^+ gradient to the oxidative phosphorylation of fuel substrates such as fatty acids or glucose. To guarantee continuous generation of ATP as the main source of energy for cellular function and viability, mitochondria operate an oxidative phosphorylation system to supply a proton gradient. In this system, PMF is generated during respiration by the passage of protons against their gradient through the electron transport chain enzyme complexes. Thus, many factors contribute directly to the regulation of PMF and thereby ATP synthesis, particularly the carrier proteins in the respiratory chain complexes in the inner membranes of mitochondria.

To maintain an appropriate level of ATP production, an uncoupling process that dissipates the proton gradient and prevents the PMF from becoming excessive can be recruited. This uncoupling process also plays a role in reducing the level of reactive oxygen species (ROS), a by-product of oxidative phosphorylation produced by the complexes of the electron transport chain.[1,2] Mitochondrial uncoupling is dominated by uncoupling proteins located in the inner mitochondrial membranes.[3] Five UCP homologues in mammals sharing distinctive UCP signature sequences are thought to be involved in fatty acid anion binding and translocation.[1,4] Based on studies of UCP1, the uncoupling activity is augmented by fatty acids and inhibited by the purine nucleotide GDP.[5,6]

Specific uncoupling proteins play a well documented, physiological role in thermogenesis regulation. UCP1, an uncoupling protein mainly expressed in brown adipose tissue, which has little ATP synthase, induces proton leak and reduces ATP production, thereby generating heat upon cold exposure and in diet-induced thermogenesis.[7,8] Activation of UCP1 stimulates respiration, fatty acid oxidation, and uncoupling activity.[9] UCP2 and UCP3 share over 50% protein sequence identity with UCP1. However, the function and regulation of these UCPs remain unclear. UCP3 is expressed in skeletal muscle and brown fat, while UCP2 is found in many tissues including heart, brain, kidney, liver, lymphocytes, pancreas, and white adipose tissue.[10] The diverse distribution of UCP2 and UCP3 led to the hypothesis that their primary physiological function involves something other than the control of thermogenesis.[11,12] Specifically, UCP2 appears to play roles in the regulation of ROS production, inflammation, and cell proliferation and death.[13,14]

Many recent reviews have discussed the possible physiological and pathophysiological roles of UCPs.[15–18] Thus, this chapter focuses on the role of UCP2 in pancreatic β cell function including the regulation of insulin secretion and cell survival.

REGULATORY FACTORS OF UCP2 EXPRESSION IN PANCREATIC β CELLS

Mitochondrial function is crucial for insulin secretion.[19] For instance, blockade of the respiratory chain by mitochondrial toxins inhibits glucose-stimulated

Islet   Islet cells

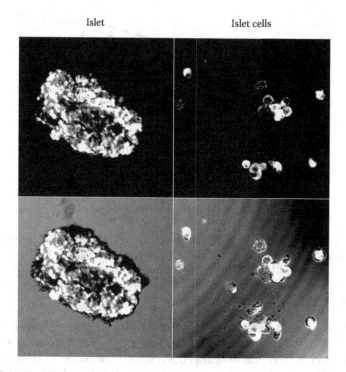

FIGURE 13.1 UCP2 protein expression in human islets. The islets were fixed with 4% paraformaldehyde, permeabilized with 0.3% Triton X-100 detergent and dispersed islet cells, and stained with goat antibody specific for UCP2 (Santa Cruz Biotechnology, Inc.). Cells were then examined using a Zeiss LSM510 confocal microscope (×40 ×1.3 magnification) and images were analyzed using LSM510 image browser software.

insulin secretion.[20] Absence of specific mitochondrial proteins such as nicotinamide nucleotide transhydrogenase (Nnt) in β cells ATP production and insulin secretion.[21] Based on the association of the UCP2 gene with obesity and hyperinsulinemia, numerous reports have demonstrated the important role of UCP2 in the development of type 2 diabetes in humans.[22]

A common −866G/A polymorphism in the promoter of the UCP2 gene contributes to diabetes susceptibility in different human populations, perhaps due to its overall effects on pancreatic islets, the immune system, and lipid metabolism.[23,24] Individuals who are heterozygous or homozygous for the −866A allele have greater UCP2 activity, which correlates with reduced glucose-induced insulin secretion and decreased oxidative stress.[25–27] Immunofluorescence staining of UCP2 revealed relatively high amounts of the UCP2 protein in isolated human islets, suggesting its importance in islet function (Figure 13.1). However, evidence linking any mutations in the UCP2 protein sequence to the pathogenesis of obesity or type 2 diabetes[28] is insufficient.

In contrast, a considerable amount is known regarding the regulation of UCP2 gene expression. The promoter region of UCP2 gene contains peroxisome

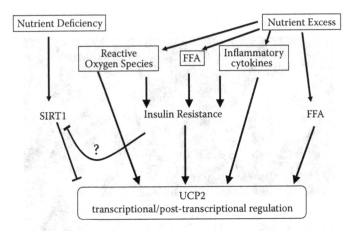

FIGURE 13.2 Proposed pathways of UCP2 regulation in β cells. UCP2 activity can be transcriptionally down-regulated by SIRT1 in response to nutrient deficiency and up-regulated by signals from ROS, FFAs, and certain pro-inflammatory cytokines due to nutrient overload and related insulin resistance. UCP2 can also be activated directly by many factors such as ROS and FFA.

proliferator response elements and a sterol regulatory element.[29–31] In pancreatic β cells, UCP2 gene expression is partially controlled by PPAR-γ coactivator PGC-1α (Peroxisome Proliferator-Activated Receptor Coactivator 1 Alpha) through sterol regulatory element binding protein isoforms (SREBPs).[32,33] Increased activity of these transcriptional activators leads to increased UCP2 expression, whereas another transcription regulator, SIRT1 (sirtuin1), represses its expression in β cells.[34]

SIRT1, a mammalian Sir2 orthologue, is a key regulator implicated in cellular stress response and survival through regulation of p53, NF-κB (Nuclear Factor KappaB) signaling, and FOXO (Forkhead Box O) transcription factors.[35] A moderate reduction in calorie intake (caloric restriction, CR) has been suggested to slow aging, reduce age-related chronic diseases, and extend lifespans. In simple model organisms, the *sir2* gene is a strong candidate to regulate CR.[36] In mammalian cells, SIRT1 deacetylase can also be induced by CR, possibly to increase the long-term function and survival of critical cell types.[37]

Interestingly, SIRT1 can directly bind to the promoter of UCP2 in response to starvation in pancreatic β cells, although it acts on PPAR-γ as well.[34,38] Therefore, a decrease in SIRT1 activity in β cells would enhance UCP2 expression and reduce insulin secretion, for instance, in response to CR (Figure 13.2). As one would expect, the levels of UCP2 are low in *sirt1* transgenic mice but high in *sirt1* knockout mice.[34,39] It is likely, under starvation conditions, that SIRT1 increases to maintain β cell survival and the cell cycle, leading to the suppression of UCP2 expression. Although food deprivation increased UCP2 mRNA in mouse pancreas,[34] cells under serum and glucose withdrawal for 15 hr increased SIRT1, possibly leading to the suppression of UCP2.[40] Conversely, under obesity or

Untreated	TNF	IL-1	Palmitate	C2-Ceramide

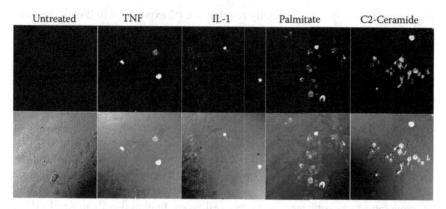

FIGURE 13.3 Induction of UCP2 protein expression in pancreatic β cell MIN6. Pancreatic clonal MIN6 β cells were treated with cytokines (TNF or IL-1, 100 ng/ml), palmitate (1 mM), or ceramide (10 μM) for 48 hr, then fixed, permeabilized, and immunofluorescently stained with goat anti-UCP2 as indicated in Figure 13.1.

caloric overload conditions, low levels of SIRT1 result in release of suppression on the UCP2 gene, leading to an increase of UCP2 expression. Clearly, more research is needed to determine the important link between UCPs and SIRT1 in islets.

Our studies have shown that UCP2 is the predominant isoform expressed in murine pancreatic islets, with UCP1 and UCP3 present only at very low levels as demonstrated by real-time PCR (unpublished data). Interestingly, in genetically obese animal models, animals fed high fat diets, and in clonal MIN6 β cells exposed to ROS, free fatty acids (FFAs), or inflammatory cytokines, UCP2 expression was up-regulated (Figure 13.3), implying that induction of UCP2 in β cells directly or indirectly correlates with insulin resistance.[41,42]

This is further supported by our unpublished results showing that the attenuation of insulin signaling by chronic silencing of the insulin receptor in pancreatic cells increases the expression of UCP2 protein and augments mitochondrial uncoupling activity. Together, these data suggest that β cells may adapt to obesity and insulin resistance by increasing UCP2 levels and thus uncoupling activity, consistent with the notion that UCP2 expression adapts to the metabolic state of an organism in a tissue-specific manner.[43] This adaptation with insulin resistance would result in the attenuation of insulin secretion from β cells in response to increased glucose (Figure 13.2).

Because UCP2 is up-regulated in β cells in response to insulin resistant states, identifying signaling pathways regulating its expression will facilitate understanding its function at the molecular level. Along these lines, we have observed that interleukin 1(IL-1), a primary regulator of inflammatory and immune responses, increases UCP2 protein expression in mouse clonal MIN6 β cells without causing significant cytotoxic effects (Figure 13.3). In contrast, in rat clonal INS-1 β cells, IL-1β down-regulated UCP2 mRNA accompanied by decreased cell viability, suggesting that IL-1 can differently regulate UCP2 expression in β cells depending on cell signaling status.[44]

IL-1 binds to cell surface-specific receptors, activating specific protein kinases such as NIK (NF-κB Induced Kinase) and MAPK (Mitogen-Activated Protein Kinase), and modulates transcription factors including NF-κB, AP-1 (Adaptor-related Protein Complex 1) and CREB (cAMP Responsive Binding Protein 1), resulting in the expression of immediate early genes central to the inflammatory response.[45] Therefore, the fact that UCP2 protein expression can be up-regulated by IL-1 suggests that UCP2 is one of the acute response genes in β cells. Consistent with this finding, IL-1β is a key inflammatory mediator causing pancreatic islet dysfunction and apoptosis through the up-regulation of inducible nitric oxide synthase and cyclooxygenase-2.[46] Although IL-1 is produced mainly by monocytes, macrophages, and other cell types, human β cells make IL-1β in response to high glucose concentrations independently of an immune-mediated process. However, the β cells also express IL-1 receptor antagonist (IL-1Ra), a naturally occurring anti-inflammatory cytokine, to protect themselves from glucotoxicity.[47,48] The pathway of IL-1-mediated β cell UCP2 regulation is therefore a potential target for controlling β cell function and survival.

UCP2 ACTION IN INSULIN SECRETION FROM PANCREATIC β CELLS: ROLE OF ATP

The β cell is essential for glucose homeostasis due to its ability to secrete insulin in response to nutrients.[49] In general, glucose metabolism in β cells leads to an elevated intracellular ATP:ADP ratio, which closes the ATP-sensitive K^+ (K_{ATP}) channel, resulting in depolarization of the cell membranes. Closure of K_{ATP} channels activates voltage-dependent Ca^{2+} channels, leading to Ca^{2+} entry and insulin exocytosis.[50] Thus, given the key role ATP plays in mediating insulin secretion and synthesis, it is not surprising that UCP2 negatively regulates insulin secretion since its uncoupling activity reduces mitochondrial ATP production.

Our laboratory and others have shown that cellular ATP levels in $UCP2^{-/-}$ animals are elevated in association with enhanced insulin secretion,[42,51] while decreased ATP levels and impaired glucose-stimulated insulin secretion occur when UCP2 is overexpressed.[42,52,53] Furthermore, it appears that increased islet UCP2 levels occur in several rodent models of type 2 diabetes and β cell dysfunction, including hyperglycemic HFD-fed mice, *ob/ob* mice, *fa/fa* rats, mice overexpressing SREBP-1c in β cells, and rodents expressing a dominant-negative IGF-1 receptor in skeletal muscle.[42,53a–57]

In addition to transcriptional regulation of UCP2 expression, regulatory mechanisms that occur more rapidly are important because they allow cells to rapidly adjust ATP levels in response to various stimuli or stresses. Post-transcriptional mechanisms regulate UCP2 activity, suggesting that UCP2 is able to respond acutely to stimuli.[58] Furthermore FFAs and superoxide radicals directly activate

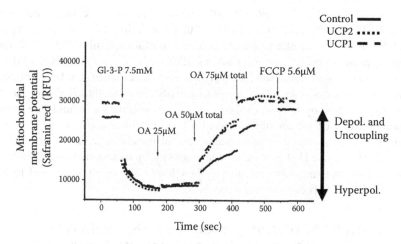

FIGURE 13.4 FFA activates mitochondrial uncoupling. Pancreatic clonal MIN6 β cells were infected with recombinant adenovirus expressing UCP1, UCP2, and control green fluorescence protein for 48 hr, then permeabilized and subjected to measurement of mitochondrial membrane potential in a buffer containing 2.5 μM safranin. Mitochondrial membrane potential as reflected by the fluorescence was monitored by a FluoroCount plate reader at excitation/emission wavelengths of 530/590 nm. Respiratory substrate glycerol-3-phosphate (Gl-3-P, 7.5 μM) was first added to induce mitochondrial hyperpolarization. To induce OA-mediated uncoupling effects, free fatty acid oleate (OA) was added to build up OA concentration from 25 to 75 μM. To confirm the depolarization of mitochondrial membrane, carbonyl cyanide p-trifluoromethoxyphenylhydrazone (FCCP) was added to a final concentration of 5.6 μM. Vertical arrows denote time points of chemical additions. (Courtesy of Vasilij Koshkin, Wheeler Laboratory.)

uncoupling. Mitochondrial superoxide activates UCP2-mediated proton leaks and influences ATP production and insulin secretion in β cells.[59–64]

Using safranin fluorescence to measure mitochondrial membrane potential, we demonstrated that 50 to 75 μM oleate acutely induced UCP2-mediated uncoupling within seconds in MIN6 cells, suggesting that UCP2 activity can be induced by FFAs directly or by an immediate consequence of ROS generation and lipid peroxidation of FFAs (Figure 13.4). Although the mechanism of FFA-induced UCP2 activation remains unclear, it presumably occurs through direct binding of FFA to UCP2.[1]

The links of UCP activators and their acute uncoupling effects on mitochondria are interesting and suggest an avenue for future investigations that may shed light on the function of UCP2. It should be noted that acute exposure of pancreatic cells to both high glucose concentrations and saturated FFAs stimulates appreciable insulin secretion, whereas chronic exposure results in desensitization and suppression of secretion.[65] Binding of FFA to its receptor GPR40 would increase intracellular cAMP levels and stimulate Ca^{2+} influx through voltage-dependent Ca^{2+} channels, resulting in an increase in insulin secretion.[66]

The effect from the FFA–GPR40 pathway may override the effect from FFA–UCP2 activity initially, considering that uncoupling has a negative impact on FFA-stimulated insulin secretion. However, it is also possible that UCP2 may initially promote insulin secretion but may ultimately attenuate it. Which of these opposing effects predominates? This is difficult to determine since UCP2 protein levels do not necessarily reflect levels of activity. Specifically, UCP2 appears to require activation by fatty acids, and studies demonstrating a quantitative correlation between UCP2 activity and insulin secretion *in vivo* have yet to be performed. Nevertheless, it is conceivable that in metabolic states where fuel is plentiful, increased UCP2 expression and activity by factors such as pro-inflammatory cytokines and high circulating FFA levels are highly correlated to the desensitization of β cells to release insulin.

UCP2 ACTION IN β CELL MASS AND SURVIVAL: ROLE OF ATP AND ROS PRODUCTION

In type 2 diabetes, reduced β cell mass is a key feature associated with defects in insulin secretion.[61] The role of UCP2 in ATP production and ROS regulation has linked this protein not only to β cell function, but also to survival and proliferation. Increased ROS are thought to induce insulin resistance in numerous settings.[68] Undoubtedly, excessive ROS production is harmful to cell viability. However, moderately elevated ROS levels can also be essential for activating defensive signaling pathways to protect cells.[69]

For example, ROS can protect against ischemia–reperfusion injury in cardiomyocytes.[2,69] The physiological role behind this phenomenon is that mitochondrial ROS is triggered to promote signaling pathways for gene transcription and cell growth in cells with low energy states. ROS may function as mediators to facilitate efficient energy transfer between mitochondria and ATPases in cells with high energy states.[70] Therefore, the beneficial effect of having lower levels of uncoupling activity in certain cell types like β cells under normal physiologic conditions may be to maintain a reasonable amount of ROS production. However, under pathological conditions when ROS production becomes excessive and harmful to cells, UCP2 may induce uncoupling of mitochondria in order to reduce ROS production.

In support of this, evidence suggests that UCP2 controls mitochondrial ROS production which, in turn, confers neuroprotection, cardioprotection, and life span extension in Drosophila.[14,71,74] In the pancreatic β cell line INS-1, overexpression of UCP2 improved cell survival when the cells were treated with H_2O_2.[75] However, under chronic metabolic stress, even modestly increased levels of UCP2 protein may promote cell death in other types of tissues and cells including liver cells and Hela cells.[76–78]

With respect to cell proliferation, based on our study, UCP2 overexpression in MIN6 cells not only reduces mitochondrial ROS formation, but also increases proliferation under both starvation and nutritional excess conditions (unpublished data). Consistent with these observations, $Ucp2^{-/-}$ mice had increased islet masses after 4.5 months of high fat diet feeding, suggesting that UCP2 may play a role in the regulation of β cell growth, probably by directly augmenting FFA-mediated ATP production.[51] The potential effect on cell growth is supported by the tumor formation that occurs upon deletion of UCP2 in association with increased NF-κB action and oxidative stress.[79]

Islets of $UCP2^{-/-}$ mice showed increased ROS levels when compared with islets from wild type (WT) mice.[80] In addition, FFAs also increased ROS production in isolated islets from WT and cultured pancreatic β cells, but not in islets from $UCP2^{-/-}$ mice,[80,81] suggesting UCP2 contributes to the regulation of ROS production in β cells. However, data from our laboratory also suggest that despite the role of UCP2 in β cells, other factors may be involved in β cell regulation of intracellular ROS. Specifically, induction of ROS formation *in vitro* using menadione, ceramide, or cytokines in both $Ucp2^{-/-}$ and WT islet cells failed to reveal any protective role of UCP2, perhaps because these mediators act beyond the UCP2 pathway (unpublished data). In addition, $Ucp2^{-/-}$ and WT islets had equal sensitivities to apoptosis induced by menadione, ceramide, and cytokines; overexpression of UCP2 did not protect MIN6 cells from either cytokine- or menadione-mediated apoptosis. Thus, these data suggest that pancreatic β cells may have a specific regulatory signaling network to maintain their own levels of ROS, which may only be partially regulated by UCP2.

CONCLUSION

Through its fundamental role in the regulation of mitochondria ATP generation and ROS production, UCP2 is a key regulator of β cell function and survival. The pharmacological targeting of UCP2 in β cells presents a potential way to modulate insulin secretion and β cell proliferation. However, β cell-specific signaling pathways mediating the induction, activation, and effects of UCP2 need to be identified to enable precise manipulation without deleterious side effects.

ACKNOWLEDGMENTS

This work was funded by Grants MOP 12898 and MOP 43987 from the Canadian Institutes of Health Research (CIHR) to M.B.W. and C.B.C, respectively. M.B.W. is supported by an investigator award from CIHR. J.D. was supported by a fellowship award from the Canadian Diabetes Association. J.W. Joseph, V. Koshkin, A.V. Gyulkhandanyan, S.C. Lee, G.T. Karaman, C. Thorn, and G. Bikopoulos are gratefully acknowledged as contributors to the work presented here. We thank D. Yau, E. Allister, and N. Wijesekara for help with editing of the manuscript.

REFERENCES

1. Garlid KD, Jaburek M, Jezek P, and Varecha M. How do uncoupling proteins uncouple? *Biochim. Biophys. Acta* 2000; 1459: 383.
2. Pain T et al. Opening of mitochondrial K(ATP) channels triggers the preconditioned state by generating free radicals. *Circ. Res.* 2000; 87: 460.
3. Sohal RS and Weindruch R. Oxidative stress, caloric restriction, and aging. *Science* 1996; 273: 59.
4. Jezek P and Urbankova E. Specific sequence of motifs of mitochondrial uncoupling proteins. *IUBMB Life* 2000; 49: 63.
5. Modriansky M, Murdza-Inglis DL, Patel HV, Freeman KB, and Garlid KD. Identification by site-directed mutagenesis of three arginines in uncoupling protein that are essential for nucleotide binding and inhibition. *J. Biol. Chem.* 1997; 272: 24759.
6. Jezek P and Freisleben HJ. Fatty acid binding site of the mitochondrial uncoupling protein: demonstration of its existence by EPR spectroscopy of 5-DOXYL-stearic acid. *FEBS Lett.* 1994; 343: 22.
7. Rothwell NJ and Stock MJ. A role for brown adipose tissue in diet-induced thermogenesis. *Nature* 1979; 281: 31.
8. Jacobsson A, Stadler U, Glotzer MA, and Kozak LP. Mitochondrial uncoupling protein from mouse brown fat: molecular cloning, genetic mapping, and mRNA expression. *J. Biol. Chem.* 1985; 260: 16250.
9. Cannon B and Nedergaard J. Brown adipose tissue: function and physiological significance. *Physiol. Rev.* 2004; 84: 277.
10. Jezek P and Garlid KD. Mammalian mitochondrial uncoupling proteins. *Int. J. Biochem. Cell Biol.* 1998; 30: 1163.
11. Dridi S, Onagbesan O, Swennen Q, Buyse J, Decuypere E, and Taouis M. Gene expression, tissue distribution and potential physiological role of uncoupling protein in avian species. *Comp. Biochem. Physiol. A Mol. Integr. Physiol.* 2004; 139: 273.
12. Vidal-Puig AJ et al. Energy metabolism in uncoupling protein 3 gene knockout mice. *J. Biol. Chem.* 2000; 275: 16258.
13. Saleh MC, Wheeler MB, and Chan CB. Uncoupling protein-2: evidence for its function as a metabolic regulator. *Diabetologia* 2002; 45: 174.
14. Arsenijevic D et al. Disruption of the uncoupling protein-2 gene in mice reveals a role in immunity and reactive oxygen species production. *Nat. Genet.* 2000; 26: 435.
15. Chan CB, Saleh MC, Koshkin V, and Wheeler MB. Uncoupling protein 2 and islet function. *Diabetes* 2004; 53, Suppl 1: S136.
16. Brand MD and Esteves TC. Physiological functions of the mitochondrial uncoupling proteins UCP2 and UCP3. *Cell Metab.* 2005; 2: 85.
17. Andrews ZB, Diano S, and Horvath TL. Mitochondrial uncoupling proteins in the CNS: in support of function and survival. *Nat. Rev. Neurosci.* 2005; 6: 829.
18. Krauss S, Zhang CY, and Lowell BB. The mitochondrial uncoupling-protein homologues. *Nat. Rev. Mol. Cell Biol.* 2005; 6: 248.
19. Lowell BB and Shulman GI. Mitochondrial dysfunction and type 2 diabetes. *Science* 2005; 307: 384.

20. Malaisse WJ, Hutton JC, Kawazu S, Herchuelz A, Valverde I, and Sener A. The stimulus-secretion coupling of glucose-induced insulin release. XXXV. The links between metabolic and cationic events. *Diabetologia* 1979; 16: 331.

21. Freeman H, Shimomura K, Horner E, Cox RD, and Ashcroft FM. Nicotinamide nucleotide transhydrogenase: a key role in insulin secretion. *Cell Metab.* 2006; 3: 35-45.

22. Fleury C et al. Uncoupling protein-2: a novel gene linked to obesity and hyperinsulinemia. *Nat. Genet.* 1997; 15: 269.

23. Gable DR, Stephens JW, Cooper JA, Miller GJ, and Humphries SE. Variation in the UCP2-UCP3 gene cluster predicts the development of type 2 diabetes in healthy middle-aged men. *Diabetes* 2006; 55: 1504.

24. Bulotta A et al. The common -866G/A polymorphism in the promoter region of the UCP-2 gene is associated with reduced risk of type 2 diabetes in Caucasians from Italy. *J. Clin. Endocrinol. Metab.* 2005; 90: 1176.

25. Vogler S et al. Association of a common polymorphism in the promoter of UCP2 with susceptibility to multiple sclerosis. *J. Mol. Med.* 2005; 83: 806.

26. Vogler S et al. Uncoupling protein 2 has protective function during experimental autoimmune encephalomyelitis. *Am. J. Pathol.* 2006; 168: 1570.

27. Yamasaki H et al. Uncoupling protein 2 promoter polymorphism -866G/A affects peripheral nerve dysfunction in Japanese type 2 diabetic patients. *Diabetes Care* 2006; 29: 888.

28. Dalgaard LT and Pedersen O. Uncoupling proteins: functional characteristics and role in the pathogenesis of obesity and type 2 diabetes. *Diabetologia* 2001; 44: 946.

29. Roduit R et al. Glucose down-regulates the expression of the peroxisome proliferator-activated receptor-alpha gene in the pancreatic beta cell. *J. Biol. Chem.* 2000; 275: 35799.

30. Patane G, Anello M, Piro S, Vigneri R, Purrello F, and Rabuazzo AM. Role of ATP production and uncoupling protein-2 in the insulin secretory defect induced by chronic exposure to high glucose or free fatty acids and effects of peroxisome proliferator-activated receptor-gamma inhibition. *Diabetes* 2002; 51: 2749.

31. Medvedev AV et al. Regulation of the uncoupling protein-2 gene in INS-1 beta-cells by oleic acid. *J. Biol. Chem.* 2002; 277: 42639.

32. Oberkofler H, Klein K, Felder TK, Krempler F, and Patsch W. Role of peroxisome proliferator-activated receptor-gamma coactivator-1alpha in the transcriptional regulation of the human uncoupling protein 2 gene in INS-1E cells. *Endocrinology* 2006; 147: 966.

33. Medvedev AV, Snedden SK, Raimbault S, Ricquier D, and Collins S. Transcriptional regulation of the mouse uncoupling protein-2 gene: double E-box motif is required for peroxisome proliferator-activated receptor-gamma-dependent activation. *J. Biol. Chem.* 2001; 276: 10817.

34. Bordone L et al. Sirt1 regulates insulin secretion by repressing UCP2 in pancreatic beta cells. *PLoS. Biol.* 2006; 4: e31.

35. Wolf G. Calorie restriction increases life span: a molecular mechanism. *Nutr. Rev.* 2006; 64: 89.

36. Lin SJ, Defossez PA, and Guarente L. Requirement of NAD and SIR2 for lifespan extension by calorie restriction in *Saccharomyces cerevisiae*. *Science* 2000; 289: 2126.

37. Cohen HY et al. Calorie restriction promotes mammalian cell survival by inducing the SIRT1 deacetylase. *Science* 2004; 305: 390.

38. Motta MC et al. Mammalian SIRT1 represses forkhead transcription factors. *Cell* 2004; 116: 551.
39. Moynihan KA et al. Increased dosage of mammalian Sir2 in pancreatic beta cells enhances glucose-stimulated insulin secretion in mice. *Cell Metab.* 2005; 2: 105.
40. Nemoto S, Fergusson MM, and Finkel T. Nutrient availability regulates SIRT1 through a forkhead-dependent pathway. *Science* 2004; 306: 2105.
41. Joseph JW et al. Uncoupling protein 2 knockout mice have enhanced insulin secretory capacity after a high-fat diet. *Diabetes* 2002; 51: 3211.
42. Zhang CY et al. Uncoupling protein-2 negatively regulates insulin secretion and is a major link between obesity, beta cell dysfunction, and type 2 diabetes. *Cell* 2001; 105: 745.
43. Samec S, Seydoux J, and Dulloo AG. Post-starvation gene expression of skeletal muscle uncoupling protein 2 and uncoupling protein 3 in response to dietary fat levels and fatty acid composition: a link with insulin resistance. *Diabetes* 1999; 48: 436.
44. Li LX, Yoshikawa H, Egeberg KW, and Grill V. Interleukin-1-beta swiftly down-regulates UCP-2 mRNA in beta-cells by mechanisms not directly coupled to toxicity. *Cytokine* 2003; 23: 101.
45. Stylianou E and Saklatvala J. Interleukin-1. *Int. J. Biochem. Cell Biol.* 1998; 30: 1075.
46. Hohmeier HE, Tran VV, Chen G, Gasa R, and Newgard CB. Inflammatory mechanisms in diabetes: lessons from the beta cell. *Int. J. Obes. Relat. Metab. Disord.* 2003; 27: S12.
47. Maedler K et al. Glucose-induced beta cell production of IL-1-beta contributes to glucotoxicity in human pancreatic islets. *J. Clin. Invest.* 2002; 110: 851.
48. Maedler K et al. Leptin modulates beta cell expression of IL-1 receptor antagonist and release of IL-1-beta in human islets. *Proc. Natl. Acad. Sci. USA* 2004; 101: 8138.
49. Baetens D, Malaisse-Lagae F, Perrelet A, and Orci L. Endocrine pancreas: three-dimensional reconstruction shows two types of islets of Langerhans. *Science* 1979; 206: 1323.
50. Koster JC, Permutt MA, and Nichols CG. Diabetes and insulin secretion: the ATP-sensitive K^+ channel (K_{ATP}) connection. *Diabetes* 2005; 54: 3065.
51. Joseph JW et al. Uncoupling protein 2 knockout mice have enhanced insulin secretory capacity after a high-fat diet. *Diabetes* 2002; 51: 3211.
52. Chan CB et al. Increased uncoupling protein-2 levels in beta cells are associated with impaired glucose-stimulated insulin secretion: mechanism of action. *Diabetes* 2001; 50: 1302.
53. Wang MY et al. Adenovirus-mediated overexpression of uncoupling protein-2 in pancreatic islets of Zucker diabetic rats increases oxidative activity and improves beta-cell function. *Diabetes* 1999; 48: 1020.
53a. Saleh MC, Wheeler MB, and Chan CB. Endogenous islet uncoupling protein-2 expression and loss of glucose homeostasis in ob/ob mice. *J. Endocrinol.* 2006; 190: 657.
54. Briaud I, Kelpe CL, Johnson LM, Tran PO, and Poitout V. Differential effects of hyperlipidemia on insulin secretion in islets of Langerhans from hyperglycemic versus normoglycemic rats. *Diabetes* 2002; 51: 662.
55. Kassis N et al. Correlation between pancreatic islet uncoupling protein-2 (UCP2) mRNA concentration and insulin status in rats. *Int. J. Exp. Diab. Res.* 2000; 1: 185.

56. Takahashi A et al. Transgenic mice overexpressing nuclear SREBP-1c in pancreatic beta-cells. *Diabetes* 2005; 54: 492.

57. Asghar Z, Yau D, Chan F, Leroith D, Chan CB, and Wheeler MB. Insulin resistance causes increased beta-cell mass but defective glucose-stimulated insulin secretion in a murine model of type 2 diabetes. *Diabetologia* 2006; 49: 90.

58. Pecqueur C et al. Uncoupling protein 2, *in vivo* distribution, induction upon oxidative stress, and evidence for translational regulation. *J. Biol. Chem.* 2001; 276: 8705.

59. Jarmuszkiewicz W et al. Redox state of endogenous coenzyme q modulates the inhibition of linoleic acid-induced uncoupling by guanosine triphosphate in isolated skeletal muscle mitochondria. *J. Bioenerg. Biomembr.* 2004; 36: 4932.

60. Koshkin V, Wang X, Scherer PE, Chan CB, and Wheeler MB. Mitochondrial functional state in clonal pancreatic beta-cells exposed to free fatty acids. *J. Biol. Chem.* 2003; 278: 19709.

61. Hirabara SM et al. Acute effect of fatty acids on metabolism and mitochondrial coupling in skeletal muscle. *Biochim. Biophys. Acta* 2006; 1757: 57.

62. Krauss S, Zhang CY, and Lowell BB. A significant portion of mitochondrial proton leak in intact thymocytes depends on expression of UCP2. *Proc. Natl. Acad. Sci. USA* 2002; 99: 118.

63. Krauss S et al. Superoxide-mediated activation of uncoupling protein 2 causes pancreatic beta cell dysfunction. *J. Clin. Invest.* 2003; 112: 1831.

64. Echtay KS et al. Superoxide activates mitochondrial uncoupling proteins. *Nature* 2002; 415: 96.

65. Haber EP, Procopio J, Carvalho CR, Carpinelli AR, Newsholme P, and Curi R. New insights into fatty acid modulation of pancreatic beta-cell function. *Int. Rev. Cytol.* 2006; 248: 1.

66. Gromada J. The free fatty acid receptor GPR40 generates excitement in pancreatic beta-cells. *Endocrinology* 2006; 147: 672.

67. Donath MY and Halban PA. Decreased beta-cell mass in diabetes: significance, mechanisms and therapeutic implications. *Diabetologia* 2004; 47: 581.

68. Houstis N, Rosen ED, and Lander ES. Reactive oxygen species have a causal role in multiple forms of insulin resistance. *Nature* 2006; 440: 944.

69. Sen CK and Packer L. Antioxidant and redox regulation of gene transcription. *FASEB J.* 1996; 10: 709.

70. Garlid KD, Dos SP, Xie ZJ, Costa AD, and Paucek P. Mitochondrial potassium transport: role of the mitochondrial ATP-sensitive K(+) channel in cardiac function and cardioprotection. *Biochim. Biophys. Acta* 2003; 1606: 1.

71. Bai Y et al. Persistent nuclear factor-kappa B activation in Ucp2–/– mice leads to enhanced nitric oxide and inflammatory cytokine production. *J. Biol. Chem.* 2005; 280: 19062.

72. Mattiasson G et al. Uncoupling protein-2 prevents neuronal death and diminishes brain dysfunction after stroke and brain trauma. *Nat. Med.* 2003; 9: 1062.

73. Teshima Y, Akao M, Jones SP, and Marban E. Uncoupling protein-2 overexpression inhibits mitochondrial death pathway in cardiomyocytes. *Circ. Res.* 2003; 93: 192.

74. Fridell YW, Sanchez-Blanco A, Silvia BA, and Helfand SL. Targeted expression of the human uncoupling protein 2 (hUCP2) to adult neurons extends life span in the fly. *Cell Metab.* 2005; 1: 145.

75. Li LX, Skorpen F, Egeberg K, Jorgensen IH, and Grill V. Uncoupling protein-2 participates in cellular defense against oxidative stress in clonal beta-cells. *Biochem. Biophys. Res. Commun.* 2001; 282: 273.

76. Chavin KD et al. Obesity induces expression of uncoupling protein-2 in hepatocytes and promotes liver ATP depletion. *J. Biol. Chem.* 1999; 274: 5692.

77. Gnaiger E, Mendez G, and Hand SC. High phosphorylation efficiency and depression of uncoupled respiration in mitochondria under hypoxia. *Proc. Natl. Acad. Sci. USA* 2000; 97: 11080.

78. Mills EM, Xu D, Fergusson MM, Combs CA, Xu Y, and Finkel T. Regulation of cellular oncosis by uncoupling protein 2. *J. Biol. Chem.* 2002; 277: 27385.

79. Derdak Z et al. Enhanced colon tumor induction in uncoupling protein-2 deficient mice is associated with NF-kappa-B activation and oxidative stress. *Carcinogenesis* 2006; 27: 956.

80. Joseph JW et al. Free fatty acid-induced beta-cell defects are dependent on uncoupling protein 2 expression. *J. Biol. Chem.* 2004; 279: 51049.

81. Koshkin V, Wang X, Scherer PE, Chan CB, and Wheeler MB. Mitochondrial functional state in clonal pancreatic beta-cells exposed to free fatty acids. *J. Biol. Chem.* 2003; 278: 19709.

Section II

Influence of Dietary Factors, Micronutrients, and Metabolism

14 Nutritional Modulation of Inflammation in Metabolic Syndrome

Uma Singh, Sridevi Devaraj, and Ishwarlal Jialal

CONTENTS

ABSTRACT

Metabolic syndrome (MetS) is a disorder composed of central adiposity, dyslip-idemia, abnormal glucose tolerance, and hypertension. It confers an increased risk for diabetes and cardiovascular disease (CVD). Inflammation plays a pivotal role in atherosclerosis and is involved in abnormalities associated with MetS such as insulin resistance (IR) and adiposity. The various biomarkers of inflammation such as inflammatory cytokines (TNF-α, IL-6, and IL-1β), chemokines (MCP-1 and IL-8), and C-reactive protein (CRP) are increased in obesity and correlate with IR and CVD. IR is also associated with endothelial dysfunction (ED).

Inflammation, IR, and ED amplify the cascade of metabolic and vascular derangements. The etiology of this syndrome is largely unknown but presumably represents a complex interaction of genetic and environmental factors. Since MetS is associated with chronic low grade inflammation, strategies are being explored to ameliorate its pro-inflammatory status. MetS has been identified also as a target for diet therapy to reduce risk of CVD. Weight loss appears to be the best modality to reduce inflammation. Intervention trials convincingly demonstrate that weight loss and/or increased physical activity reduce biomarkers of inflammation such as CRP and IL-6. Thus, therapeutic lifestyle change (TLC) is strongly suggested for MetS subjects both for weight reduction and also to reduce the inflammation associated with this syndrome. Much more research is needed to define the roles of individual dietary factors on the biomarkers of inflammation and the mechanisms of the anti-inflammatory effects of weight loss in MetS.

METABOLIC SYNDROME

MetS comprises a cluster of abnormalities with insulin resistance and adiposity as its central features.[1-3] Five diagnostic criteria were identified by the Adult Treatment Panel III in the executive summary of a report of the National Cholesterol Education Program (NCEP). The presence of any three of the five features is considered sufficient to diagnose MetS.[4] These include central obesity, high triglycerides, low HDL, hypertension, and impaired fasting glucose. Using the NCEP definition on a representative sample of 8814 men and women from the United States, the age-adjusted prevalence of MetS was 24% in men and 23.4% in women.[5] Applying this figure to the U.S. population in the year 2000 means about 47 million residents have MetS.

Recently, a worldwide consensus for the MetS defined by the International Diabetes Federation (IDF) includes central obesity as a main component and the other factors such as hypertension, dyslipidemia, and impaired fasting glucose as other components, all of which lead to insulin resistance and independently predict cardiovascular events.[6] The National Heart, Lung, and Blood Institute/American Heart Association statement defining MetS has just been published. It encompasses the recent IDF guidelines, but includes central obesity, dyslipidemia, hypertension, and impaired fasting glucose as the features.[7] This chapter will focus on biomarkers of inflammation in relation to MetS and its nutritional modulation. Particular emphasis will be given to nutritional insights derived from therapeutic interventions with diet and exercise in subjects with metabolic abnormalities.

METABOLIC SYNDROME AND CARDIOVASCULAR DISEASE

Subjects with MetS have increased burdens of cardiovascular disease. In the Kuopio Ischemic Heart Disease Study, Lakka et al.[8] convincingly showed that men with MetS, even in the absence of baseline CAD or diabetes, had significantly increased mortality from CAD. In the Botnia Study,[9] MetS was defined as the presence of at least two of the following risk factors: obesity, hypertension, dyslipidemia, and microalbuminuria. Cardiovascular mortality was assessed in 3606 subjects with a median follow-up of 6.9 years. In women and men, respectively, MetS was seen in 10 and 15% of subjects with normal glucose tolerance (NGT), 42 and 64% of those with impaired fasting glucose/impaired glucose tolerance (IFG/IGT), and 78 and 84% of those with type 2 diabetes mellitus. The risk for coronary heart disease and stroke was increased three-fold in subjects with MetS (p <0.001) and cardiovascular mortality was increased six-fold (12.0 versus 2.2%, p <0.001).

Using data from NHANES III, Alexander et al.[10] also reported that MetS is very common, with 44% of the U.S. population over 50 years of age meeting NCEP criteria. Those who had MetS without diabetes had higher CHD prevalence (13.9%). Those with both MetS and diabetes had the highest prevalence of CHD (19.2%) compared with those with neither condition. MetS was a significant univariate predictor of prevalent CHD. Recently, the Hoorn Study[11] reported that the NCEP definition of MetS was associated with about a two-fold increase in age-adjusted risk of fatal CVD in men and nonfatal CVD in women.

METABOLIC SYNDROME AND DIABETES

Besides its effects on cardiovascular morbidity and mortality, the components of MetS have been associated with diabetes. Factor analysis was used to identify the components of MetS present in 1918 Pima Indians.[12] Insulin resistance factor was strongly associated with diabetes in a 4-year follow-up. Also, body size and lipid factor predicted diabetes whereas the blood pressure factor did not.

In the WOSCOPS study,[13] MetS increased the risk for CHD events and for diabetes. Subjects with four or five features of the syndrome showed 3.7-fold increases in risk for CHD and 24.5-fold increased risks for diabetes compared with those who had none. The PROCAM study[14] also reported a 2.3-fold increased incidence of CVD in subjects with MetS and these effects persisted after adjustment for conventional risk factors. Thus, MetS per se confers increased propensities to both diabetes and cardiovascular disease. Low grade chronic inflammation appears to be a central feature for the increased risks of CVD and diabetes in MetS.

INFLAMMATION, METABOLIC SYNDROME, AND ACUTE PHASE PROTEINS

The relationship between inflammation and MetS is supported by the demonstration[15,16] that IL-6 and hsCRP are elevated in persons with diabetes having features of MetS. Several studies collectively record the universal observation that CRP is elevated in subjects with features of MetS.[17,18] Furthermore, a linear relationship exists between the number of components of MetS and increasing levels of hsCRP.

Pickup et al.[19] were among the first researchers to show that subjects with more than two features of MetS in fact had greater inflammation (increased serum CRP and serum IL-6 levels) compared to those who had fewer than two features. Yudkin et al.[20] reported a very significant correlation between inflammatory markers and several features of MetS by conducting Z-score analyses in 107 non-diabetic subjects. CRP levels were shown to be strongly associated with insulin resistance calculated from the HOMA model, blood pressure, low HDL, triglycerides, and levels of the IL-6 and TNF pro-inflammatory cytokines. Body mass index (BMI) and insulin resistance were the strongest determinants of the inflammatory state.

Festa et al.[15] who participated in the Insulin Resistance and Atherosclerosis Study (IRAS) showed that hsCRP was positively correlated with BMI, waist circumference, and other individual risk factors of MetS. Multivariate linear regression models reported the strongest correlation of CRP with BMI, systolic blood pressure, and insulin sensitivity index. All these studies record the universal observation that CRP is elevated in subjects with features of MetS and indicate a linear relationship between CRP levels and number of MetS features. Furthermore, the strongest associations are observed between CRP levels and central adiposity and insulin resistance.

PRO-INFLAMMATORY CYTOKINES, MONOCYTES, AND METABOLIC SYNDROME

Cytokines, in particular IL-1, TNF, and IL-6, are the main inducers of the acute phase response (APR).[21] Several lines of evidence indicate that IL-6 is the central mediator of the inflammatory response and promotes insulin resistance.[22] In addition, IL-6 administered to humans subcutaneously induces an acute inflammatory response.[23] In fact, IL-6 is believed to be the main driver of CRP release from hepatocytes.[24]

A very tight correlation exists between IL-6 and CRP; such a strong correlation does not exist for other cytokines. Furthermore, IL-6 and hsCRP are associated with visceral adiposity and 30% of circulating IL-6 is derived from human adipose tissue.[25] Moreover, baseline IL-6 levels independently predict future CVD.[26] Pickup et al.[19] demonstrated increased levels of IL-6 in subjects with more than two features of MetS. Our group also showed that monocyte release of IL-6 in type 2 diabetes is significantly increased when compared to non-diabetic controls.[16] Several studies have shown that TNF secreted by monocytes–macrophages, endothelial cells, and also to a large extent by adipose tissue, is an important regulator of insulin sensitivity.[27]

Elevated IL-18 level has been recently reported as an independent risk predictor for MetS in the absence of a history of type 2 diabetes in a large community-based population sample.[28] IL-18 functions as a pleiotropic pro-inflammatory cytokine and stimulates the production of TNF and secondarily IL-6. IL-18 may form a link between MetS and atherosclerosis because it is highly expressed in atherosclerotic plaques, and its role in plaque destabilization has been suggested.[29] Additionally, data from the recently reported MONICA/KORA study show that elevated levels of IL-18 are associated with considerably increased risks of type 2 diabetes.[30]

Chemokines also play an important role in leukocyte trafficking. Circulating levels of chemokines have been shown to increase inflammatory processes including obesity-related pathologies. Kim et al.[31] reported that circulating levels of MCP-1 and IL-8 are related to obesity-related parameters such as BMI, waist circumference, CRP, IL-6, HOMA, and low HDL-cholesterol. These findings suggest that the circulating MCP-1 and/or IL-8 may be a potential candidate linking obesity with obesity-related metabolic complications such as atherosclerosis and diabetes.

Since low grade inflammation is a feature of MetS, it is important to state that the pro-inflammatory status appears to result from an imbalance between pro- and anti-inflammatory cytokines. Importantly, in the Leiden 85-Plus Study, subjects who developed diabetes and had more than three MetS features had lower levels of LPS-stimulated whole blood release of IL-10 (a potent anti-inflammatory cytokine) than those who did not.[32] Esposito et al.[33] also reported that in both obese and non-obese women, IL-10 levels were significantly lower in women with MetS.

ADIPOSE TISSUE AND INFLAMMATION

Changes in adipose tissue (AT) mass are associated with altered endocrine and metabolic functions of the adipose tissue.[34] Most importantly, adipose tissue is a depot of inflammatory molecules, collectively called the adipocytokines or adipokines, including tumor necrosis factor (TNF), angiotensinogen, PAI-1, leptin, and adiponectin. Unlike adipose tissues of lean individuals, adipose tissues of obese subjects secreted increased amounts of TNF, IL-6, iNOS, TGF-β, CRP, sICAM, MCP-1, procoagulant proteins such as PAI-1, and tissue factor.[35] Elevated PAI-1 levels, the principal inhibitors of fibrinolysis, have been reported in many clinical and population studies in obese subjects and correlate with abdominal patterns of obesity[36] and other components of MetS. Furthermore, it appears to be even a better predictor of the MetS status than CRP.

A large body of evidence documents low levels of adiponectin in subjects with MetS.[37] Adiponectin belongs to the soluble defense collagen family and has been shown to suppress expression of adhesion molecules by endothelial cells, lipid accumulation, TNF release from monocytes, and vascular smooth muscle cell proliferation; it also reduces atherosclerotic vascular lesions *in vivo*. Adiponectin is an atypical adipokine because in contrast to the dramatic increase in plasma levels of all other adipokines, the circulating concentration of adiponectin is paradoxically decreased in obesity.

A strong correlation between low levels of adiponectin and increased insulin resistance is well established in humans.[38] Adiponectin also improves insulin resistance *in vivo*. Among Asian Indians, plasma levels of adiponectin in subjects with IGT are strong predictors of the development of diabetes.[39] In a step-wise regression model of a study examining the association between CRP and adiponectin levels, hsCRP was independently associated inversely with levels of

TABLE 14.1
Metabolic Syndrome: Evidence for a
Pro-Inflammatory State

↑ Acute phase proteins (HsCRP and SAA)
↓ Adiponectin
↑ Pro-inflammatory status (IL-6, IL-18, and TNF-α)
↑ Leptin
↑ Chemokines (MCP-1 and IL-8)
↓ Anti-inflammatory cytokines (IL-10)

adiponectin and positively with levels of leptin.[40] Adiponectin exerts anti-athero-genic properties by suppressing the endothelial inflammatory response, inhibiting TNF and monocyte adhesion, and suppresses transformation of macrophages to foam cells.

THERAPEUTIC MODULATION OF INFLAMMATION IN METABOLIC SYNDROME

Inflammation can be reduced by a variety of approaches including diet, exercise, and pharmacotherapy (statins, PPAR-α and -γ agonists, and the endocannabinoid receptor-1 blocker, rimonabant). As reviewed previously[41] and summarized in Table 14.1, people with MetS typically manifest pro-thrombotic and pro-inflammatory states. Therapeutic lifestyle changes (TLCs), including diet and exercise, serve as cornerstone therapies for the treatment of MetS. Thus, the first step in reducing the excess cardiovascular risk associated with MetS is the adoption of a healthier lifestyle (particularly reducing body weight and increasing physical activity).

WEIGHT LOSS, HYPOCALORIC DIETS, INFLAMMATION, AND METABOLIC SYNDROME

Weight losses ranging from 3 to 15 kg achieved through different diet programs (low fat, high protein, or hypocaloric diets) result in concomitant reductions of CRP levels by 7 to 48% as shown in several studies. We recently reviewed the literature[42,43] and documented a significant positive correlation ($r^2 = 0.8693$; p <0.05) between weight loss and percentage reduction of CRP levels from pooled data accumulated from several dietary intervention trials. The few studies that focus on nutritional modulation of biomarkers of inflammation in subjects with MetS are discussed in Table 14.2.

Esposito et al.[44] emphasized meal modulation of cytokines (IL-6, -8, -18, and adiponectin) as a therapeutic approach toward attenuating atherogenic inflammatory activities in obese women. In an effort to determine the effects of lifestyle changes on markers of systemic vascular inflammation and IR, they conducted a randomized single-blind trial in 120 premenopausal (aged 20 to 46 years) obese women (BMI ≥30 kg/m²) without diabetes, hypertension, or hyperlipidemia. The intervention group (n = 60) adhered to a low energy Mediterranean-style diet (foods rich in complex carbohydrates, monounsaturated fat, and fiber; lower ratios of omega-6 to omega-3 fatty acids) and increased physical activity. The control group (n = 60) was given general information about healthy food choices and exercise. BMI decreased significantly more in the intervention group than in controls, as did serum concentrations of IL-6, IL-18, and CRP, while adiponectin levels increased significantly. In multivariate analyses, changes in FFA and adiponectin levels were independently associated with changes in insulin sensitivity

The same group of authors[45] investigated the effects of Mediterranean-style diets on endothelial function and vascular inflammatory markers in 180 patients with MetS. The 99 men and 81 women with MetS, as defined by the ATP III, were randomized equally in the intervention and placebo groups for 2 years. The patients in the intervention group were instructed to follow a Mediterranean-style diet and received detailed advice about how to increase daily consumption of whole grains, fruits, vegetables, nuts, and olive oil. Patients in the control group followed a prudent diet (50 to 60% carbohydrates, 15 to 20% proteins, total fat <30%). Compared with patients consuming the control diet, patients adhering to the intervention diet exhibited significantly reduced serum concentrations of hsCRP and IL-6 and decreased insulin resistance, whereas adiponectin levels increased significantly. However, IL-18 was non-significantly decreased. Endothelial function scores improved in the intervention group but remained stable in the control group. A 2-year follow-up revealed that only 40 patients in the intervention group still had features of MetS, compared with 78 in the control group. Hence, it was concluded that a Mediterranean-style diet may be effective in reducing the prevalence of MetS and its associated cardiovascular risk.

Tchernof et al.[46] carried out a caloric restriction-induced weight loss program in 61 obese (BMI = 35.6 kg/m²) postmenopausal women (56.4 ± 5.2 years) and reported that plasma CRP levels were positively associated with dual x-ray absorptiometry-measured total body fatness and CT-measured intra-abdominal body fat area. Significant correlations were also found between plasma CRP and triglyceride levels and glucose disposal measured by the hyperinsulinemic–euglycemic clamp technique. The average weight loss was 15.6%, with losses of 25% fat mass and 6% fat-free mass. Visceral and subcutaneous fat areas were reduced by 36.4% and 23.7%, respectively. Plasma CRP levels were significantly reduced by weight loss: average 32.3%, from 3.06 to 1.63 µg/mL. The changes in body weight and in total body fat mass were positively associated with plasma CRP level reductions.

There is a definite relation between an optimal pace of exercise combined with weight loss. Lifestyle intervention achieved by the Diabetes Prevention Program (DPP)[47] included various components: low calorie, low fat diet (<15% total fat and <7% saturated fat), physical activity involving brisk walking or bicycling at least 150 minutes weekly, and a 16-lesson curriculum covering diet,

TABLE 14.2
Randomized Clinical Trials Examining Effects of Lifestyle Interventions (Diet and Exercise) on Biomarkers of Inflammation

Study Investigators, Duration, Reference	Subjects	Lifestyle Intervention Type	Changes in Biomarkers of Inflammation
Esposito et al. 2003 (2 years)[44]	120 premenopausal obese women (20 to 46 years, BMI ≥30 kg/m²)	Mediterranean-style diet (rich in complex carbohydrates, MUSFAs and fiber, lower ratio of omega 6/omega 3) and increased physical activity versus control subjects informed of healthy food choices and exercise	Decreased CRP, IL-6, and IL-18 and increased APN compared with controls
Esposito et al. 2004 (2 years)[45]	180 MetS patients (99 men and 81 women)	Mediterranean style diet versus prudent diet (50 to 60% carbohydrates, 15 to 20% proteins, <30% total fat)	Increased CRP, decreased IR, and increased APN; non-significant decrease of IL-18; improved endothelial function
Tchernof et al. 2002 (14 months)[46]	61 obese, postmenopausal women (mean age 56.4 years, BMI-35.6 kg/m²)	Supervised weight loss program aimed at reducing body weight to <120% of ideal value; 1200 kcal/d; American Heart Association Step 2 Diet	Significantly reduced CRP
Orchard et al. 2005 (3.2 years)[48]	Participants had IGT; 657 controls and 638 subjected to intervention	Lifestyle intervention designed to achieve and maintain 7% weight loss and 150 minutes of exercise per week versus controls	Greater reduction in CRP levels in patients with impaired glucose tolerance in intervention group versus placebo
Seshadari et al. 2004 (6 months)[50]	78 severely obese patients, 86% with diabetes and/or MetS	Low carbohydrate diet versus conventional diet (fat and calorie restricted)	Decreased CRP in both groups, greater decrease in CRP in those with high risk baseline levels (>3 mg/dl, n = 48 on low carbohydrate diet)
Kopp et al 2003 (14 months after surgery)[51]	37 morbidly obese patients (BMI = 49 kg/m²)	Elective gastroplasty for induction of weight loss	Decreased CRP and IL-6; TNF-α unchanged

TABLE 14.2 (CONTINUED)
Randomized Clinical Trials Examining Effects of Lifestyle Interventions
(Diet and Exercise) on Biomarkers of Inflammation

Study Investigators, Duration, Reference	Subjects	Lifestyle Intervention Type	Changes in Biomarkers of Inflammation
Troseid et al. 2004 (12 weeks)[54]	15 MetS subjects in exercise (n = 9) and non-exercise (n = 6) groups	Endurance exercise (walking or jogging on treadmill) and strength training 45 to 60 minutes three times weekly in training studio under supervision	Significant reduction in MCP-1 and IL-8
Troseid et al. 2005 (12 weeks)[55]	32 subjects with MetS	2 × 2 randomized factorial trial, physical exercise	No change in CAMs
Roberts et al. 2005 (3 weeks)[56]	31 obese patients; 15 with MetS	High fiber, low fat diet; 3-week residential program (ad libitum food and daily aerobic exercise)	Significant reductions in CRP, sICAM, sP-selectin, MIP-1, MMP-9; increased NO; significant reduction in monocyte adhesion and chemotactic activity
Brinkworth et al. 2004 (12 weeks energy restriction + 4 weeks energy balance + 52 weeks with minimal professional support)[57]	58 (13 male, 45 female) obese non-diabetic subjects with hyperinsulinemia; mean age = 50 years, BMI = 34 kg/m²	Randomization to standard protein (15% protein, 55% carbohydrate) or high protein (30% protein, 40% carbohydrate) diet	Significant (<0.001) fat loss with no diet effect; decreased sICAM and CRP (p <0.05); increased HDL (p <0.001) with both diets

IGT = impaired glucose tolerance. APN = adiponectin.

exercise, and behavior modification taught through individual and group sessions. The goal was to lose 7% of body weight. Orchard et al.[48] reported that the baseline incidence of MetS was 51% in the DPP and that the incidence of MetS was reduced by 41% in subjects in the lifestyle intervention group. They also reported that lifestyle interventions in both men and women resulted in significantly greater reduction in CRP levels in patients with IGT relative to placebo. The reduction in CRP could be explained by the greater weight loss. Meydani[49] critically reviewed the importance of moderate increases in physical activity along with a detailed and tailored Mediterranean-style diet to reduce the prevalence of MetS and associated cardiovascular risks through reducing systemic inflammation and endothelium dysfunction, particularly in patients who failed to lose weight.

In obese women, hypocaloric diets have been associated with weight reduction as well as inflammation. In this context, Seshadari et al.[50] found overall favorable effects on inflammation in 78 severely obese subjects from a low carbohydrate diet (LCD) versus a conventional (fat- and calorie-restricted) diet for a period of 6 months; 86% of the subjects were either diabetic or had features of MetS. Overall, CRP levels decreased modestly in both diet groups. However, patients with high risk baseline levels (CRP >3 mg/dL, n = 48) experienced greater decreases in CRP on a low carbohydrate diet, independent of weight loss.

Kopp et al.[51] examined the cross-sectional and longitudinal relationships of CRP, IL-6, and TNF with features of the IRS in 37 morbidly obese patients (BMI = 49 kg/m^2) with different stages of glucose tolerance before and 14 months after gastric surgery. Weight loss after gastric surgery induced a significant shift from diabetes (37 versus 3%) to IGT (40 versus 33%) and NGT (23 versus 64%). Concentrations of CRP and IL-6 decreased after weight loss, whereas serum levels of TNF-alpha remained unchanged. It was concluded that weight loss results in a marked decrease in circulating levels of inflammatory markers in association with a reversal of diabetes in morbidly obese individuals after gastroplastic surgery. However, long-term studies are needed to show whether this improvement in cardiovascular risk factors will eventually translate into a significant clinical benefit with regard to cardiovascular morbidity and mortality.

Cellular adhesion molecules (CAMs) such as E-selectin, intercellular adhesion molecule-1 (ICAM-1), and vascular cell adhesion molecule-1 (VCAM-1) are involved in the rolling, adhesion, and extravasation of monocytes and T lymphocytes into the atherosclerotic plaque. Serum concentrations of CAMs are higher in patients with coronary artery disease than in healthy control subjects.[52] Moreover, male participants in the Physician's Health Study who had ICAM-1 levels in the highest quartile are at greater cardiovascular risk than men in the lowest quartile.[53] In line with these findings, Troseid et al.[54] published a study conducted as an unmasked randomized 2 × 2 factorial trial involving an intensive exercise protocol of 12 weeks' duration. In the combined exercise groups, significant reductions in MCP-1 and IL-8 as compared to the combined non-exercise groups were noted along with a significant reduction versus baseline for both chemokines. The changes in MCP-1 were significantly correlated to changes in visceral fat.

However, the same group of investigators[55] recently reported a negative study with regard to the effect of physical exercise on serum levels of CAMs. These authors explored the possible role of adipose tissue in regulating serum levels of CAMs. No significant changes in CAMs were observed in the intervention group. On examination of the whole study population regardless of intervention, changes in serum E-selectin were significantly correlated to changes in body mass index, waist circumference, fasting glucose, and HbA1c, but not to changes in visceral fat, subcutaneous fat, TNF, or adiponectin. Thus this study highlights changes in glycemic control and obesity rather than regional fat distribution to influence E-selectin levels in subjects with MetS.

Additionally, Roberts et al.[56] recently reported the results of examining the effects of lifestyle modification on various key contributing factors to atherogenesis, including oxidative stress, inflammation, chemotaxis, and cell adhesion. Obese men (n = 31), 15 of whom had MetS, were placed on a high fiber, low fat diet in a 3-week residential program where food was provided ad libitum and daily aerobic exercise (45 to 60 min) was performed. After 3 weeks, significant reductions in BMI, CRP, sICAM-1, sP-selectin, MIP-1α, MMP-9, and biomarkers of oxidative stress with concomitant increases in NO production were noted. Additionally, both monocyte adhesion and monocyte chemotactic activity (MCA) significantly decreased. Nine of 15 subjects were no longer positive for MetS post-intervention. This study postulated the amelioration of CAD risk factor by intensive lifestyle modification in men with MetS factors prior to reversal of obesity.

Comparative effects of two low fat diets differing in carbohydrate-to-protein ratio for biomarkers of CVD risk in obese subjects with hyperinsulinemia were examined.[57] Two groups totaling 58 obese, non-diabetic subjects with hyperinsulinemia (13 males and 45 females, mean age 50.2 years, mean BMIs of 34.0 kg/m[2], mean fasting insulin levels of 17.8 mU/l) were randomly assigned to either a standard protein (SP; 15% protein, 55% carbohydrate) or high protein (HP; 30% protein, 40% carbohydrate) diet during 12 weeks of energy restriction (approximately 6.5 MJ/day) and 4 weeks of energy balance (approximately 8.3 MJ/day). They were subsequently asked to maintain the same dietary pattern for the succeeding 52 weeks with minimal professional support. The measurements included fasting blood lipids, glucose, insulin, and CRP and sICAM-1 at baseline and at weeks 16 and 68. In total, 43 subjects completed the study with similar dropouts in each group. At week 68, there was net weight loss (SP –2.9 ± 3.6%; HP –4.1 ± 5.8%; p <0.44) due entirely to fat loss (p <0.001) with no diet effect. Both diets significantly increased HDL-C (p <0.001) and decreased fasting insulin, insulin resistance, sICAM-1, and CRP (p <0.05). Nonetheless, both dietary patterns achieved net weight loss and improvements in cardiovascular risk.

PHARMACOLOGICAL THERAPIES WITH POTENTIAL TO PREVENT OR TREAT METABOLIC SYNDROME

Although, lifestyle intervention remains the cornerstone therapy, research continues to focus on other options such as pharmacotherapy and bariatric surgery to decrease the inflammatory responses. Pharmacological therapy is a critical step in the management of patients with MetS when lifestyle modifications fail to achieve therapeutic goals. Basic research has ushered in numerous potential new molecular drug targets for treating MetS.[58]

Chief among these are statins, insulin sensitizers such as metformin and thiazolidinediones (rosiglitazone and pioglitazone), fibrates, and a newly discovered drug named rimonabant. In various large prospective studies, statins have been shown to reduce hsCRP.[59] Rosiglitazone has also been shown to significantly reduce hsCRP levels,[60] suggesting that PPAR-γ agonists may reduce markers for subclinical inflammation to the levels seen in statin trials. Furthermore, therapy with fenofibrate, a fibric acid derivative, has been shown to lower IL-6 and hsCRP in borderline hyperlipidemic subjects.[61] In addition, the endocannabinoid system appears to play a key role in metabolism and weight gain. The investigational rimonabant is a cannabinoid receptor type 1 blocker that has been employed in trials involving more than 6500 patients. It has resulted in reduction in the prevalence of MetS as well as biomarkers of inflammation along with improved glycemic and lipid profiles.[58,62]

CONCLUSION

It appears from the available literature that therapeutic lifestyle change remains the cornerstone in modulating inflammation and CVD. If lifestyle change is not sufficient, then drug therapies for abnormalities in individual risk factors are to be indicated. To date, we have insufficient evidence for primary use of drugs that target the underlying causes of MetS. Weight loss and hypocaloric diets are definitely associated with reduced inflammation and overall risk reduction of CVD.

Also, most studies reported effects on CRP, whereas a few focused on pro-inflammatory cytokines such as IL-6, IL-18 or TNF-α and chemokines (IL-8 and MCP-1). Future research should focus on the roles of specific nutritional components in various diets on biomarkers of inflammation, as they may modulate inflammation through different mechanisms. Despite the fact that precise mechanisms have yet to be established, both diet and physical activity play pivotal roles in improving many factors associated with MetS including modulation of various adipocytokines. A more thorough understanding of the clustering of metabolic abnormalities and their underlying etiology will help to define diet and physical activity guidelines for preventing and treating MetS — an important aspect of CVD prevention.

REFERENCES

1. Reaven, G.M. (2005) The insulin resistance syndrome: definition and dietary approaches to treatment. *Ann Rev Nutr* 25, 391.
2. Eckel, R.H., Grundy, S. M., and Zimmet, P.Z. (2005) The metabolic syndrome. *Lancet* 365, 1415.
3. Haffner, S. and Cassells, H.B. (2003) Metabolic syndrome: a new risk factor of coronary heart disease? *Diabetes Obes Metab* 5, 359.

4. Expert Panel on Detection, Evaluation, and Treatment of High Blood Cholesterol in Adults. (2001) Executive Summary of Third Report of the National Cholesterol Education Program (NCEP) Expert Panel on Detection, Evaluation, and Treatment of High Blood Cholesterol in Adults (Adult Treatment Panel III). (2001) *JAMA* 285, 2486.

5. Ford, E.S., Giles, W.H., and Dietz, W.H. (2002) Prevalence of the metabolic syndrome among U.S. adults: findings from the third National Health and Nutrition Examination Survey. *JAMA* 287, 356.

6. Alberti, K.G, Zimmet, P., and Shaw, J. (2005) IDF Epidemiology Task Force Consensus Group: the metabolic syndrome — a new worldwide definition. *Lancet* 366, 1059.

7. Grundy, S.M. et al. (2005) Diagnosis and management of the metabolic syndrome: an American Heart Association/National Heart, Lung, and Blood Institute scientific statement. *Circulation* 112, 2735.

8. Lakka, H.M. et al. (2002) The metabolic syndrome and total and cardiovascular disease mortality in middle-aged men. *JAMA* 288, 2709.

9. Isomaa, B. et al. (2001) Cardiovascular morbidity and mortality associated with the metabolic syndrome. *Diabetes Care* 24, 683.

10. Alexander, C.M., Landsman, P.B., Teutsch, S.M., and Haffner, S.M. (2003) Third National Health and Nutrition Examination Survey (NHANES III); National Cholesterol Education Program (NCEP). NCEP-defined metabolic syndrome, diabetes, and prevalence of coronary heart disease among NHANES III participants age 50 years and older. *Diabetes* 52, 1210.

11. Dekker, J.M. et al. (2005) Metabolic syndrome and 10-year cardiovascular disease risk in the Hoorn Study. *Circulation* 112, 666.

12. Hanson, R.L, Imperatore, G., Bennett, P.H., and Knowler, W.C. (2002) Components of the "metabolic syndrome" and incidence of type 2 diabetes. *Diabetes* 51, 3120.

13. Wallace, A.M. et al. (2001) Plasma leptin and the risk of cardiovascular disease in the West of Scotland Coronary Prevention Study (WOSCOPS). *Circulation* 104, 3052.

14. Assmann, G., Nofer, J.R., and Schulte, H. (2004) Cardiovascular risk assessment in metabolic syndrome: view from PROCAM. *Endocrinol Metab Clin North Am* 33, 377.

15. Festa, A. et al. (2000) Chronic subclinical inflammation as part of the insulin resistance syndrome: the Insulin Resistance Atherosclerosis Study (IRAS). *Circulation* 102, 42.

16. Devaraj, S. and Jialal, I. (2000) Alpha-tocopherol supplementation decreases serum C-reactive protein and monocyte interleukin-6 levels in normal volunteers and type 2 diabetic patients. *Free Radic Biol Med* 29, 790.

17. Ridker, P.M., Buring, J.E., Cook, N.R., and Rifai, N. (2003) CRP, the metabolic syndrome, and risk of incident cardiovascular events: an 8-year follow-up of 14719 initially healthy American women. *Circulation* 107, 391.

18. Ford, E.S. (2003) The metabolic syndrome and C-reactive protein, fibrinogen, and leukocyte count: findings from the Third National Health and Nutrition Examination Survey. *Atherosclerosis* 168, 351.

19. Pickup, J.C., Mattock, M.B., Chusney, G.D., and Burt, D. (1997) NIDDM as a disease of the innate immune system: association of acute-phase reactants and interleukin-6 with metabolic syndrome X. *Diabetologia* 40, 1286.

20. Yudkin, J.S., Kumari, M, Humphries, S.E., and Mohamed-Ali, V. (2000) Inflammation, obesity, stress and coronary heart disease: is interleukin-6 the link? *Atherosclerosis* 148, 209.
21. Baumann, H. and Goldie, J. (1994) The acute phase response. *Immunol Today* 15, 74.
22. Heinrich, P.C., Castell, J.V., and Andus T. (1990) Interleukin-6 and the acute phase response. *Biochem J* 265, 621.
23. Banks, R.E. et al. (1995) The acute phase protein response in patients receiving subcutaneous IL-6. *Clin Exp Immunol* 102, 217.
24. Li, S.P. and Goldman, N.D. (1996) Regulation of human C-reactive protein gene expression by two synergistic IL-6 responsive elements. *Biochemistry* 35, 9060.
25. Aldhahi, W. and Hamdy, O. (2003) Adipokines, inflammation, and the endothelium in diabetes. *Curr Diab Rep* 3, 293.
26. Ridker, P.M., Hennekens, C.H., Buring, J.E., and Rifai, N. (2000) C-reactive protein and other markers of inflammation in the prediction of cardiovascular disease in women. *New Engl J Med* 342, 836.
27. Hotamisligil, G.S. and Spiegelman, B.M. (1994) Tumor necrosis factor-alpha: a key component of the obesity-diabetes link. *Diabetes* 43, 1271.
28. Hung, J., McQuillan, B.M., Chapman, C.M., Thompson, P.L., and Beilby, J.P. (2005) Elevated interleukin-18 levels are associated with the metabolic syndrome independent of obesity and IR. *Arterioscler Thromb Vasc Biol* 25, 1268.
29. SoRelle, R. (2003) Interleukin-18 predicts coronary events. *Circulation* 108, e9051.
30. Thorand, B. et al. (2005) Elevated levels of interleukin-18 predict the development of type 2 diabetes: results from the MONICA/KORA Augsburg Study, 1984–2002. *Diabetes* 54, 2932.
31. Kim, C.S., Park, H.S., Kawada, T., Kim, J.H., and Erickson, K.L. (2006) Circulating levels of MCP-1 and IL-8 are elevated in human obese subjects and associated with obesity-related parameters. *Int J Obes* [epub ahead of print].
32. van Exel, E., Gussekloo, J., and de Craen, A.J. (2002) Low production capacity of interleukin-10 associates with the metabolic syndrome and type 2 diabetes: the Leiden 85-Plus Study. *Diabetes* 51, 1088.
33. Esposito, K., Pontillo, A., and Giugliano, F. (2003). Association of low interleukin-10 levels with the metabolic syndrome in obese women. *J Clin Endocrinol Metab* 88, 1055.
34. Lau, D.C., Dhillon, B., Yan, H., Szmitko, P.E., and Verma S. (2005) Adipokines: molecular links between obesity and atherosclerosis. *Am J Physiol Heart Circ Physiol* 288, H2031.
35. Rajala, M.W. and Scherer, P.E. (2003) Minireview: the adipocyte at the crossroads of energy homeostasis, inflammation, and atherosclerosis. *Endocrinology* 144, 3765.
36. Sakkinen, P.A, Wahl, P., Cushman, M., Lewis, M.R., and Tracy, R.P. (2000) Clustering of procoagulation, inflammation, and fibrinolysis variables with metabolic factors in insulin resistance syndrome. *Am J Epidemiol* 152, 897.
37. Kondo, H., Shimomura, I., and Matsukawa, Y. (2002) Association of adiponectin mutation with type 2 diabetes: a candidate gene for the insulin resistance syndrome. *Diabetes* 51, 2325.

38. Weyer, C., Funahashi, T., and Tanaka, S. (2001) Hypoadiponectinemia in obesity and type 2 diabetes: close association with insulin resistance and hyperinsulinemia. *J Clin Endocrinol Metab* 86, 1930.
39. Snehalatha, C., Mukesh, B., and Simon, M. (2003) Plasma adiponectin is an independent predictor of T2DM in Asian Indians. *Diabetes Care* 26, 3226.
40. Ouchi, N. et al. (2003) Reciprocal association of C-reactive protein with adiponectin in blood stream and adipose tissue. *Circulation* 107, 671.
41. Devaraj, S., Rosenson, R.S., and Jialal, I. (2004) Metabolic syndrome: an appraisal of the pro-inflammatory and procoagulant status. *Endocrinol Metab Clin North Am* 33, 431.
42. Basu, A., Devaraj, S., and Jialal, I. (2006) Dietary factors that promote or retard inflammation. *Arterioscler Thromb Vasc Biol* 26, 995.
43. Dietrich, M. and Jialal, I. (2005) The effect of weight loss on a stable biomarker of inflammation, C-reactive protein. *Nutr Rev* 63, 22.
44. Esposito, K. et al. (2003) Effect of weight loss and lifestyle changes on vascular inflammatory markers in obese women: a randomized trial. *JAMA* 289, 1799.
45. Esposito, K. et al. (2004) Effect of a Mediterranean-style diet on endothelial dysfunction and markers of vascular inflammation in metabolic syndrome: a randomized trial. *JAMA* 292, 1440.
46. Tchernof, A., Nolan, A., Sites, C.K., Ades, P.A., and Poehlman, E.T. (2002) Weight loss reduces C-reactive protein levels in obese postmenopausal women. *Circulation.* 105, 564.
47. Diabetes Prevention Program (DPP) Research Group. (2002) The Diabetes Prevention Program (DPP): description of lifestyle intervention. *Diabetes Care* 25, 2165.
48. Orchard, T.J. et al. (2005) The effect of metformin and intensive lifestyle intervention on the metabolic syndrome: the Diabetes Prevention Program randomized trial. *Ann Intern Med* 142, 611.
49. Meydani, M. (2005). A Mediterranean-style diet and metabolic syndrome. *Nutr Rev* 63, 312.
50. Seshadri, P. et al. (2004) A randomized study comparing the effects of a low-carbohydrate diet and a conventional diet on lipoprotein subfractions and C-reactive protein levels in patients with severe obesity. *Am J Med* 117, 398.
51. Kopp, H.P. et al. (2003) Impact of weight loss on inflammatory proteins and their association with the insulin resistance syndrome in morbidly obese patients. *Arterioscler Thromb Vasc Biol* 23, 1042.
52. Haught, W.H., Mansour, M., and Rothlein, R. (1996) Alterations in circulating intercellular adhesion molecule-1 and L-selectin further evidence for chronic inflammation in ischemic heart disease. *Am Heart J* 132, 1.
53. Ridker, P.M., Hennekens, H., Roitman-Johnson, M.B., Stampfer, J., and Allen, J. (1998) Plasma concentration of soluble intercellular adhesion molecule 1 and risks of future myocardial infarction in apparently healthy men, *Lancet* 351, 88.
54. Troseid, M., Lappegard, K.T., and Claudi T. (2004) Exercise reduces plasma levels of chemokines MCP-1 and IL-8 in subjects with the metabolic syndrome. *Eur Heart J* 25, 349.
55. Troseid, M., Lappegard, K.T., Mollnes, T.E., Arnesen, H., and Seljeflot, I. (2005) Changes in serum levels of E-selectin correlate to improved glycaemic control and reduced obesity in subjects with the metabolic syndrome. *Scand J Clin Lab Invest* 65, 283.

56. Roberts, C.K. et al. (2005) Effect of a diet and exercise intervention on oxidative stress, inflammation, MMP-9 and monocyte chemotactic activity in men with metabolic syndrome factors. *J Appl Physiol* 100, 1657.

57. Brinkworth, G.D., Noakes, M., Keogh, J.B., Wittert, G.A., and Clifton, P.M. (2004) Long-term effects of a high-protein, low-carbohydrate diet on weight control and cardiovascular risk markers in obese hyperinsulinemic subjects. *Int J Obes Relat Metab Disord* 28, 661.

58. Davis, S.N. (2006) Contemporary strategies for managing cardiometabolic risk factors. *J Manag Care Pharm* 12, S4.

59. Devaraj, S. and Jialal, I. (2004) Effects of statins on C-reactive protein: are all statins similar? In *Statins: Understanding Clinical Use,* Elsevier Science, Amsterdam, p. 189.

60. Esposito, K. et al. (2006) Effect of rosiglitazone on endothelial function and inflammatory markers in patients with the metabolic syndrome. *Diabetes Care* 29, 1071.

61. Zambon, A., Gervois, P., Pauletto, P., Fruchart, J.C., and Staels, B. (2006) Modulation of hepatic inflammatory risk markers of cardiovascular diseases by PPAR-alpha activators: clinical and experimental evidence. *Arterioscler Thromb Vasc Biol* 26, 977.

62. Despres, J.P., Golay, A., and Sjostrom, L. (2005) Rimonabant in Obesity-Lipids Study Group: effects of rimonabant on metabolic risk factors in overweight patients with dyslipidemia. *New Engl J Med* 353, 2121.

15 Dietary Fatty Acids and Metabolic Syndrome

Helen M. Roche

CONTENTS

INTRODUCTION: DIETARY FAT INTAKE, INFLAMMATION, AND METABOLIC SYNDROME —DOUBLE HIT HYPOTHESIS

Nutrition is a key environmental factor that is particularly involved in the pathogenesis and progression of several polygenic, diet-related diseases. Metabolic syndrome is a very common condition, characterized by insulin resistance, abdominal obesity, dyslipidemia [increased triacylglycerol (TAG) or reduced high density lipoprotein (HDL) cholesterol concentrations], hypertension, and urinary microalbuminuria.[1–3] It often precedes type 2 diabetes mellitus (T2DM) and is associated with a greater risk of cardiovascular disease (CVD).[4,5]

The development of metabolic syndrome can be attributed to both genetic and environmental factors, as reviewed elsewhere.[6] Figure 15.1 illustrates that insulin resistance is the most important metabolic defect that leads to the development of the metabolic syndrome — a state that may be triggered by excessive metabolic stressors [high fat diet, obesity, elevated plasma non-esterified fatty acid (NEFA) levels, etc.].[7,8]

The link between dietary fat, obesity, and T2DM is partly due to elevated fatty acid concentrations on systemic responsiveness to insulin, resulting in impaired insulin action in several peripheral tissues including the liver, skeletal muscles, and adipose tissues and dysregulated carbohydrate and lipid metabolism, which leads to a compensatory hyperinsulinemia.[9] Interestingly, insulin resistance

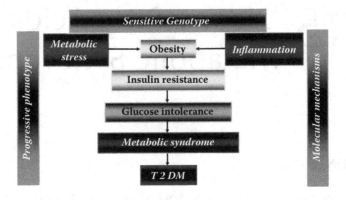

FIGURE 15.1 Development and progression of metabolic syndrome. (From Roche, H.M., *Proc Nutr Soc* 64, 23, 2005. With permission.)

often co-exists with a subacute chronic pro-inflammatory state.[10–12] Indeed, recent research suggests that pro-inflammatory cytokines derived from adipose tissue play a key role in the development of insulin resistance.[13,14] The model in Figure 15.1 suggests that an individual with a genetic predisposition to metabolic syndrome cannot sustain a "double hit" from metabolic and inflammatory stressors. Also, the pathway between obesity and insulin resistance toward the metabolic syndrome and T2DM represents a progressive phenotype. Therefore, attenuating the impacts of metabolic stressors through dietary fat modification to improve insulin sensitivity would be advantageous.

DIETARY FATTY ACIDS, INFLAMMATION, AND INSULIN SIGNALLING: CELLULAR PERSPECTIVE

Several lines of evidence suggest that the combination of excessive nutrient-derived metabolic stressors and pro-inflammatory stressors plays an important role in the development of insulin resistance and metabolic syndrome. High fat diets and excessive adipose tissue TAG storage result in increased circulating plasma NEFA flux to peripheral tissues that promotes insulin resistance. Fatty acids induce insulin resistance through inhibition of insulin signaling.[15–17]

Evidence also indicates that saturated fatty acids (SFAs) may play a particular role in attenuating peripheral tissue responsiveness to insulin[3] while some poly-unsaturated fatty acids (PUFAs) counteract this effect. Adipose tissue may be the source of insulin de-sensitization of pro-inflammatory molecules collectively known as adipocytokines or adipokines, which predispose to insulin resistance.[13,14] Tumor necrosis factor (TNF)-α, interleukin (IL)-6, resistin, and adiponectin (acrp30) have all been shown to influence insulin sensitivity.[18–20] The insulin de-sensitizing effects of TNF-α are probably best characterized. TNF-α inhibits autophosphorylation of tyrosine residues of the insulin receptor, promotes serine phosphorylation of insulin receptor substrate (IRS)-1, and reduces transcription of key targets in the insulin signaling cascade, all of which impede transduction of the insulin signal.[21]

Knocking out TNF-α or the TNF-α receptor improves insulin resistance in animal models of obesity-induced insulin resistance.[22,23] Other groups have proposed that other components of the inflammatory response, IκB kinase-β (IKK-β) or c-Jun amino-terminal kinases (JNKs), are central to the interplay of dietary fatty acids, obesity, and insulin resistance.[24,25] In summary, these studies suggest that down-regulation of several components of the inflammatory response affords substantial protection from obesity-induced insulin resistance.

Consistent with our hypothesis that dietary fatty acids can modulate the inflammatory profiles of adipocytes, two recent studies determined the effects of other fatty acids *in vitro*. The first showed that the palmitate SFA activated nuclear factor-κB (NF-κB) activity and induced TNF-α and IL-6 expression in 3T3-L1 adipocytes.[26] The second extensive investigation demonstrated that a mixture of SFA and PUFA NEFA treatments impaired insulin signalling at multiple sites, decreased insulin-stimulated glucose transporter (GLUT)-4 translocation and glucose transport, and activated the stress/inflammatory kinase JNK pathway in 3T3-L1 adipocytes.[27] Thus it may be possible to attenuate the pro-inflammatory phenotype associated with obesity-induced insulin resistance by altering the composition of circulating NEFAs through dietary fatty acid modification.

Therefore, in terms of manipulating dietary factors to attenuate the inflammatory response in adipose tissue to improve insulin sensitivity, the most obvious treatment is to reduce adipose tissue mass. Nevertheless the prevalence of obesity is increasing and due to poor compliance, current therapies are largely ineffective. Therefore other strategies to attenuate the impact of insulin resistance in the presence of obesity are required.

Our group demonstrated that a sub-group of fatty acids known as conjugated linoleic acids and, in particular, the cis-9, trans-11 CLA isomer (c9, t11-CLA), may have the potential to improve lipid metabolism and insulin sensitivity within the context of obesity.[28,29] This effect was ascribed to differential sterol regulatory element-binding protein (SREBP)-1c gene expression, a key regulatory transcription factor involved in lipogenesis and glucose metabolism. Feeding a c9, t11-CLA-rich diet produced divergent tissue-specific effects on SREBP-1c expression, significantly reducing hepatic SREBP-1c and increasing adipose tissue SREBP-1c expression, both of which could contribute to improved lipid and glucose metabolism.[28] Interestingly, this study also showed that TNF-α regulated SREBP-1c expression in human adipocytes, but not in hepatocytes, thus supporting the hypotheses that crosstalk exists between molecular markers of insulin sensitivity and adipocytokines which in turn can be modified by fatty acids.

Further work showed that the insulin sensitizing effect of the c9, t11-CLA-rich diet was associated with a marked reduction in adipose tissue TNF-α expression that may be related to lower NF-κB DNA binding, which has been attributed to lower nuclear P65 levels and increased cytosolic inhibitor of κBα (IκBα) expression.[29] This study suggests that the fatty acid composition of the diet can be adjusted to attenuate the pro-inflammatory insulin de-sensitizing effect of obesity-induced insulin resistance. Indeed there was a significant reduction in the adipose tissue macrophage population observed in the c9, t11-CLA fed mice. This is an extremely important observation, in that it shows that dietary modification can alter the cellular profile of adipose tissue in obesity, which was associated with positive metabolic effects.

WHOLE BODY METABOLIC PERSPECTIVE: EVIDENCE FROM HUMAN DIETARY FATTY ACID INTERVENTION STUDIES

Several studies have shown a consistent relationship between plasma fatty acid composition and insulin resistance. A prospective cohort study investigated the interaction between serum fatty acid composition and the development of T2DM in a cohort of middle-aged normoglycemic men.[30] Baseline serum esterified and non-esterified SFA levels were significantly higher and PUFA levels were lower in the men who developed T2DM after 4 years.

Recent evidence from the Nurses' Health Study showed that higher intake of saturated fat and a low dietary P:S ratio were related to increased CVD risk among women with T2DM.[31] This study estimated that replacement of 5% of energy from saturated fat with equivalent energy from carbohydrates or MUFAs was associated with 22 and 37% lower risks of CVD, respectively. This finding suggests that dietary fatty acid modification may also determine secondary outcomes associated with metabolic syndrome and T2DM.

Relatively few human dietary intervention studies have determined the relationship between dietary fatty acid composition and insulin sensitivity as the primary metabolic end point. For the purpose of this review, studies that investigated the effect of isocaloric substitution of dietary fatty acids will be reviewed, to exclude confounding effects of reduced energy intake on body weight. The KANWU study, a controlled multicenter, isoenergetic dietary intervention study involving 162 individuals, showed that decreasing dietary SFA and increasing MUFA improved insulin sensitivity but had no effect on insulin secretion.[32] Interestingly the favorable effect of substituting SFA for MUFA was only seen when total fat intake was below 37% energy. Within each dietary group a second assignment of n-3 PUFA supplementation or placebo was completed, but the n-3 PUFA intervention had no effect on insulin sensitivity despite reduced TAG concentrations.

A smaller randomized crossover dietary intervention study in 59 young healthy subjects randomly assigned to isoenergetic carbohydrate- and MUFA-rich diets for 28 days significantly improved insulin sensitivity compared to a high-SFA diet.[33] Also ex vivo analysis showed that both the carbohydrate- and MUFA-rich diets significantly increased basal and insulin-stimulated glucose uptake in monocytes. In contrast, another randomized, double-blind, crossover study comparing the effect of MUFA, SFA, and trans-fatty acid diets failed to show any significant effects on insulin sensitivity or secretion.[34] When the group members were subdivided according to body mass index (BMI), insulin sensitivity was 24% lower in overweight individuals (BMI >25 kg/m^2) after the SFA diet compared to the MUFA diet. It is interesting to note that these diets were very low in fat (28% energy), Therefore the effects of dietary fat composition may be more obvious without the background low fat diet.

Overall, human dietary intervention studies suggest that the removal of dietary saturated fat, as verified by alterations in plasma fatty acid composition, can have a direct effect on insulin sensitivity. This effect has also been confirmed in a metabolic study that showed that altering the compositions of infused free fatty acids affected insulin sensitivity. A SFA-rich lipid infusion significantly reduced insulin sensitivity indices (40 to 50%) to a much greater extent than a PUFA-rich lipid infusion (20 to 27%) in healthy subjects.[35]

Recent studies have attempted to determine whether alterations in dietary fat intake or endogenous fatty acid synthesis accounted for altered fatty acid composition associated with metabolic syndrome. Cross-sectional data suggest that insulin-resistant states are associated with high levels of activity of stearoyl-CoA desaturates (SCD-1) and Δ6-desaturase (D6D) and low Δ5-desaturase (D5D) activity.[36]

In a prospective study, Warensjo et al.[37] evaluated serum cholesteryl ester fatty acid composition and estimated SCD-1, D6D, and D5D activities as precursors to fatty acid ratios. The study showed that baseline fatty acid profiles predicted the development of metabolic syndrome 20 years later, SFA levels were significantly higher and linoleic acid levels were lower in subjects who subsequently developed metabolic syndrome by age 70. In addition SCD-1 and D6D activities were significantly higher and D5D activity was lower in those who developed metabolic syndrome during follow-up. The clinical relevance of altered desaturase activity in the development of the metabolic syndrome requires further study.

Long chain n-3 PUFAs have a number of positive health benefits relevant to metabolic syndrome, particularly with respect to TAG metabolism.[36] However, we have relatively little evidence that n-3 PUFA supplementation improves insulin sensitivity in humans despite several studies showing that feeding n-3 PUFA had positive effects on glucose and insulin metabolism in different animal models of T2DM and metabolic syndrome.[39,40]

Also human epidemiological studies suggest that habitual dietary fish intake is inversely associated with the incidence of impaired glucose tolerance and T2DM.[41,42] Some studies have reported positive effects of n-3 PUFA supplementation on insulin sensitivity in individuals with impaired glucose tolerance and diabetes.[43,44] Other studies have not shown positive effects[32,45] even though n-3 PUFA supplementation improved TAG metabolism. Clearly the putative effects of n-3 PUFA on human insulin resistance and impaired glucose tolerance require further clarification.

FUTURE PERSPECTIVES

The increasing prevalence of obesity requires us to reduce the impact of the adverse health effects, particularly T2DM and CVD. By 2010, some 31 million people in Europe and an estimated 239 million worldwide will require treatment for T2DM and its related complications.[46,47] The incidence of the metabolic syndrome is increasing exponentially as a consequence of the sharp rise in obesity — the key etiological factor in the development and severity of metabolic syndrome. To date, public health strategies have been largely unsuccessful at reducing the prevalence of obesity. Therefore dietary interventions that attenuate the severity of metabolic syndrome within the context of obesity are required and attenuating the impact of environmental causes through dietary fatty acid modification to improve insulin sensitivity would be advantageous.

REFERENCES

1. Alberti, K., Zimmet, P., and Consultation, W., Definition, diagnosis and classification of dibetes mellitus and its complications. Part 1: diagnosis and classification of diabetes mellitus, provisional report of a WHO consultation. *Diabetic Med.* 15, 539, 1998.
2. Executive Summary, Third Report of the National Cholesterol Education Programme (NCEP) Expert Panel on Detection, Evaluation and Treatment of High Blood Cholesterol in Adults (Adult Treatment Panel III). *JAMA* 285, 2486, 2001.
3. Roche, H.M., Fatty acids and the metabolic syndrome. *Proc. Nutr. Soc.* 64, 23, 2005.
4. Magliano, J.D., Shaw, J.E., and Zimmet, P.Z., How best to define the metabolic syndrome. *Ann. Med.* 58, 34, 2006.
5. Phinney, S.D., Fatty acids, inflammation and the metabolic syndrome. *Am. J. Clin. Nutr.* 82, 1151, 2005.
6. Phillips, C.P. et al., Genetic and nutrient determinants of the metabolic syndrome, *Curr. Op. Cardiol.* 21, 185, 2006.
7. Kahn, B.B. and Flier, J.S., Obesity and insulin resistance. *J. Clin. Invest.* 106, 473, 2000.
8. Roche, H.M., Phillips, C., and Gibney, M.J., The metabolic syndrome: the crossroads of diet and genetics. *Proc. Nutr. Soc.* 64, 371, 2005.
9. Saltiel, A.R., The molecular and physiological basis of insulin resistance: emerging implications for metabolic and cardiovascular diseases. *J. Clin. Invest.* 106, 163, 2000.
10. Ghamin, H., Circulating mononuclear cells in the obese are in a proinflammatory state. *Circulation* 110, 1564, 2004.
11. Dandona, P., Ajada, A., and Bandyopadhyay, A., Inflammation: the link between insulin resistance, obesity and diabetes. *Trends Immunol.* 25, 4, 2004.
12. Roche, H.M., Dietary lipids and gene expression. *Biochem. Soc. Trans.* 32, 999, 2004.
13. Weisberg, S.P. et al., Obesity is associated with macrophage accumulation in adipose tissue. *J. Clin. Invest.* 112, 1796, 2004.

14. Xu, H. et al., Chronic inflammation in fat plays a crucial role in the development of obesity-related insulin resistance. *J. Clin. Invest.* 112, 1821, 2004.

15. Griffin, M.E. et al., Free fatty acid-induced insulin resistance is associated with activation of protein kinase C-theta and alterations in the insulin signaling cascade. *Diabetes* 48, 1270, 1999.

16. Shulman, G.I., Cellular mechanisms of insulin resistance. *J. Clin. Invest.* 106, 171, 2000.

17. Le Marchand-Brustel, Y. et al., Fatty acid-induced insulin resistance: role of insulin receptor substrate 1 serine phosphorylation in the retroregulation of insulin signalling. *Biochem. Soc. Trans.* 31, 1152, 2003.

18. Wellen, K.E. and Hotamisigil, G.S., Inflammation, stress and diabetes. *J. Clin. Invest.* 115, 1111, 2005.

19. Senn, J.J., et al., Suppressor of cytokine signaling-3 (SOCS-3), a potential mediator of interleukin-6-dependent insulin resistance in hepatocytes. *J. Biol. Chem.* 278, 13740, 2003.

20. Steppan, C.M. et al., The hormone resistin links obesity to diabetes. *Nature* 409, 307, 2001.

21. Hotamisligil, G.S., Inflammatory pathways and insulin action. *Int. J. Obesity Rel. Metab Disord.* 27, S53, 2003.

22. Peraldi, P. and Spiegelman, B., TNF-alpha and insulin resistance: summary and future prospects. *Mol. Cell. Biochem.* 182, 169, 1998.

23. Uysal, K.T., Weisbrock, S.M., and Hotamisligil, G.S., Functional analysis of tumor necrosis factor (TNF) receptors in TNF-alpha-mediated insulin resistance in genetic obesity. *Endrocrinology* 139, 4832, 1998.

24. Kim, K.H. et al., A cysteine-rich adipose tissue-specific secretory factor inhibits adipocyte differentiation. *J. Biol. Chem.* 276, 11252, 2001.

25. Hirosumi, J. et al., A central role for JNK in obesity and insulin resistance. *Nature* 420, 333, 2002.

26. Ajuwon, K.M. and Spurlock, M.E., Palmitate activates the NF-κB transcription factor and induces IL-6 and TNFα expression in 3T3-L1 adipocytes. *J. Nutr.* 135, 1841, 2005.

27. Nguyen, M.T. et al., JNK and tumor necrosis factor-α mediate free fatty acid-induced insulin resistance in 3T3-L1 adipocytes. *J. Biol. Chem.* 280, 35361, 2005.

28. Roche, H.M. et al., Isomer-dependent metabolic effects of conjugated linoleic acid (CLA): insights from molecular markers SREBP-1c and LXRα. *Diabetes* 51, 2037, 2002.

29. Moloney, F. et al., Anti-diabetic effects of cis-9, trans-11 conjugated linoleic acid may be mediated via anti-inflammatory effects in white adipose tissue. *Diabetes* 56(3), 574, 2007.

30. Laaksonen, D.E. et al., Serum fatty acid composition predicts development of impaired fasting glycaemia and diabetes in middle-aged men. *Diabetes Med.* 19, 456, 2002.

31. Tanasescu, M. et al., Dietary fat and cholesterol and the risk of cardiovascular disease among women with type 2 diabetes. *Am. J. Clin. Nutr.* 79, 99, 2004.

32. Vessby, B. et al., Substituting dietary saturated for monounsaturated fat impairs insulin sensitivity in healthy men and women: the KANWU study. *Diabetologia* 44, 312, 2001.

33. Perez-Jimenez, E. et al., A Mediterranean and a high-carbohydrate diet improve glucose metabolism in healthy young persons. *Diabetologia*, 44, 2038, 2001.

34. Lovejoy, J.C. et al., Effects of diets enriched in saturated (palmitic), monounsaturated (oleic), or trans (eladic) fatty acids in insulin sensitivity and substrate oxidation in healthy adults. *Diabetes Care* 25, 1283, 2002.

35. Stefan, N. et al., Effect of the pattern of elevated free fatty acids on insulin sensitivity and insulin secretion in healthy humans. *Hormone Metab. Res.* 33, 432, 2001.

36. Vessby, B. et al., Desaturation and elongation of fatty acids and insulin action. *Ann. NY Acad. Sci.* 967, 183, 2002.

37. Warensjo, E., Riserus, U., and Vessby, B., Fatty acid composition of serum lipids predicts the development of the metabolic syndrome in men. *Diabetologia,* 48, 1999, 2005.

38. Roche, H.M. and Gibney, M.J., The effect of long-chain n-3 PUFA on fasting and postprandial triacylglycerol metabolism. *Am. J. Clin. Nutr.* 71, 232, 2000.

39. Storlien, L.H. et al., Fish oil prevents insulin resistance induced by high fat feeding in rats. *Science* 237, 885, 1987.

40. Aguilera, A.A. et al., Effects of fish oil on hypertension, plasma lipids and tumor necrosis factor-a in rats with sucrose-induced metabolic syndrome. *J. Nutr. Biochem.* 15, 350, 2004.

41. Feskens, E.J. et al., Dietary factors determining diabetes and impaired glucose tolerance: a 20-year follow-up of the Finnish and Dutch cohorts of the Seven Countries Study. *Diabetes Care* 18, 1104, 1985.

42. Feskens, E.J., Bowles, C.H., and Kromhout, D., Inverse association between fish intake and risk of glucose intolerance in normoglycemic elderly men and women. *Diabetes Care,* 14, 935, 1991.

43. Fasching, P. et al., Metabolic effects of fish-oil supplementation in patients with impaired glucose tolerance. *Diabetes,* 40, 583, 1991.

44. Popp-Snijders, C. et al., Dietary supplementation of omega-3 polyunsaturated fatty acids improves insulin sensitivity in non-insulin-dependent diabetes. *Diabetes Res.* 4, 141, 1987.

45. Brady, L.M. et al., Increased n-6 polyunsaturated fatty acids do not attenuate the effects of long-chain n-3 polyunsaturated fatty acids on insulin sensitivity or triacylglycerol reduction in Indian Asians. *Am. J. Clin. Nutr.* 79, 983, 2004.

46. King, H., Aubert, R.E., and Herman, W.H., Global burden of diabetes 1995–2025. *Diabetes Care* 21, 1414, 1998.

47. Zimmet, P., Alberti, K.G.M.M., and Shaw, J., Global and societal implications of the diabetes epidemic. *Nature* 414, 782, 2001.

16 Lipid-Induced Death of Macrophages: Implication for Destabilization of Atherosclerotic Plaques

Oren Tirosh and Anna Aronis

CONTENTS

ABSTRACT

In recent decades, the incidence of atherosclerosis in Western society has been on the rise. Lipid particles, of which low density lipoprotein (LDL) is the most studied, accumulate in the intima of blood vessels and lead to inflammatory processes. Leukocytes, mostly macrophages, are recruited from circulating blood to the vessel walls. After they are overexposed to lipids and transform to foam cells, cell death processes take place. In macrophages, the exposure to lipids is known to trigger cell deaths of both apoptotic and necrotic types. External, receptor-dependent, and internal mitochondria-affecting apoptotic death pathways are involved in these processes. Although oxidized LDL is known as a major factor for plaque formation, triacylglycerols (TGs) are increasingly believed to be independently associated with coronary heart disease. Accumulation of intracellular TGs causes elevation of reactive oxygen species (ROS), probably of mitochondrial sources, and cell death. In this chapter, the role of macrophage cell death in atherogenic plaque instability and the specific effects of TG in macrophage lipotoxicity are discussed.

INTRODUCTION

Coronary heart disease (CHD) is the largest cause of morbidity and mortality in the Western world.[1] Three cellular components of the circulation, monocytes, platelets, and T lymphocytes, together with two cell types of the artery wall cells, endothelial and smooth muscle cells (SMC), interact in multiple ways in concert with lipoprotein particles in generating atherosclerotic lesions. Accumulation and oxidation of LDL in vessel walls promotes up-regulation of adhesion molecules in endothelial cells and leads to recruitment of blood monocytes and lymphocytes to the intima.[2]

As shown by epidemiological and clinical studies, the Western way of life and consumption of an unbalanced diet lead to a high incidence of atherosclerosis and conditions significantly increasing the risk of CHD such as obesity, metabolic syndrome, and diabetes.[3] An atherogenic shift in blood lipid spectrum profile that may manifest by increases of serum levels of cholesterol, LDLs, TGs, and very low density lipoproteins (VLDLs), and decreases of serum levels of anti-atherogenic HDLs has become a frequent problem in developed countries.

Recruitment of immune cells to the intima of blood vessels indicates an inflammation reaction caused by lipid particles. Indeed, obese persons have increased serum or plasma concentrations of acute-phase proteins or pro-inflammatory cytokines such as C-reactive protein (CRP), interleukin (IL)-6, IL-8, and tumor necrosis factor (TNF). Circulating levels of these immune mediators can be lowered by weight loss[4,5] or other plasma lipid-lowering interventions, for example, short-term diet and exercise.[6,7] CRP, which has been reported to be expressed by the liver, macrophages and SMC-like cells, correlates with macrophage accumulation in coronary arteries.[8] Therefore, a reduction of CRP by lipid-lowering measures can contribute to prevention of macrophage accumulation in atherosclerotic plaques.

LDL cholesterol is currently defined as a major risk factor of CHD. However, the role of TG, which is increasingly believed to be independently associated with CHD, is not yet well studied. A high blood level of TG often occurs together with low HDL. Such an effect on the blood lipid profile often occurs with normal levels of LDL-C. These abnormalities of the TG-HDL axis are characteristic of patients with metabolic syndrome.[9] Recent clinical studies have also revealed that increased serum triglyceride (TG) levels are closely related to atherosclerosis independently of serum levels of HDL and LDL.

Among TG-rich lipoproteins (TRLs), remnant lipoproteins (RLPs) are considered atherogenic and independent coronary risk factors. It was previously reported[10] that monocytes cultured in the presence of RLPs increased their adhesion capability to vascular endothelial cells. In the Third Report of the Adult Treatment Panel (ATP III) establishing the National Cholesterol Education

Program (NCEP) in the United States, the attitude toward elevated plasma triglyceride levels has changed, especially as related to moderately elevated and borderline high levels. TGs were cited as independent risk factors for CHD.

MACROPHAGE CELL DEATH

PROGRAMMED CELL DEATH

Apoptosis is often associated with morphological and biochemical changes.[11,12] During apoptosis, the nucleus and cytoplasm are condensed and the dying cells disintegrate into membrane-bound apoptotic bodies. Nucleases are activated and cause the degradation of chromosomal DNA, at first into large chromosomal DNA (50 to 300 kb) and ultimately into very small oligonucleosomal fragments.[11,12] The signaling events leading to apoptosis can be divided into two distinct pathways involving either mitochondria or death receptors.[13,14]

In the mitochondrial pathway, death signals led to changes in mitochondrial membrane permeability and the subsequent release of pro-apoptotic factors involved in various aspects of apoptosis.[13] The released factors included cytochrome c (cyt c),[15] apoptosis inducing factor (AIF),[16] second mitochondria-derived activator of caspase (Smac/DIABLO),[17,18] and endonuclease G.[19] Cytosolic cyt c forms an essential part of the *apoptosome* apoptosis complex composed of cyt c, Apaf-1, and procaspase-9. Formation of the apoptosome leads to the activation of caspase-9, which then processes and activates other caspases to orchestrate the biochemical executions of cells. Smac/DIABLO is also released from the mitochondria along with cyt c during apoptosis, and it functions to promote activation of caspases by inhibiting IAP (inhibitor of apoptosis) family proteins.[17,18]

In the death receptor pathway, the apoptotic events are initiated by engaging the tumor necrosis factor (TNF) family receptors including eight different death domain (DD)-containing receptors (TNFR1 also called DR1; Fas, also called DR2; DR3, DR4, DR5, DR6, NGFR, and EDAR).[20,21] Upon ligand binding or when overexpressed in cells, DD receptors aggregate, resulting in the recruitment of various adapter proteins that mediate both cell death and proliferation[20,21] and can activate the caspase system. The human genome encodes 12 to 13 distinct caspases that function in cytokine processing and inflammation, and at least seven (caspases 2, 3, 6, 7, 8, 9, and 10) contribute to cell death.[22] To date more than 280 caspase targets have been identified.[23] Some of these proteins may be cleaved very late and less completely during apoptosis or may not be cleaved in all cell types. The functional consequences of the cleavage of many of the identified substrates are unknown.[23] Most of the protein substrates can be categorized into a few functional groups: apoptosis regulator, cell adhesion, cytoskeletal and structural, etc. Rucci et al. noted that proteins of the electron transfer chains of mitochondria are targets for caspase-dependent degradation during apoptosis.[24]

Active caspase-9 and caspase-3 have been observed in the mitochondria, but their origins are unclear. Theoretically, procaspase-9 may be activated in the mitochondria in a cytochrome c/Apaf-1-dependent manner, or activated caspase-9 and -3 may translocate to the mitochondria as suggested by Chandra and Tang.[25] Using a system of positive staining of the mitochondria with calcein and cobalt as a fluorescent quencher, Poncet et al. showed that during induction of apoptosis, the inner membrane of the mitochondria losses its barrier function and becomes permeable.[26] These data suggest that caspase activity can translocate to the mitochondrial matrix. Indeed strong support for this idea is based on detection of mitochondrial caspase activity in real time, *in situ*, in live cells. Using a mitochondrially targeted CFP-caspase 3 substrate-YFP construct (mC3Y), a caspase-3-like activity in the mitochondrial matrices of some cells during apoptosis was demonstrated.[27]

CELL DEATH AND ATHEROSCLEROSIS

One of the earliest events in atherosclerosis is the entry of monocytes into focal areas of the arterial subendothelium. Once inside the arterial intima, monocytes differentiate into macrophages.[28] Macrophages are constantly exposed to excess lipids following overnutrition and they are also involved in the development of atherosclerosis — an inflammatory disease process.[29-31] Macrophages are considered to constitute one of the most important factors in its initiation and progression. Macrophages accumulate large amounts of intracellular cholesterol via accumulation of lipoproteins. The presence of cholesterol-loaded macrophages in atherosclerotic lesions is a prominent feature throughout the lives of the lesions and these cells exert major impacts on lesion progression.

Dead macrophages are frequently observed in human atherosclerotic lesions, and are considered to be involved in atherosclerotic plaque instability.[32] Unstable plaques demonstrate a greater portion of apoptotic cells than stable ones.[33,34] Immunohistochemical staining of ruptured plaques has shown that apoptotic nuclei in plaque rupture sites are essentially those of macrophages and much less frequently of smooth muscle and T lymphocytes.[33] Fibrous caps of ruptured plaques have more macrophages and also contain fewer SMCs than those of unruptured ones. The death of macrophages and SMCs in blood vessels is a complicated process owing to their mutual interactions and the presence of monocytes and macrophages in plaques increases also the rate of SMC apoptosis.[35]

Accumulated evidence suggests that plaque macrophage deaths are due to three possible pathways: (1) Fas-mediated death and promotion of inflammation, (2) oxidized LDL which at least *in vitro* was proven to be an effective cell death inducer following cellular uptake, and (3) excess accumulation of free cholesterol that may lead to caspase-dependent or independent cell death. The level of apoptotic cell death is strongly related to the stage of development of the atherosclerotic plaque.[36] It has been suggested that in the early stages of lesions, the induction of macrophages apoptosis actually prevents the growth and development of the lesion. This effect is probably attributed to controlling the number of macrophages in the lesion. The experimental evidence of this effect of apoptosis includes a number of genetic alterations in mouse models of atherosclerosis that resulted in early increases or decreases of macrophage apoptosis. An inverse relationship was found between early lesional macrophage apoptosis and lesion area.

When bone marrow taken from P53–/– (a pro-apoptotic protein) mice was transplanted to APOE*3-Leiden mice, a significant 2.3-fold increase was observed in early lesion size. When bone marrow of Bax–/– mice was transplanted to LDLR–/– mice, the same effect was observed, indicating a negative effect of apoptosis on plaque growth. On the other hand, in an advanced atherosclerotic plaque region, the apoptotic processes may lead to massive macrophage cell deaths. Impaired capacities of other macrophages to clean and clear the plaque regions from cell debris may lead to the development of necrotic cores and accelerated inflammatory processes in the intima. Macrophages express multiple metalloproteinases (e.g., stromelysin) and serine proteases (e.g., urokinase) that degrade the extracellular matrix, weakening the plaque and making it rupture-prone.

Most studies report apoptotic macrophage deaths in ruptured plaques on the basis of immunohistochemical staining methods. However, standard TUNEL DNA staining lacks specificity, especially in late phases of cell death when most DNA passes degradation.[36] Moreover, because of variable uptake of lipids and differences in cell size, apoptosis-characterizing cell shrinkage is not always apparent.[33] This is why an indication of the active form of "committing-to-die" caspase-3 is necessary for final determination of apoptotic cell death. Some researchers indicate caspase-3 independent cell deaths in LDL-exposed macrophage cell cultures or histological cuts of plaques,[33,37] supposing an alternative of oncotic (necrotic) cell death.

EFFECTS OF TGS ON MACROPHAGE CELL DEATHS

Previous studies have suggested that TGs are the main signaling components of endocytosed VLDL in macrophages.[38] Our study showed death pathways in macrophages resulting from exposure to TGs — a mechanism that may be relevant to the development of atherosclerosis. Murine J774.2 macrophages in culture were exposed to a commercial soybean oil-based TG lipid emulsion (0.1 to 1.5 mg lipid/ml) known to be taken up by macrophages via a coated pit-dependent mechanism mediated by macrophage secretion of apolipoprotein E.[40] The exposure of the cells to the lipid emulsion led to intracellular change in fatty acid profile and high accumulation of TGs as measured with gas chromatography and Nile red staining, respectively (Figure 16.1). The TG effect culminated in cell death, with no caspase-3 activation. Dual staining with propidium iodide and annexin V followed by flow cytometric analysis showed that TG facilitated cell deaths with clear necrotic characteristics.

Reactive oxygen species (ROS) are involved in macrophage activation and death. While ROS may induce macrophage activation, macrophage foam cells contain potent oxidant-generating lipid-targeting systems such as inducible NO

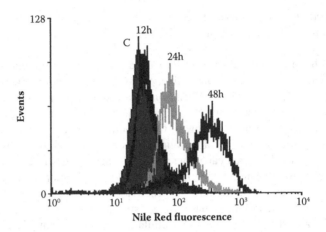

FIGURE 16.1 Accumulation of intracellular lipids following exposure of J774.2 macrophages to lipid treatment. Nile red staining of the cells was carried out after treatment with soybean oil-based lipid emulsion, 1 mg/ml.

synthase and 15-lipoxygenase, allowing increased recognition and uptake by macrophages and creating a positive feedback loop.[41]

Following 24 hr of exposure of macrophages to 1 mg/ml TGs, cellular ROS levels were strongly elevated. In contrast, after 48 hr, when 50% of the macrophages underwent cell death, ROS production was arrested. Most of the TG-mediated ROS production was demonstrated to be via mitochondrial complex 1 of the electron transfer chain, as demonstrated by the use of rotenone, a mitochondrial complex 1 inhibitor that significantly attenuated cellular ROS levels in TG-treated cells (Scheme 16.1).

To elucidate whether the cell death process was indeed oxidant dependent, antioxidant protection was studied. Treatment with 0.5 mM N-acetyl-cysteine (NAC), 0.05 mM ascorbic acid, and 0.2 mM resveratrol protected against the TG lipotoxic effect, while lipophilic antioxidants surprisingly did not. For further study, the combination of NAC, ascorbic acid, and resveratrol was used at much lower concentrations (one-tenth of original concentrations), which led to the appearance of a synergistic protective effect.

Exposure of J774.2 macrophages to increased levels of TG leads to the induction of oxidative stress-mediated lipotoxicity. Decreased GSH levels and the protective role of NAC are strong indications of the pivotal role of ROS in TG-induced lipotoxicity. In our model, the NAC, ascorbic acid, and resveratrol antioxidants played a protective role in TG-induced lipotoxicity. The protective effect of NAC, which is a precursor of glutathione, suggests an interaction in the level of water-soluble antioxidants. Moreover, TG caused ROS-dependent G1 arrest in the macrophage cell cycle. The exposure of untreated cells to an inhibitor of glutathione synthesis BSO mimicked TG-induced G1 arrest, reinforcing an important role of GSH in prevention of TG-induced mechanisms of lipotoxicity.

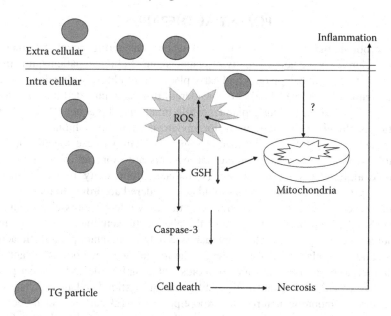

SCHEME 16.1 Mechanism for TG-induced cell death in macrophages. Lipid droplets accumulate inside the macrophages, in time generating foam cells. Oxidative stress develops as a result of the lipid stress and leads to a necrotic type of cell death. Overall, the effect of macrophage disintegration in atherogenic plaques will result in inflammation and plaque rupture due to destabilization.

It is possible that prolonging ROS production from endogenous cellular sources such as mitochondria may attenuate caspase activity and shift apoptosis to necrosis. We therefore evaluated the effect of TG on apoptotic cells showing high caspase activity. Indeed, TG induced elevated ROS levels and suppressed caspase-3 in apoptotic cells pretreated for 24 hr with cycloheximide. These results indicate that exposure to TG can directly regulate lipotoxicity in macrophages by inducing mitochondria-mediated prolonged oxidative stress; this, in turn, can inactivate the apoptotic caspase system, resulting in necrotic cell death which can be prevented by specific antioxidants.

Accumulation of TG in macrophages is a unique phenomenon of these cells since they are capable of taking up large amounts of this type of lipid. We suggest that rapid accumulation of fat may result in oxidative stress-dependent suppression of the caspase system which may suppress activation of type 1 programmed cell death and affect other mechanisms related to cellular signal transduction and cell cycle arrest. The TG treatment may promote characteristics of necrotic cell death or caspase-independent types of programmed cell death. This, in fact, may result in disintegration of lipid-loaded macrophages (foam cells) and release of their enzymatic cytosolic and lysosomal contents to their surroundings, leading to lesion erosion and rupture of the plaques with the formation of a thrombus.

POTENTIAL THERAPIES

One problem that must be encountered when considering novel therapies in atherosclerosis is the heterogeneity of plaque status. In human patients, it is most probable that advanced plaques and early plaques are relatively abundant. Therefore, a careful approach should be considered. It is recommended that apoptosis will be promoted in the early plaques while in advanced plaques where inflammation is already a considerable factor, apoptosis should be inhibited.[42]

One approach for blocking macrophage cell death is to use specific antioxidants. We found that water-soluble antioxidants are more efficient than lipid-soluble antioxidants in preventing TG-induced lipotoxicity in macrophages. Therefore the use of antioxidants should be considered according to their relative death-preventing effects. Only compounds that may affect lipotoxicity should be used. Another approach is to enhance the phagocytic activities of macrophages in atherosclerotic plaques. This approach will help to maintain phagocytic activities in advance plaques and will prevent the formation of necrotic cores and the inflammatory responses. Use of eicosanoids including LXA4, LXB4, and aspirin-triggered 15-epi-LXB4, and their stable analogues stimulated macrophages to phagocytose apoptotic neutrophils. Apoplipoprotein E3 can correct defects in macrophage phagocytic effects and finally the use of glucocorticoids may boost phagocytic activity.[42]

REFERENCES

1. Shah, P.K. (2003) Pathophysiology of plaque rupture and the concept of plaque stabilization, *Cardiol. Clin.* 21, 303.
2. Osterud, B. and Bjorklid, E. (2003) Role of monocytes in atherogenesis, *Physiol. Rev.* 83, 1069.
3. Gasbarrini, A. and Piscaglia, A.C. (2005) A natural diet versus modern Western diets? A new approach to prevent "well-being syndromes," *Dig. Dis. Sci.* 50, 1.
4. Herder, C. et al. (2006) Systemic monocyte chemoattractant protein-1 concentrations are independent of type 2 diabetes or parameters of obesity: results from the Cooperative Health Research in the Region of Augsburg Survey S4 (KORA S4), *Eur. J. Endocrinol.* 154, 311.
5. Heilbronn, L.K. and Clifton, P.M. (2002) C-reactive protein and coronary artery disease: influence of obesity, caloric restriction and weight loss, *J. Nutr. Biochem.* 13, 316.
6. Roberts, C.K. et al. (2006) Effect of a short-term diet and exercise intervention on oxidative stress, inflammation, MMP-9, and monocyte chemotactic activity in men with metabolic syndrome factors, *J. Appl. Physiol.* 100, 1657.
7. Kasim-Karakas, S.E., Tsodikov, A., Singh, U., and Jialal, I. (2006) Responses of inflammatory markers to a low-fat, high-carbohydrate diet: effects of energy intake, *Am. J. Clin. Nutr.* 83, 774.
8. Turk, J.R. et al. (2003) C-reactive protein correlates with macrophage accumulation in coronary arteries of hypercholesterolemic pigs, *J. Appl. Physiol.* 95, 1301.

9. Szapary, P.O. and Rader, D.J. (2004) The triglyceride-high-density lipoprotein axis: an important target of therapy? *Am. Heart J.* 148, 211.

10. Kawakami, A. and Yoshida, M. (2005) Remnant lipoproteins and atherogenesis, *J. Atheroscler. Thromb.* 12, 73.

11. Steller, H. (1995) Mechanisms and genes of cellular suicide, *Science* 267, 1445.

12. Nagata, S. (1997) Apoptosis by death factor, *Cell* 88, 355.

13. Green, D.R. and Reed, J.C. (1998) Mitochondria and apoptosis, *Science* 281, 1309.

14. Ashkenazi, A. and Dixit, V.M. (1998) Death receptors: signaling and modulation, *Science* 281, 1305.

15. Liu, X., Kim, C.N., Yang, J., Jemmerson, R., and Wang, X. (1996) Induction of apoptotic program in cell-free extracts: requirement for dATP and cytochrome c, *Cell* 86, 147.

16. Susin, S.A. et al. (1999) Molecular characterization of mitochondrial apoptosis-inducing factor, *Nature* 397, 441.

17. Du, C., Fang, M., Li, Y., Li, L., and Wang, X. (2000) Smac, a mitochondrial protein that promotes cytochrome c-dependent caspase activation by eliminating IAP inhibition, *Cell* 102, 33.

18. Verhagen, A.M. et al. (2000) Identification of DIABLO, a mammalian protein that promotes apoptosis by binding to and antagonizing IAP proteins, *Cell* 102, 43.

19. Li, L.Y., Luo, X., and Wang, X. (2001) Endonuclease G is an apoptotic DNase when released from mitochondria, *Nature* 412, 95.

20. Bhardwaj, A. and Aggarwal, B.B. (2003) Receptor-mediated choreography of life and death, *J. Clin. Immunol.* 23, 317.

21. Singh, A., Ni, J., and Aggarwal, B.B. (1998) Death domain receptors and their role in cell demise, *J. Interferon Cytokine Res.* 18, 439.

22. Kroemer, G. and Martin, S.J. (2005) Caspase-independent cell death, *Nat. Med.* 11, 725.

23. Fischer, U., Janicke, R.U., and Schulze-Osthoff, K. (2003) Many cuts to ruin: a comprehensive update of caspase substrates, *Cell Death Differ.* 10, 76.

24. Ricci, J.E. et al. (2004) Disruption of mitochondrial function during apoptosis is mediated by caspase cleavage of the p75 subunit of complex I of the electron transport chain, *Cell* 117, 773.

25. Chandra, D. and Tang, D.G. (2003) Mitochondrially localized active caspase-9 and caspase-3 result mostly from translocation from the cytosol and partly from caspase-mediated activation in the organelle: lack of evidence for Apaf-1-mediated procaspase-9 activation in the mitochondria, *J. Biol. Chem.* 278, 17408.

26. Poncet, D., Boya, P., Metivier, D., Zamzami, N., and Kroemer, G. (2003) Cyto-fluorometric quantitation of apoptosis-driven inner mitochondrial membrane per-meabilization, *Apoptosis* 8, 521.

27. Zhang, Y. et al. (2004) Detection of mitochondrial caspase activity in real time *in situ* in live cells, *Microsc. Microanal.* 10, 442.

28. Linton, M.F. and Fazio, S. (2003) Macrophages, inflammation, and atherosclerosis, *Int. J. Obes. Relat. Metab. Disord.* 27, Suppl. 3, S35.

29. Hansson, G.K., Zhou, X., Tornquist, E., and Paulsson, G. (2000) The role of adaptive immunity in atherosclerosis, *Ann. NY Acad. Sci.* 902, 53.

30. Patrick, L. and Uzick, M. (2001) Cardiovascular disease: C-reactive protein and the inflammatory disease paradigm: HMG-CoA reductase inhibitors, alpha-toco-pherol, red yeast rice, and olive oil polyphenols: a review of the literature, *Altern. Med. Rev.* 6, 248.

31. Isomaa, B. (2003) A major health hazard: the metabolic syndrome, *Life Sci.* 73, 2395.

32. Feng, B., Zhang, D., Kuriakose, G., Devlin, C.M., Kockx, M., and Tabas, I. (2003) Niemann-Pick C heterozygosity confers resistance to lesional necrosis and macrophage apoptosis in murine atherosclerosis, *Proc. Natl. Acad. Sci. USA* 100, 10423.

33. Kolodgie, F.D. et al. (2000) Localization of apoptotic macrophages at the site of plaque rupture in sudden coronary death, *Am. J. Pathol.* 157, 1259.

34. Moreno, P.R., Falk, E., Palacios, I.F., Newell, J.B., Fuster, V., and Fallon, J.T. (1994) Macrophage infiltration in acute coronary syndromes: implications for plaque rupture, *Circulation* 90, 775.

35. Seshiah, P.N. et al. (2002) Activated monocytes induce smooth muscle cell death: role of macrophage colony-stimulating factor and cell contact, *Circulation* 105, 174.

36. Kockx, M.M. (1998) Apoptosis in the atherosclerotic plaque: quantitative and qualitative aspects, *Arterioscler. Thromb. Vasc. Biol.* 18, 1519.

37. Baird, S.K., Reid, L., Hampton, M.B., and Gieseg, S.P. (2005) OxLDL-induced cell death is inhibited by the macrophage synthesised pterin, 7,8-dihydroneopterin, in U937 cells but not THP-1 cells, *Biochim. Biophys. Acta* 1745, 361.

38. Chawla, A. et al. (2003) PPARdelta is a very low-density lipoprotein sensor in macrophages, *Proc. Natl. Acad. Sci. USA* 100, 1268.

39. Aronis, A., Madar, Z., and Tirosh, O. (2005) Mechanism underlying oxidative stress-mediated lipotoxicity: exposure of J774.2 macrophages to triacylglycerols facilitates mitochondrial reactive oxygen species production and cellular necrosis, *Free Radic. Biol. Med.* 38, 1221.

40. Carvalho, M.D. et al. (2002) Macrophages take up triacylglycerol-rich emulsions at a faster rate upon co-incubation with native and modified LDL: investigation on the role of natural chylomicrons in atherosclerosis, *J. Cell. Biochem.* 84, 309.

41. Bennett, M.R. (2001) Reactive oxygen species and death: oxidative DNA damage in atherosclerosis, *Circ. Res.* 88, 648.

42. Tabas, I. (2005) Consequences and therapeutic implications of macrophage apoptosis in atherosclerosis: the importance of lesion stage and phagocytic efficiency, *Arterioscler. Thromb. Vasc. Biol.* 25, 2255.

17 α-Lipoic Acid Prevents Diabetes Mellitus and Endothelial Dysfunction in Diabetes-Prone Obese Rats

Woo Je Lee, Ki-Up Lee, and Joong-Yeol Park

CONTENTS

INTRODUCTION

Recent evidence suggests that oxidative stress plays a causative role in many chronic diseases such as diabetes and cardiovascular disease. Many efforts have been made to prevent the development and progression of diabetes and cardiovascular disease by using materials with antioxidative properties. Among a variety of antioxidants, α-lipoic acid (α-LA) has recently received much research attention. This chapter will focus on the roles and mechanisms of α-LA in the prevention of diabetes mellitus and endothelial dysfunction in obese animals.

α-LIPOIC ACID

α-LA is a naturally occurring compound that is widely distributed in plants and animals. In humans, α-LA can be supplied by diet or obtained through *de novo* synthesis by the liver and other tissues. Only the *R*-isomer of α-LA is synthesized naturally. Conventional chemical synthesis of α-LA results in a 50/50 (racemic) mixture of two optical isomers, *R*-α-LA and *S*-α-LA.

Free α-LA is rapidly taken up by cells and reduced to dihydrolipoic acid (DHLA) intracellularly. In most cells containing mitochondria, α-LA is reduced by an NADH-dependent reaction with lipoamide dehydrogenase to form DHLA. In cells that lack mitochondria, α-LA can be reduced to DHLA via NADPH with glutathione and thioredoxin reductase.[1]

α-LA contains two sulfur molecules that can be oxidized or reduced (Figure 17.1). This feature allows α-LA to function as a cofactor for several important enzymes as well as a potent antioxidant. α-LA functions as a cofactor of mitochondrial key enzymes such as pyruvate dehydrogenase and α-keto-glutarate dehydrogenase and thus has been shown to be required for the oxidative decarboxylation of pyruvate to acetyl-CoA, the critical step leading to the production of cellular energy (ATP).[2,3] In addition, when large amounts of free α-LA are available (e.g., with supplementation), α-LA is also able to function as an antioxidant.[4] α-LA and its reduced form, DHLA, are powerful antioxidants.[5] DHLA can then be transported easily out of the interiors of cells and function effectively in the extracellular spaces. Thus, in this form, it is believed to function directly as an antioxidant and possess its greatest antioxidant potential.[6] However, because DHLA is also rapidly eliminated from cells, the extents to which its antioxidant effects can be sustained remain unclear.

Many reports describe the functions of α-LA[4,5,7–10] and the functions include (1) quenching of reactive oxygen species, (2) regeneration of exogenous and endogenous antioxidants such as vitamins C and E and glutathione, (3) chelation of metal ions, and (4) prevention of membrane lipid peroxidation and reparation of oxidized proteins. Because of its antioxidant properties, α-LA is known to be effective in both prevention and treatment of oxidative stress in a number of

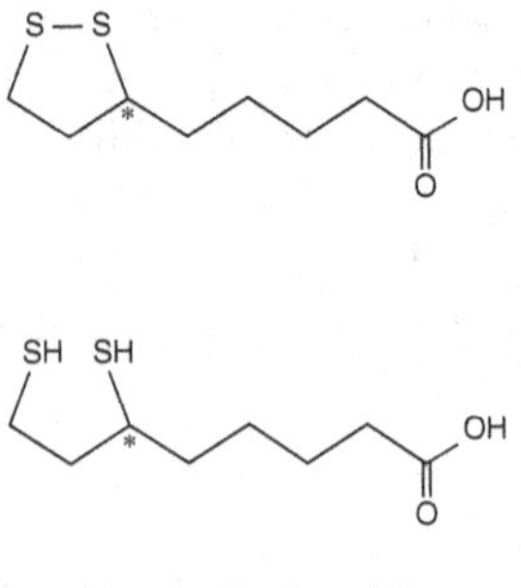

FIGURE 17.1 Structures of lipoic acid and dihydrolipoic acid.

models or clinical conditions including diabetes,[11–14] vascular dysfunction,[15] and neurodegenerative diseases.[16]

α-LIPOIC ACID PREVENTS DIABETES MELLITUS IN DIABETES-PRONE OBESE RATS

Increased oxidative stress is a widely accepted participant in the development and progression of diabetes and its complications.[17–19] Diabetes is usually accompanied by increased production of free radicals[18–21] or impaired antioxidant defenses.[22,23] Hence, it is likely that a substance known to reduce oxidative stress would reduce the progression of cell damage in diabetes.

α-LA, an essential cofactor in oxidative metabolism, has been reported to exert a number of potentially beneficial effects in oxidative stress-related conditions. Experimental and clinical studies have indicated that α-LA treatment reduces the development of diabetic complications.[7,24–29] Furthermore, α-LA facilitates glucose metabolism and increases glucose uptake leading to improved glucose utilization *in vitro* and *in vivo*.[30] These features may enable α-LA to be used as a potential agent for the prevention and treatment of diabetes mellitus.

Experimental and clinical studies have indicated that α-LA improved insulin sensitivity. α-LA administration improved insulin-stimulated whole body and skeletal muscle glucose utilization in insulin-resistant obese rats[30,31] and type 2 diabetic humans.[32,33] Moreover, several studies using cultured muscle cells,[34] isolated rat diaphragm,[35] and perfused preparations of normal and diabetic rabbits[36] demonstrated that α-LA activates glucose transport and stimulates glycolysis.

Obesity is the most important risk factor for type 2 diabetes mellitus.[37] Many possible mechanisms link obesity and the development of diabetes mellitus. One mechanism is that oxygen free radicals derived from excessive triglycerides and long chain fatty acyl CoA (LCAC) in muscles and pancreatic β cells cause functional defects in these tissues, leading to the development of type 2 diabetes.[38–40] α-LA has been shown to protect against oxidative stress-induced insulin resistance *in vitro*[41,42] and improve insulin sensitivity in human and animal models.[30,43–47] In addition, it was recently reported that administration of α-LA to rodents reduced body weight by suppressing food intake and visceral fat mass.[48] Thus, it is conceivable that α-LA may have a preventive role in the development of diabetes.

A recent study (14) demonstrated that α-LA could prevent the development of diabetes in an obese animal model. The authors used Otsuka Long-Evans Tokushima Fatty (OLETF) rats as obese animal models, because these rats are obese from a young age and become diabetic after 18 weeks of age.[49] These features in OLETF rats resemble those of human type 2 diabetes. To investigate the preventive effect of α-LA on the development of diabetes, 9-week old OLETF rats were fed standard rat chow with or without racemic α-LA (200 mg/kg of body weight/day) for 3 weeks and the development of diabetes was evaluated. Approximately 80% of rats fed rat chow without α-LA developed diabetes at 40 weeks of age, but none of the rats fed rat chow with α-LA developed diabetes. In addition, administration of α-LA protected pancreatic β cell destruction and reduced triglyceride accumulations in skeletal muscles and pancreatic β cells. These data indicate that α-LA prevents the development of diabetes in diabetes-prone obese rats by decreasing lipid accumulation in skeletal muscle and exerting beneficial effects on pancreatic β cells.

This report seems to indicate that the mechanism of preventive effect of α-LA on the development of diabetes may be multifarious. First, because oxidative stress has been suggested to be associated with insulin resistance[50,51] and α-LA has potent antioxidative capacities, its preventive effects may be due to its anti-oxidant action. Indeed, α-LA reduced plasma levels of oxidative stress markers such as malondialdehyde and 8-hydroxy-deoxyguanosine. However, this relationship between protective effect and antioxidant action of α-LA was not investigated directly.

A second possible mechanism of the preventive role of α-LA is its effects on lipids in skeletal muscles and pancreatic β cells. Skeletal muscle is the major tissue responsible for 70 to 80% of whole body glucose uptake. Indeed, insulin resistance in skeletal muscle is a common characteristic of type 2 diabetes, and many previous studies in animal models and human subjects have shown that lipid accumulation in skeletal muscle is associated with insulin resistance in obesity. In addition, lipid accumulation in pancreatic β cells also affects β cell damage. LCAC accumulation in pancreatic β cells may induce apoptosis of the cells. Increased plasma free fatty acid by lipid–heparin infusion induced β cell hypertrophy. Since α-LA reduces visceral fat mass and decreases plasma free fatty acid and triglyceride concentrations and tissue triglyceride contents in OLETF rats, insulin resistance in skeletal muscle and β cell apoptosis can be prevented.

While the possible mechanisms of the effects of α-LA on the development of diabetes in obese rats could be inferred from this report, we cannot exclude the possibility that α-LA prevented the development of diabetes due to its weight reducing effects. In addition, the exact molecular mechanism by which α-LA reduces lipid contents in skeletal muscle was not investigated in this report. Thus, other studies[15,52] were performed to minimize the effect of α-LA on body weight change and investigate the mechanism of α-LA in reducing lipid content in skeletal muscle. The authors fed rats for a short period (200 mg/kg body weight) via intravenous infusion for 2 hr or 0.5% (wt/wt) racemic α-LA (mixed in food for 3 days) to minimize the effect of α-LA on body weight change. In addition, since AMPK was known to be responsible for the effect of α-LA in the hypo-thalamus and because it increases both glucose uptake and fatty acid oxidation, the authors focused on AMPK as a molecular target of α-LA. AMPK is known as a cellular "energy sensor" because its activity is sensitively changed by its cellular energy state. AMPK controls a number of metabolic processes to help restore energy depletion in peripheral tissues.[53,54] For example, AMPK activation in skeletal muscle enhanced glucose uptake and mitochondrial fatty acid oxidation.[55] AMPK is also expressed in vascular endothelium,[56] and AMPK dysregulation has been suggested to contribute to endothelial dysfunction.[57]

Administration of α-LA to OLETF rats improved insulin-stimulated whole body glucose uptake and whole body glycogen synthesis. Insulin-stimulated glucose uptake and glycogen synthesis in skeletal muscle were also improved in OLETF rats treated with α-LA. In skeletal muscles of OLETF rats, AMPK (α2-AMPK, not α1-AMPK) activity was decreased compared to control rats. The reason for the decreased AMPK activity is not yet known. However, because AMPK is enzyme activated when the cellular energy state is low, abundant fuels such as elevated plasma free fatty acid and/or glucose in OLETF rats may reduce AMPK activity.

Administration of α-LA increased α2-AMPK and fatty acid oxidation and decreased triglyceride contents in skeletal muscles of OLETF rats. Overexpression of the dominant negative α2-AMPK gene in skeletal muscles of OLETF rats reversed the effects of α-LA on triglyceride contents, fatty acid oxidation, and insulin-stimulated glucose uptake, suggesting that α-LA exerted its effects via AMPK activation. Because AMPK is known to increase fatty acid oxidation and because lipid accumulation in muscle is associated with insulin resistance, we could deduce from this report that α-LA improved insulin sensitivity by activating AMPK and increasing fatty acid oxidation, and subsequently by reducing lipid accumulation in skeletal muscle. However, further study is necessary to explain the precise molecular mechanism by which α-LA prevents the development of diabetes and improves insulin sensitivity.

α-LIPOIC ACID PREVENTS ENDOTHELIAL DYSFUNCTION IN DIABETES-PRONE OBESE RATS

Oxidative stress and impaired bioactivity of vascular nitric oxide (NO) play important roles in the pathogenesis of microvascular and macrovascular complications in diabetes mellitus. Antioxidative properties and the capacity of increasing NO synthesis and activity may contribute to the protective role of α-LA in the development of vascular complications in diabetes. It was reported that O_2^- production was increased in aortic tissues of insulin-resistant rats,[43,58] and α-LA supplementation in chronically glucose-fed rats prevents the increase in aortic basal O_2^- production. In addition, α-LA treatment prevented increases of thiobarbituric acid-reactive substances (indirect evidence of intensified free radical production and uniformly increased substances in streptozotocin (STZ)-induced diabetic rats).[27,28]

α-LA improved NO-mediated vasodilation in diabetic patients but not in healthy control subjects. This effect seems to be associated with antioxidative properties of α-LA because the effects of α-LA were positively related to plasma levels of malondialdehyde.[59] In an in vitro study, however, α-LA increased NO synthesis and bioactivity in human aortic endothelial cells by mechanisms that appear to be independent of antioxidative properties, because cellular GSH levels and GSH:GSSG ratios were not significantly changed by α-LA treatment.[60]

α-LA improved impaired endothelium-dependent vasorelaxation in diabetic rats.[61] In STZ-induced diabetic rats, hyperglycemia induces decreased eNOS expression and increased iNOS expression in tissues such as heart, aorta, sciatic nerve, and kidney. α-LA treatment increased eNOS expression and decreased iNOS expression.[62] In addition, the decline in eNOS phosphorylation — a possible mechanism involved in vascular dysfunction in aging — can be partially restored by treating old rats with α-LA.[63]

Moreover, α-LA treatment inhibited adhesion molecule expression in HAEC and TNF-α-induced monocyte adhesion.[64] α-LA also inhibited NF-κB-mediated transcription and expression of endothelial genes such as tissue factor and endothelin-1[65] and reduced advanced glycation end product (AGE)-induced endothelial expression of VCAM-1 and monocyte binding to the endothelium.[66]

In addition to these mechanisms, α-LA may exhibit diverse protective effects in vascular disease. It may lower blood pressure in rats. α-LA supplementation decreased systolic blood pressure in spontaneously hypertensive rats,[67] in high salt-treated rats,[68] and in fructose-induced hypertensive WKY rats.[69] Similarly, hypertension induced by the addition of a 10% D-glucose drink to their diet was attenuated by α-LA in Sprague-Dawley rats.[70]

α-LA has exhibited lipid-lowering capacity. In studies of rabbits[71,72] and Japanese quail,[73] α-LA decreased levels of cholesterol and lipoprotein in serum. It also reduced serum triglyceride levels in STZ-induced diabetic rats.[24]

α-LA has been shown to decrease LDL oxidation. It was reported that DHLA inhibited Cu^{2+}-dependent LDL peroxidation by chelating copper in *in vitro* experiments.[74] In addition, in human studies, α-LA (600 mg/day) significantly increased the lag time of LDL lipid peroxide formation for both copper-catalyzed and 2,2′-azobis (2-amidinopropane) hydrochloride (AAPH)-induced LDL oxidation.[75]

Central obesity is an important risk factor for cardiovascular disease. Endothelial dysfunction, generally considered a prerequisite for atherosclerosis, is a frequent finding in obesity. A recent report indicates that administration of α-LA to patients with metabolic syndrome improved endothelial function[76] but the mechanism is not yet known. It was suggested that lipid accumulation in vascular tissue and a consequent increase in oxidative stress lead to endothelial dysfunction in obesity.[39]

As described in the previous section, α-LA enhances fatty acid oxidation and reduces lipid accumulation by activating AMPK in skeletal muscles of obese rats. Because α-LA produces these effects and because AMPK is expressed in vascular endothelial cells,[77] it is conceivable that α-LA may improve endothelial function in obesity by activating AMPK. One recent study[15] demonstrated that endothelium-dependent vasorelaxation was impaired and AMPK activity in endothelium of obese animals was decreased compared to control (lean) rats. In obese rats, α-LA administration improved the impaired endothelium-dependent vasorelaxation and increased AMPK activity and phosphorylation independent of the effect of α-LA on body weight. In addition, in an *in vitro* study using human aortic endothelial cells (HAECs) treated with a fatty acid (linoleic acid), α-LA prevented linoleic acid-induced decreases in AMPK phosphorylation. This effect was associated with normalization of endothelial apoptosis and ROS generation in the presence of linoleic acid. These results suggest that the mechanism of endothelial dysfunction in obesity is reduced AMPK activity in endothelial cells, and that α-LA may improve endothelial dysfunction in obese rats by activating AMPK in endothelial cells.

SUMMARY

A growing body of evidence suggests that α-LA exerts beneficial effects on both prevention and treatment of oxidative stress in a number of models and clinical conditions including diabetes and vascular dysfunction. However, before α-LA can be used in clinical practice to prevent diabetes or vascular dysfunction, more experimental research to elicit the mechanism and clinical trials to determine the proper dose, formula, and route of administration are needed.

ACKNOWLEDGMENT

This work was supported by National Research Laboratory grant M1040000000804J000000810 from the Ministry of Science and Technology of the Republic of Korea.

REFERENCES

1. Jones, W., Li, X., Qu, Z.C., Perriott, L., Whitesell, R.R., and May, J.M. (2002) Uptake, recycling, and antioxidant actions of alpha-lipoic acid in endothelial cells. *Free Radic Biol Med* 33, 83.
2. Packer, L., Roy, S., and Sen, C.K. (1997) Alpha-lipoic acid: a metabolic antioxidant and potential redox modulator of transcription. *Adv Pharmacol* 38, 79.
3. Reed, L.J. (1998) From lipoic acid to multi-enzyme complexes. *Protein Sci* 7, 220.
4. Biewenga, G.P., Haenen, G.R., and Bast, A. (1997) The pharmacology of the antioxidant lipoic acid. *Gen Pharmacol* 29, 315.
5. Packer, L., Witt, E.H., and Tritschler, H.J. (1995) Alpha-lipoic acid as a biological antioxidant. *Free Radic Biol Med* 19, 227.
6. Kramer, K.P., Hoppe, P., and Packer, L., Eds. (2001) *Nutraceuticals in Health and Disease Prevention*, Marcel Dekker, New York, p. 129.
7. Obrosova, I., Cao, X., Greene, D.A., and Stevens, M.J. (1998) Diabetes-induced changes in lens antioxidant status, glucose utilization and energy metabolism: effect of DL-alpha-lipoic acid. *Diabetologia* 41, 1442.
8. Scott, B.C. et al. (1994) Lipoic and dihydrolipoic acids as antioxidants: a critical evaluation. *Free Radic Res* 20, 119.
9. Packer, L. (1998) Alpha-lipoic acid: a metabolic antioxidant which regulates NF-kappa B signal transduction and protects against oxidative injury. *Drug Metab Rev* 30, 245.

10. Evans, J.L. and Goldfine, I.D. (2000) Alpha-lipoic acid: a multifunctional antioxidant that improves insulin sensitivity in patients with type 2 diabetes. *Diabetes Technol Ther* 2, 401.

11. Suzuki, Y.J., Tsuchiya, M., and Packer, L. (1992) Lipoate prevents glucose-induced protein modifications. *Free Radic Res Commun* 17, 211.

12. Roy, S., Sen, C.K., Tritschler, H.J., and Packer, L. (1997) Modulation of cellular reducing equivalent homeostasis by alpha-lipoic acid: mechanisms and implications for diabetes and ischemic injury. *Biochem Pharmacol* 53, 393.

13. Romero, F.J., Ordonez, I., Arduini, A., and Cadenas, E. (1992) The reactivity of thiols and disulfides with different redox states of myoglobin: redox and addition reactions and formation of thiyl radical intermediates. *J Biol Chem* 267, 1680.

14. Song, K.H. et al. (2005) Alpha-lipoic acid prevents diabetes mellitus in diabetes-prone obese rats. *Biochem Biophys Res Commun* 326, 197.

15. Lee, W.J. et al. (2005) Alpha-lipoic acid prevents endothelial dysfunction in obese rats via activation of AMP-activated protein kinase. *Arterioscler Thromb Vasc Biol* 25, 2488.

16. Packer, L., Tritschler, H.J., and Wessel, K. (1997) Neuroprotection by the metabolic antioxidant alpha-lipoic acid. *Free Radic Biol Med* 22, 359.

17. Ceriello, A. (2000) Oxidative stress and glycemic regulation. *Metabolism* 49, 27.

18. Baynes, J.W. and Thorpe, S.R. (1999) Role of oxidative stress in diabetic complications: a new perspective on an old paradigm. *Diabetes* 48, 1.

19. Baynes, J.W. (1991) Role of oxidative stress in development of complications in diabetes. *Diabetes* 40, 405.

20. Chang, K.C. et al. (1993) Possible superoxide radical-induced alteration of vascular reactivity in aortas from streptozotocin-treated rats. *J Pharmacol Exp Ther* 266, 992.

21. Young, I.S., Tate, S., Lightbody, J.H., McMaster, D., and Trimble, E.R. (1995) The effects of desferrioxamine and ascorbate on oxidative stress in the streptozotocin diabetic rat. *Free Radic Biol Med* 18, 833.

22. Saxena, A.K., Srivastava, P., Kale, R.K., and Baquer, N.Z. (1993) Impaired antioxidant status in diabetic rat liver: effect of vanadate. *Biochem Pharmacol* 45, 539.

23. McLennan, S.V. et al. (1991) Changes in hepatic glutathione metabolism in diabetes. *Diabetes* 40, 344.

24. Ford, I., Cotter, M.A., Cameron, N.E., and Greaves, M. (2001) The effects of treatment with alpha-lipoic acid or evening primrose oil on vascular hemostatic and lipid risk factors, blood flow, and peripheral nerve conduction in the streptozotocin-diabetic rat. *Metabolism* 50, 868.

25. Kishi, Y. et al. (1999) Alpha-lipoic acid: effect on glucose uptake, sorbitol pathway, and energy metabolism in experimental diabetic neuropathy. *Diabetes* 48, 2045.

26. Ziegler, D. and Gries, F.A. (1997) Alpha-lipoic acid in the treatment of diabetic peripheral and cardiac autonomic neuropathy. *Diabetes* 46, Suppl 2, S62.

27. Obrosova, I.G., Fathallah, L., and Greene, D.A. (2000) Early changes in lipid peroxidation and antioxidative defense in diabetic rat retina: effect of DL-alpha-lipoic acid. *Eur J Pharmacol* 398, 139.

28. Kocak, G. et al. (2000) Alpha-lipoic acid treatment ameliorates metabolic parameters, blood pressure, vascular reactivity and morphology of vessels already damaged by streptozotocin-diabetes. *Diabetes Nutr Metab* 13, 308.

29. Morcos, M. et al. (2001) Effect of alpha-lipoic acid on the progression of endothelial cell damage and albuminuria in patients with diabetes mellitus: an exploratory study. *Diabetes Res Clin Pract* 52, 175.

30. Jacob, S. et al. (1996) The antioxidant alpha-lipoic acid enhances insulin-stimulated glucose metabolism in insulin-resistant rat skeletal muscle. *Diabetes* 45, 1024.

31. Henriksen, E.J. et al. (1997) Stimulation by alpha-lipoic acid of glucose transport activity in skeletal muscle of lean and obese Zucker rats. *Life Sci* 61, 805.

32. Jacob, S. et al. (1995) Enhancement of glucose disposal in patients with type 2 diabetes by alpha-lipoic acid. *Arzneimittelforschung* 45, 872.

33. Jacob, S., Henriksen, E.J., Tritschler, H.J., Augustin, H.J., and Dietze, G.J. (1996) Improvement of insulin-stimulated glucose-disposal in type 2 diabetes after repeated parenteral administration of thioctic acid. *Exp Clin Endocrinol Diabetes* 104, 284.

34. Estrada, D.E. et al. (1996) Stimulation of glucose uptake by the natural coenzyme alpha-lipoic acid/thioctic acid: participation of elements of the insulin signaling pathway. *Diabetes* 45, 1798.

35. Haugaard, N. and Haugaard, E.S. (1970) Stimulation of glucose utilization by thioctic acid in rat diaphragm incubated *in vitro. Biochim Biophys Acta* 222, 583.

36. Singh, H.P. and Bowman, R.H. (1970) Effect of DL-alpha-lipoic acid on the citrate concentration and phosphofructokinase activity of perfused hearts from normal and diabetic rats. *Biochem Biophys Res Commun* 41, 555.

37. Lieberman, L.S. (2003) Dietary, evolutionary, and modernizing influences on the prevalence of type 2 diabetes. *Annu Rev Nutr* 23, 345.

38. Unger, R.H. and Zhou, Y.T. (2001) Lipotoxicity of beta-cells in obesity and in other causes of fatty acid spillover. *Diabetes* 50, Suppl 1, S118.

39. Bakker, S.J. et al. (2000) Cytosolic triglycerides and oxidative stress in central obesity: the missing link between excessive atherosclerosis, endothelial dysfunction, and beta-cell failure? *Atherosclerosis* 148, 17.

40. Evans, J.L., Goldfine, I.D., Maddux, B.A., and Grodsky, G.M. (2002) Oxidative stress and stress-activated signaling pathways: a unifying hypothesis of type 2 diabetes. *Endocr Rev* 23, 599.

41. Rudich, A., Tirosh, A., Potashnik, R., Khamaisi, M., and Bashan, N. (1999) Lipoic acid protects against oxidative stress induced impairment in insulin stimulation of protein kinase B and glucose transport in 3T3-L1 adipocytes. *Diabetologia* 42, 949.

42. Maddux, B.A., See, W., Lawrence, J.C., Jr., Goldfine, A.L., Goldfine, I.D., and Evans, J.L. (2001) Protection against oxidative stress-induced insulin resistance in rat L6 muscle cells by mircomolar concentrations of alpha-lipoic acid. *Diabetes* 50, 404.

43. El Midaoui, A. and de Champlain, J. (2002) Prevention of hypertension, insulin resistance, and oxidative stress by alpha-lipoic acid. *Hypertension* 39, 303.

44. Jacob, S. et al. (1999) Oral administration of RAC-alpha-lipoic acid modulates insulin sensitivity in patients with type-2 diabetes mellitus: a placebo-controlled pilot trial. *Free Radic Biol Med* 27, 309.

45. Eason, R.C., Archer, H.E., Akhtar, S., and Bailey, C.J. (2002) Lipoic acid increases glucose uptake by skeletal muscles of obese-diabetic ob/ob mice. *Diabetes Obes Metab* 4, 29.

46. Khamaisi, M. et al. (1999) Lipoic acid acutely induces hypoglycemia in fasting nondiabetic and diabetic rats. *Metabolism* 48, 504.

47. Khamaisi, M. et al. (1997) Lipoic acid reduces glycemia and increases muscle GLUT4 content in streptozotocin-diabetic rats. *Metabolism* 46, 763.

48. Kim, M.S. et al. (2004) Anti-obesity effects of alpha-lipoic acid mediated by suppression of hypothalamic AMP-activated protein kinase. *Nat Med* 10, 727.

49. Kawano, K., Hirashima, T., Mori, S., Saitoh, Y., Kurosumi, M., and Natori, T. (1992) Spontaneous long-term hyperglycemic rat with diabetic complications: Otsuka Long-Evans Tokushima fatty (OLETF) strain. *Diabetes* 41, 1422.

50. Hansen, L.L., Ikeda, Y., Olsen, G.S., Busch, A.K., and Mosthaf, L. (1999) Insulin signaling is inhibited by micromolar concentrations of H_2O_2: evidence for a role of H_2O_2 in tumor necrosis factor alpha-mediated insulin resistance. *J Biol Chem* 274, 25078.

51. Tirosh, A., Potashnik, R., Bashan, N., and Rudich, A. (1999) Oxidative stress disrupts insulin-induced cellular redistribution of insulin receptor substrate-1 and phosphatidylinositol 3-kinase in 3T3-L1 adipocytes: a putative cellular mechanism for impaired protein kinase B activation and GLUT4 translocation. *J Biol Chem* 274, 10595.

52. Lee, W.J. et al. (2005) Alpha-lipoic acid increases insulin sensitivity by activating AMPK in skeletal muscle. *Biochem Biophys Res Commun* 332, 885.

53. Kemp, B.E. et al. (2003) AMP-activated protein kinase, super metabolic regulator. *Biochem Soc Trans* 31, 162.

54. Hardie, D.G., Salt, I.P., Hawley, S.A., and Davies, S.P. (1999) AMP-activated protein kinase: an ultrasensitive system for monitoring cellular energy charge. *Biochem J* 338 (Pt 3), 717.

55. Kahn, B.B., Alquier, T., Carling, D., and Hardie, D.G. (2005) AMP-activated protein kinase: ancient energy gauge provides clues to modern understanding of metabolism. *Cell Metab* 1, 15.

56. Dagher, Z., Ruderman, N., Tornheim, K., and Ido, Y. (2001) Acute regulation of fatty acid oxidation and amp-activated protein kinase in human umbilical vein endothelial cells. *Circ Res* 88, 1276.

57. Ruderman, N. and Prentki, M. (2004) AMP kinase and malonyl-CoA: targets for therapy of the metabolic syndrome. *Nat Rev Drug Discov* 3, 340.

58. Kashiwagi, A. et al. (1999) Free radical production in endothelial cells as a pathogenetic factor for vascular dysfunction in the insulin resistance state. *Diabetes Res Clin Pract* 45, 199.

59. Heitzer, T. et al. (2001) Beneficial effects of alpha-lipoic acid and ascorbic acid on endothelium-dependent, nitric oxide-mediated vasodilation in diabetic patients: relation to parameters of oxidative stress. *Free Radic Biol Med* 31, 53.

60. Visioli, F., Smith, A., Zhang, W., Keaney, J.F., Jr., Hagen, T., and Frei, B. (2002) Lipoic acid and vitamin C potentiate nitric oxide synthesis in human aortic endothelial cells independently of cellular glutathione status. *Redox Rep* 7, 223.

61. Cameron, N.E., Jack, A.M., and Cotter, M.A. (2001) Effect of alpha-lipoic acid on vascular responses and nociception in diabetic rats. *Free Radic Biol Med* 31, 125.

62. Bojunga, J., Dresar-Mayert, B., Usadel, K.H., Kusterer, K., and Zeuzem, S. (2004) Antioxidative treatment reverses imbalances of nitric oxide synthase isoform expression and attenuates tissue-cGMP activation in diabetic rats. *Biochem Biophys Res Commun* 316, 771.

63. Smith, A.R. and Hagen, T.M. (2003) Vascular endothelial dysfunction in aging: loss of Akt-dependent endothelial nitric oxide synthase phosphorylation and partial restoration by (R)-alpha-lipoic acid. *Biochem Soc Trans* 31, 1447.

64. Zhang, W.J. and Frei, B. (2001) Alpha-lipoic acid inhibits TNF-alpha-induced NF-kappaB activation and adhesion molecule expression in human aortic endothelial cells. *FASEB J* 15, 2423.

65. Bierhaus, A. et al. (1997) Advanced glycation end product-induced activation of NF-kappa-B is suppressed by alpha-lipoic acid in cultured endothelial cells. *Diabetes* 46, 1481.

66. Kunt, T. et al. Alpha-lipoic acid reduces expression of vascular cell adhesion molecule-1 and endothelial adhesion of human monocytes after stimulation with advanced glycation end products. *Clin Sci (Lond)* 96, 75.

67. Vasdev, S., Ford, C.A., Parai, S., Longerich, L., and Gadag, V. (2000) Dietary alpha-lipoic acid supplementation lowers blood pressure in spontaneously hypertensive rats. *J Hypertens* 18, 567.

68. Vasdev, S., Gill, V., Longerich, L., Parai, S., and Gadag, V. (2003) Salt-induced hypertension in WKY rats: prevention by alpha-lipoic acid supplementation. *Mol Cell Biochem* 254, 319.

69. Vasdev, S., Ford, C.A., Parai, S., Longerich, L., and Gadag, V. (2000) Dietary lipoic acid supplementation prevents fructose-induced hypertension in rats. *Nutr Metab Cardiovasc Dis* 10, 339.

70. Midaoui, A.E., Elimadi, A., Wu, L., Haddad, P.S., and de Champlain, J. (2003) Lipoic acid prevents hypertension, hyperglycemia, and the increase in heart mitochondrial superoxide production. *Am J Hypertens* 16, 173.

71. Angelucci, L. and Mascitelli-Coriandoli, E. (1958) Anticholesterol activity of alpha-lipoic acid. *Nature* 181, 911.

72. Ivanov, V.N. (1974) Effect of lipoic acid on tissue respiration in rabbits with experimental atherosclerosis. *Cor Vasa* 16, 141.

73. Shih, J.C. (1983) Atherosclerosis in Japanese quail and the effect of lipoic acid. *Fed Proc* 42, 2494.

74. Lodge, J.K., Traber, M.G., and Packer, L. (1998) Thiol chelation of Cu^{2+} by dihydrolipoic acid prevents human low density lipoprotein peroxidation. *Free Radic Biol Med* 25, 287.

75. Marangon, K., Devaraj, S., Tirosh, O., Packer, L., and Jialal, I. (1999) Comparison of the effect of alpha-lipoic acid and alpha-tocopherol supplementation on measures of oxidative stress. *Free Radic Biol Med* 27, 1114.

76. Sola, S. et al. (2005) Irbesartan and lipoic acid improve endothelial function and reduce markers of inflammation in the metabolic syndrome: results of the Irbesartan and Lipoic Acid in Endothelial Dysfunction (ISLAND) study. *Circulation* 111, 343.

77. Dagher, Z., Ruderman, N., Tornheim, K., and Ido, Y. (1999) The effect of AMP-activated protein kinase and its activator AICAR on the metabolism of human umbilical vein endothelial cells. *Biochem Biophys Res Commun* 265, 112.

18 Lipoic Acid Blocks Obesity through Reduced Food Intake, Enhanced Energy Expenditure, and Inhibited Adipocyte Differentiation

Jong-Min Park and An-Sik Chung

CONTENTS

INTRODUCTION

Obesity is rapidly increasing throughout the world, substantially shortening life expectancy, and closely correlated with the prevalence of diabetes and cardiovascular diseases. Plasma levels of leptin, tumor necrosis factor (TNF)-α and nonesterified fatty acids are elevated in obesity and substantially contribute to the development of insulin resistance.[1]

Although obesity research including studies of leptin[2] and the leptin receptor gene[3] has been extensively pursued for more than two decades, the molecular mechanisms of obesity are not yet completely understood. Finding target molecules of weight regulatory mechanisms will contribute to the development of safe and effective pharmaceuticals for blocking obesity and preventing diabetes and cardiovascular diseases. α-lipoic acid (LA) and its reduced dihydrolipoic acid (DHLA) are considered antioxidants (Figure 18.1). LA has components of α-keto dehydrogenases including pyruvate dehydrogenases that catalyze various redox-based reactions.[4] Recent studies demonstrated that LA facilitates glucose transport and utilization in fully differentiated adipocytes as well as animal models of diabetes.[5–7] LA also dramatically reduced rodent weights by suppressing food intake and increasing energy expenditures.[8]

Obesity is caused by both hypertrophy of adipose tissue and by adipose tissue hyperplasia, which triggers the transformation of pre-adipocytes into adipocytes.[9] LA decreases hypothalamic AMP-activated protein kinase (AMPK) activity and induces preformed weight losses in rodents by reducing food intake and enhancing energy expenditures.[8] This chapter will focus on antiobesity mechanisms of LA: blocking adipocyte differentiation and reducing hypothalamic AMPK.

OBESITY RELATED TO FATTY CELL BIOLOGY, ENERGY EXPENDITURE, AND FEEDING REGULATION

Adipose tissue is the major storage organ for surplus energy. It is now clear that adipose tissue is a complex and highly active metabolic and endocrine organ.

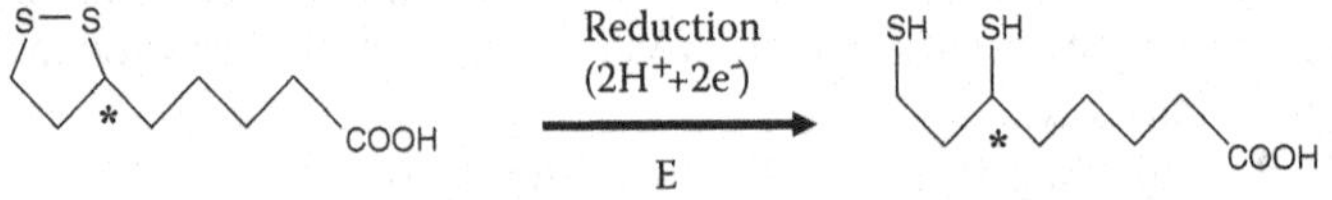

α-Lipoic acid
(1,2-dithiolane-3-pentanoic acid)

Dihydrolipoic acid (DHLA)

Reduction
(2H⁺+2e⁻)
E

E: Lipoamide dehydrogenases
Glutathione reductases
Thioredoxin reductases

FIGURE 18.1 Structures of α-lipoic acid and dihyrolipoic acid.

Leptin is a representative adipocyte-derived hormone that signals information about the body's energy status from the adipose tissue to the brain.

Although it is well known that leptin-deficient ob/ob and leptin receptor deficient db/db mice develop severe obesity, leptin deficiency is very rare in humans. Most obese individuals have increased plasma leptin concentrations, suggesting that they are leptin-resistant. In addition to secreting leptin, adipose tissue secretes a variety of bioactive peptides, collectively called adipocytokines, including TNF-α, adiponectin, plasminogen activator inhibitor (PAI)-1, interleukin (IL)-6, and angiotensinogen. The expression profiles of adipocytokines in subcutaneous adipose tissues and those in visceral adipose tissues are different. Adiponectin, PAI-1, IL-6, and angiotensinogen are mainly shown in the latter.

The change in these important adipocytokines by visceral obesity is regarded as the cause of the detrimental metabolic effects. Recently, the physiologic role of adiponectin has received considerable attention. Adiponectin has been shown to reduce insulin resistance and atherosclerotic processes and to increase fatty acid oxidation rates.[10] The actions of adiponectin are considered to arise from the activation of AMPK through the recently cloned adiponectin receptor.[11] Unlike other adipocytokines, plasma adiponectin levels are reduced in accordance with body (visceral) fat mass.[10] TNF-α, which increases in obesity and inhibits insulin signals, is considered a key factor in the regulation of adiponectin production.

Body weight results from a balance between food intake and energy expenditure. The sympathetic nervous system is activated in response to excess energy or a cold environment. Mice with deletions of the genes encoding the three adrenergic receptor subtypes[12] developed severe obesity as a result of their inability to increase energy expenditure in response to a high caloric diet. In rodents, uncoupling protein (UCP)-1 in brown adipose tissue (BAT) is the main regulator of basal energy expenditure and the expression of this protein is increased by adrenergic stimulation. In humans, however, the main regulator of energy expenditure is not yet known. The UCP-1 homologues known as UCP-2 and UCP-3 that are expressed in human tissues were formerly thought to be the main regulators of energy expenditure. Indeed, hyperexpression of UCP-2 and UCP-3 was shown to increase energy expenditure and reduce body fat. These proteins, however, may not be the major regulators of whole body energy expenditure, inasmuch as mice deficient in either protein did not develop obesity.[13]

The neural system that regulates body weight and appetite is centered in the hypothalamus, which coordinates both afferent sensing and efferent action signals. Long-term afferent signals such as leptin and insulin sense the long-term status of body energy stores, whereas short-term (meal-related) afferent signals derived from the gut are involved in regulating the onset or termination of individual meals. Neuronal cells, which sense nutrient availability, trigger feeding behavior.[13]

Intracerebroventricular (ICV) administration of glucose or long chain fatty acid has been found to inhibit food intake,[14] whereas central administration of 2-deoxyglucose (2-DG), a non-metabolizable glucose analogue, or mercaptoacetate, an inhibitor of fatty acid oxidation, elicits feeding behavior.[15]

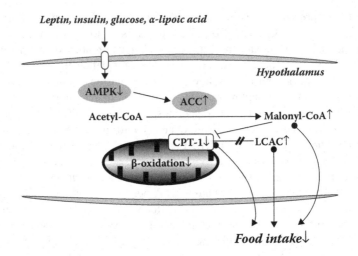

FIGURE 18.2 Possible mechanism by which decreased hypothalamic AMP-activated protein kinase (AMPK) activity reduces food intake. The reduction in hypothalamic AMPK activity in response to feeding-inhibiting factors such as leptin, insulin, glucose, and α-LA increases acetyl-CoA carboxylase (ACC) activity. Increased ACC activity leads to increase in malonyl-CoA levels, which in turn inhibits carnitine palmitoyltransferase-1 (CPT-1) and mitochondrial β oxidation of long chain fatty acyl-CoA (LCAC). Recent studies suggest that increases in malonyl-CoA and/or LCAC levels in the hypothalamus decrease food intake.[16,18,28]

Several signaling pathways are thought to be involved in mediating nutrient-induced feeding. For example, central administration of the fatty acid synthase (FAS) inhibitors, cerulenin and C75, reduces food intake, but it can be prevented by the coadministration of the acetyl-CoA carboxylase (ACC) inhibitor, TOFA.[16] This result suggests that malonyl-CoA, an intermediate metabolite between ACC and FAS, may be an anorexigenic signal (Figure 18.2).

Patients with metabolic syndrome (MS) are characterized by insulin resistance, obesity, hyperlipidemia, premature atherosclerosis, and type-2 diabetes. It has been proposed that the common feature linking this syndrome is dysregulation of the AMPK/malonyl-CoA fuel-sensing and signaling network.[17] Both fuel surfeit and reduced AMPK activity result in increased cellular malonyl-CoA levels, which reduces fat oxidation and favors abnormal tissue accumulation of lipids. Some evidential factors indicate that activated AMPK and/or reduced malonyl-CoA levels may reverse these abnormalities or prevent them from occurring. In contrast, inhibition of hypothalamic carnitine palmitoyl transferase-1 (CPT-1) reduces food intake.[18]

These findings led to the suggestion that increased cytosolic concentrations of long chain fatty acyl-CoA and malonyl-CoA levels may serve as anorexigenic signals. Like pancreatic β cells, some neurons in the ventromedial and arcuate nuclei of the hypothalamus have glucose-sensing machinery that includes GLUT2, glucokinase, and the ATP-sensitive potassium (KATP) channel.[19] The anorexic hormones leptin and insulin can activate the KATP channel in glucose-responsive hypothalamic neurons.[20]

ANTIOBESITY FUNCTION THROUGH SUPPRESSION OF HYPOTHALAMIC AMPK

REGULATION OF ENERGY EXPENDITURE AND FOOD INTAKE

Chronic LA treatment significantly reduced body weight gain and visceral fat mass in obese Otsuta Long-Evans Tokushima fatty (OLETF) rats.[21] Male Sprague-Dawley rats given a standard chow containing LA for 2 weeks reduced food intake and body weight in a dose-dependent manner. Acute administration of LA by intraperitoneal injection (50, 75, or 100 mg/kg of body weight) also suppressed food intake. In addition, ICV injection of a small amount of LA reduced food intake, suggesting that the CNS is the primary site of the anorexic effect.

Dietary administration of LA (0.5%, w/w) for 14 weeks also decreased body weight and visceral fat mass in genetically OLETF rats. This effect was accompanied by reductions in plasma glucose, insulin, free fatty acids, and leptin. It has been further demonstrated that LA improved endothelial dysfunction in obese rats by activating AMPK in endothelial cells by reducing glyceride and lipid peroxide and increasing NO synthesis.[22]

LA also increases whole body energy expenditure. One study compared body weight changes in rats given dietary LA (0.5%) and in pair-fed rats given the same amount of food. The LA-treated rats weighed significantly less than the pair-fed rats, indicating enhanced use of ingested energy. Energy expenditure measured by indirect calorimetry was higher in rats given LA than in control or pair-fed rats. UCP-1 in BAT is a chief regulator of energy expenditures in rodents.[23] UCP-1 is located in the mitochondrial inner membranes and dissipates proton electrochemical energy as heat. The pair-fed rats show reduced expression of UCP-1 in BAT and ectopic expression of UCP-1 in white adipose tissue. These results suggest that weight loss induced by LA is due in part to an enhancement of energy expenditure.

The effects of LA on food intake and energy metabolism are similar to the reported effects of leptins 1 and 2. To determine whether the action of LA is mediated by leptin or leptin receptor signaling, leptin-deficient (Lep–/–) or leptin receptor-deficient (Lepr–/–) mice were fed a diet containing LA (0.5%). LA reduced food intake and caused weight losses in both strains of mice, indicating that leptin and its receptor are not essential for LA-induced anorexia. Leptin exerts its anorexic effect through the regulation of hypothalamic neuropeptides.[24] Intraperitoneal (i.p.) administration of leptin (1 mg/kg) to Sprague-Dawley rats reduced the expression of hypothalamic neuropeptide Y (NPY) mRNA and increased that of pro-opiomelanocortin (Pomc) and corticotropin releasing hormone (Crh) mRNA. By contrast, the i.p. administration of LA (75 mg/kg) did not cause any acute changes in the levels of hypothalamic NPY, Pomc, Crh, pro-melanin concentrating hormone, or hypocretin mRNA.

ROLE OF HYPOTHALAMIC AMPK

AMPK is an enzyme that acts as an intracellular energy sensor.[25] It is activated when cell energy status is low. When activated, AMPK inhibits ATP-consuming pathways (e.g., fatty acid synthesis) and activates ATP-generating pathways (e.g., fatty acid oxidation and glycolysis), thus maintaining energy balance within cells. AMPK activation in skeletal muscle enhances glucose uptake and mitochondrial fatty acid oxidation.[25]

In the liver, AMPK activation suppresses endogenous glucose production.[25] In pancreatic β cells, AMPK seems to antagonize the effect of glucose on insulin secretion and induce β cell apoptosis.[26] Activation of AMPK has been reported to play a favorable role in preserving β cell function under lipid overloading conditions.[27]

Recent studies[21,28,29] have demonstrated that AMPK activity in hypothalamic neurons is altered by various factors and mediates their feeding effects. Hypothalamic AMPK activity is regulated by nutritional availability in hypothalamic neurons. Administration of 2-DG increases hypothalamic AMPK activity, while co-administration of an AMPK inhibitor, compound C, inhibited the 2-DG-induced glycolysis activity.[21] Conversely, ICV administration of glucose or restoration of food intake has been found to decrease hypothalamic AMPK activity.[29] AMPK activity is reduced by ICV administration of anorexigenic hormones such as insulin and leptin, but increased by ICV administration of ghrelin, an orexigenic hormone.[28,29] In the hypothalamic paraventricular nucleus, AMPK activity is decreased by the MT-II melanocortin receptor agonist but increased by the melanocortin receptor antagonist, agouti-related protein (AgRP).

Taken together, these findings indicate that AMPK is part of the common signaling pathway by which various factors regulate feeding behavior, that is, hypothalamic AMPK activity is reduced by feeding inhibiting factors and increased by feeding stimulating factors. Although the mechanism by which AMPK activity in hypothalamic neurons affects feeding behavior is still not fully understood, the leptin-induced reduction in hypothalamic AMPK activity is shown to decrease feeding by down-regulating expression of the NPY and AgRP orexigenic hormones.[29] Alternatively, changes in AMPK activity may affect feeding via changes in intracellular malonyl-CoA concentrations and CPT-1 activity.[16,18]

REDUCTION OF FOOD INTAKE AND INCREASE OF ENERGY EXPENDITURE VIA SUPPRESSION OF HYPOTHALAMIC AMPK

It was recently demonstrated that LA has anti-obesity effects mediated by the suppression of hypothalamic AMPK activity.[21] LA is an essential cofactor of mitochondrial pyruvate dehydrogenase and α-ketoacid dehydrogenase. In addition, LA enhances glucose metabolism in insulin-resistant rat skeletal muscle, and showed potent antioxidant activity by chelating transition metal ions and increasing cytosolic glutathione and vitamin C levels.

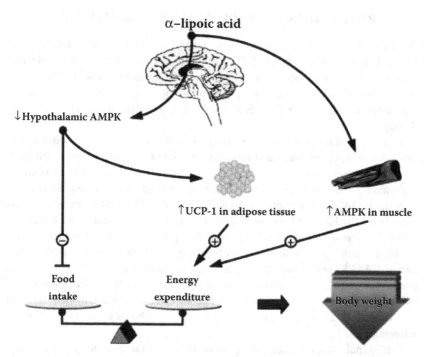

FIGURE 18.3 Mechanism of body weight regulation by LA. LA reduces food intake and increases energy expenditure by suppressing hypothalamic AMP-activated protein kinase (AMPK) activity. It increases energy expenditure by increasing uncoupling protein (UCP)-1 expression in adipose tissue and by activating AMPK in skeletal muscle.

Administration of LA to rodents reduces food intake and body weight as well as stimulating whole body energy expenditure. Central administration of very small amounts of LA increased UCP-1 mRNA in BATs, whereas co-administration of an AMPK activator, 5-aminoimidazole-4-carboxamide ribonucleoside, prevented the LA-induced enhancement of energy expenditure and UCP-1 expression.[21] These results indicate that food intake, energy expenditure, and UCP-1 are mediated via hypothalamic AMPK inhibition by treatment with LA (Figure 18.3). Finally, in contrast to its effects on the hypothalamus, LA increased glucose uptake and fatty acid oxidation in skeletal muscle by activating AMPK.[30]

The mechanism by which LA decreases hypothalamic AMPK activity is presently unknown. However, LA has been found to stimulate glucose transport and ATP synthesis in peripheral tissues,[31,32] and it may decrease hypothalamic AMPK activity by increasing glucose uptake and/or its metabolism in the hypothalamus. LA has been shown to activate ACC by decreasing ACC phosphorylation in hypothalamics.[21] Since activation of ACC can increase intracellular malonyl-CoA in hypothalamic neurons, malonyl-CoA may be a key downstream mediator of AMPK activity responsible for the decrease in food intake by LA administration as shown by Loftus et al.[16] and Ruderman and Prentki.[17]

REGULATION OF ADIPOCYTE DIFFERENTIATION

Obesity is closely correlated with the prevalence of diabetes and cardiovascular disease. Plasma levels of leptin, TNF-α, and non-esterified fatty acids are elevated in obesity and substantially contribute to the development of insulin resistance.[1] Obesity is caused not only by hypertrophy of adipose tissue, but also by adipose tissue hyperplasia that triggers the transformation of pre-adipocytes into adipocytes.[9]

Adipocyte differentiation is a complex process that involves coordinated expression of specific genes and proteins associated with each stage of differentiation. This process is regulated by several signaling pathways.[33] Insulin, the major anabolic hormone, promotes *in vivo* accumulation of adipose tissue.[34]

Structurally unrelated inhibitors of PI3-K, LY294002 and wortmannin, have been shown to block adipocyte differentiation in a time- and dose-dependent fashion,[35] suggesting that the IR/Akt signaling pathway is important in transducing the pro-adipogenic effects of insulin. In contrast, mitogen-activated protein kinases (MAPKs) such as extracellular signal-regulated kinase (ERK) and c-Jun N-terminal kinase (JNK) suppress the process of adipocyte differentiation.[36,37] TNF-α is known to exert its anti-adipogenic effects, at least in part, through activation of the ERK pathway.[36] However, p38K is shown to promote adipocyte differentiation.[38]

The signals that regulate adipogenesis either promote or block the cascades of transcription factors that coordinate the differentiation process. CCAAT element binding proteins (C/EBPs)-β and -δ and sterol response element binding protein-1 (ADD1/ SREBP1) are active during the early stages of the differentiation process and induce the expression and/or activity of peroxisome proliferator-activated receptor-γ, (PPAR-γ), a pivotal coordinator of adipocyte differentiation. PPAR-γ, PPAR-α, and PPAR-δ are important regulators of lipid metabolism. Although they share significant structural similarities, the biological effects associated with each isotype are distinct. For example, PPAR-α and PPAR-δ regulate fatty acid metabolism, while PPAR-γ controls lipid storage and adipogenesis. Activated PPAR-γ induces exit from the cell cycle and, in cooperation with C/EBP-α, stimulates the expression of many metabolic genes such as Glut-4, lipoprotein lipase (LPL),[39] and adipocyte-specific fatty acid binding protein (aP2),[40] thus constituting a functional lipogenic adipocyte.

JNK and ERK suppress this process by phosphorylating and thereby attenuating the transcriptional activity of PPAR-γ.[36,37] Besides these integral members of the adipogenesis program, other transcription factors such as AP-1[41] and cAMP responsive element binding protein (CREB)[42] are known to promote adipogenesis, whereas nuclear factor-κB (NF-κB) suppresses adipocyte differentiation.[43] Therefore, the activity and/or expression of these transcription factors are attractive pharmacological targets for modulating adipocyte tissue formation and deposition.

INHIBITION OF ADIPOCYTE DIFFERENTIATION VIA MITOGEN-ACTIVATED PROTEIN KINASE PATHWAYS

MAPK SIGNALING PATHWAYS MEDIATE ACTIONS OF LA ON ADIPOGENESIS

Several lines of evidence indicate that pro-adipogenic transcription factors such as PPAR-γ and members of the C/EBP family can be negatively regulated by MAPKs. Epidermal growth factor, platelet-derived growth factor, lipoxygenase-1 metabolites, and prostaglandin $F_{2\alpha}$ phosphorylate and attenuate transcriptional activity of PPAR-γ by activating MAPK signaling pathways.[44–46] Similarly, LA treatment of pre-adipocytes inhibits the insulin- or hormonal cocktail-induced transcriptional activity of PPAR-γ and C/EBPα, which is accompanied by strong activation of ERK and JNK.

Inhibitors of ERK and JNK activity abolish the inhibitory effect of LA on insulin- or hormonal cocktail-induced adipogenesis. On the other hand, LA hardly stimulates phosphorylation of insulin receptor (IR) or insulin receptor substrate (IRS)-1 in pre-adipocytes and adipocytes at the early stages of differentiation. In particular, upon LA treatment, a transient Akt phosphorylation is detected in pre-adipocytes although it is not detectable in adipocytes at early stages of differentiation. In contrast, insulin strongly activates IR and IRS-1 and induces long-lasting Akt activation in pre-adipocytes and in adipocytes at early stages of differentiation. These findings exclude possible involvement of Akt activation in LA-induced inhibition of adipogenesis and demonstrate that LA down-regulates PPAR-γ and C/EBP-α through activation of MAPK signaling pathways such as ERK and JNK (Figure 18.4).

MODULATION OF AUXILIARY TRANSCRIPTION FACTORS IN ADIPOGENESIS

Transcriptional activities of AP-1 and CREB are increased in fully differentiated 3T3-L1 adipocytes as well as after 2-hr treatment with a hormonal cocktail in 3T3-L1 pre-adipocytes. AP-1 is involved in transcriptional regulation of aP2 and LPL genes.[47,48] CREB appears to stimulate transcription of several adipocyte-specific genes such as aP2, fatty acid synthetase, and phosphoenolpyruvate carboxykinase.[42] LA, however, strongly down-regulates AP-1 and CREB activities, whereas it up-regulates NF-κB activities in pre-adipocytes.

Many anti-adipogenic factors such as pro-inflammatory cytokines,[49] TNF-α,[50] and endrin[51] are also known to up-regulate NF-κB activity, whereas pro-adipogenic factors such as troglitazone display opposite effects in 3T3-L1 cells.[52] Considering that AP-1,[53] CREB,[54] and NF-κB[55] mediate major downstream effects of MAPK signaling pathways, our findings suggest that activation of the MAPK signaling pathways by LA leads to differential regulation of these transcription factors, which eventually results in decreased expression of the adipocyte-specific genes, and consequently suppresses adipogenesis.

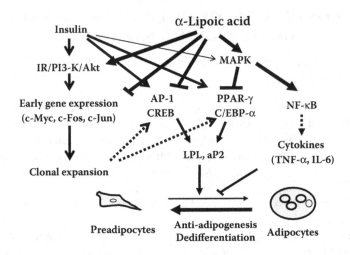

FIGURE 18.4 Regulation of adipocyte differentiation by LA. LA activates both the IR/PI3-K/Akt pathway and the MAPK pathway. Activation of the MAPK pathway mainly leads to the inhibition of pro-adipogenic transcription factors such as PPAR-γ, C/EBP-α, AP-1, and CREB, and the activation of anti-adipogenic factor such as NF-κB. The inhibition of pro-adipogenic transcription factors decreases transcription of several proteins including LPL and aP2, which are two major markers of adipogenesis. In contrast, activation of NF-κB may enhance cytokine release and inhibit adipogenesis. LA may also prevent the adipogenic process by regulating the expression of immediate early genes, which are involved in the process of clonal expansion.

MAPK SIGNALING PATHWAYS MEDIATE LA ACTIONS ON CELL CYCLE AND CLONAL EXPANSION

In the course of adipogenesis, one of the first events following hormonal induction is re-entry of growth-arrested pre-adipocytes into the cell cycle. LA has been demonstrated to inhibit the process of clonal expansion when induced by insulin or a hormonal cocktail, indicating that insulin and LA oppositely regulate cell cycle progression. This differential effect seems to be due to the potency and/or the kinetics of activating of MAPK and IR/Akt signaling pathways.

Both insulin and LA activate MAPK signaling pathways in pre-adipocytes. However, insulin, but not LA, also strongly activates the IR/Akt signaling pathway. This observation indicates that progression in the cell cycle and clonal expansion may require activation of both MAPK and IR/Akt signaling pathways. On the other hand, insulin-induced MAPK activation is transient, whereas that of LA lasts longer, indicating that duration of MAPK activation may be another important factor in determining the fate of a cell in the cell cycle. Indeed, transient activation of MAPK has been considered as a contributor to cell cycle progression whereas its prolonged activation can result in cell cycle arrest via induction of p21[Cip1/Waf1] expression and inhibition of cyclin-dependent kinase activity.[56,57]

It should be emphasized that JNK is known to activate p53, which triggers activation of several proteins involved in cell cycle arrest such as p21[Cip1/Waf1].[58] This evidence supports the notion that activation of MAPKs mediates the inhibitory effect of LA on the clonal expansion process by suppressing the expression of immediate early genes.

SUMMARY

AMPK is an enzyme that functions as an intracellular energy sensor. It is activated when the energy status of a cell is low. LA decreases hypothalamic AMPK activity and results in profound weight losses in rodents by reducing food intake and enhancing energy expenditures. However, LA increases AMPK activity in skeletal muscle and other organs, which enhances glucose uptake and mitochondrial fatty acid oxidation. Activation of hypothalamic AMPK activity reverses by LA administration, which in turn reduces food intake and increases energy expenditure. ICV administration of glucose decreases hypothalamic AMPK activity, whereas inhibition of intracellular glucose utilization through the administration of 2-DG increases hypothalamic AMPK activity and food intake. 2-DG-induced hyperphagia is reversed by inhibiting hypothalamic AMPK. These findings indicate that hypothalamic AMPK is important in the regulation of food intake and energy expenditure and that LA exerts anti-obesity effects by suppressing hypothalamic AMPK activity.

It was reported that LA inhibited differentiation of pro-adipocytes induced by a hormone mixture or troglitazone. Northern blot analysis of cells demonstrated that this inhibition was accompanied by attenuated expression of aP2 and LPL. Electrophoretic mobility shift assay and western blot analysis of cells demonstrated that LA modulates transcriptional activity and/or expression of anti- or pro-adipogenic transcriptional factors.

LA treatment of pre-adipocytes also resulted in prolonged activation of major MAPK signaling pathways. These findings suggest that LA inhibits insulin- or hormonal mixture-induced differentiation of pre-adipocytes by modulating activity and/or expression of pro- or anti-adipogenic transcriptional factors mainly by activating the MAPK pathways.

In conclusion, LA may be beneficial in obesity via two major mechanisms. It may regulate food intake and energy expenditure by decreasing hypothalamic AMPK activity and block adipocyte differentiation by activating MAPK pathways. Further studies are warranted to evaluate food intake and energy expenditure via hypothalamic AMPK activity and adipocyte differentiation by MAPK pathways in humans.

ACKNOWLEDGMENTS

We thank Drs. Ki-Up Lee and Jong-Yeol Park of the University of Ulsan College of Medicine, Asan Medical Center, Seoul, Republic of Korea, for providing their data. This study was supported by Brain Korea 21 Grant M103300.

REFERENCES

1. Leong KS and Wilding JP. Obesity and diabetes. *Baillieres Best Pract Res Clin Endocrinol Metab* 1999; 13, 221.
2. Zhang Y, Proenca R, Maffei M, Barone M, Leopold L, and Friedman JM. Positional cloning of the mouse obese gene and its human homologue. *Nature* 1994; 372, 425.
3. Tartaglia LA et al. Identification and expression cloning of a leptin receptor, OB-R. *Cell* 1995; 83, 1263.
4. Berg A and de Kok A. 2-Oxo acid dehydrogenase multienzyme complexes: the central role of the lipoyl domain. *J Biol Chem* 1997; 378, 617.
5. Czech MP and Fain JN. Cu^{++}-dependent thiol stimulation of glucose metabolism in white fat cells. *J Biol Chem* 1972; 247, 6218.
6. Krieger-Brauer HI, Medda PK, and Kather H. Insulin-induced activation of NADPH-dependent H_2O_2 generation in human adipocyte plasma membranes is mediated by G-alpha-i2. *J Biol Chem* 1997; 272, 10135.
7. Mahadev K et al. Hydrogen peroxide generated during cellular insulin stimulation is integral to activation of the distal insulin signaling cascade in 3T3-L1 adipocytes. *J Biol Chem* 2001; 276, 48662.
8. Lee WJ et al. Obesity: the role of hypothalamic AMP-activated protein kinase in body weight regulation. *Int J Biochem Cell Biol* 2005; 37, 2254.
9. Caro JF, Dohm LG, Pories WJ, and Sinha MK. Cellular alterations in liver, skeletal muscle, and adipose tissue responsible for insulin resistance in obesity and type II diabetes. *Diabetes Metab Rev* 1989; 5, 665.
10. Havel PJ. Update on adipocyte hormones: regulation of energy balance and carbohydrate/lipid metabolism. *Diabetes* 2004; 53, Suppl 1, S143.
11. Yamauchi T et al. Cloning of adiponectin receptors that mediate antidiabetic metabolic effects. *Nature* 2003; 423, 762.
12. Bachman ES et al. Beta AR signaling required for diet-induced thermogenesis and obesity resistance. *Science* 2002; 297, 843.
13. Flier JS. Obesity wars: molecular progress confronts an expanding epidemic. *Cell* 2004; 116, 337.
14. Obici S et al. Central administration of oleic acid inhibits glucose production and food intake. *Diabetes* 2002; 51, 271.
15. Sergeyev V, Broberger C, Gorbatyuk O, and Hokfelt T. Effect of 2-mercaptoacetate and 2-deoxy-D-glucose administration on the expression of NPY, AGRP, POMC, MCH and hypocretin/orexin in the rat hypothalamus. *Neuroreport* 2000; 11, 117.
16. Loftus TM et al. Reduced food intake and body weight in mice treated with fatty acid synthase inhibitors. *Science* 2000; 288, 2379.
17. Ruderman N and Prentki M. AMP kinase and malonyl-CoA: targets for therapy of the metabolic syndrome. *Nat Rev Drug Discov* 2004; 3, 340.
18. Obici S, Feng Z, Arduini A, Conti R, and Rossetti L. Inhibition of hypothalamic carnitine palmitoyltransferase-1 decreases food intake and glucose production. *Nat Med* 2003; 9, 756.
19. Yang XJ, Kow LM, Funabashi T, and Mobbs CV. Hypothalamic glucose sensor: similarities to and differences from pancreatic beta-cell mechanisms. *Diabetes* 1999; 48, 1763.

20. Spanswick D, Smith MA, Mirshamsi S, Routh VH, and Ashford ML. Insulin activates ATP-sensitive K⁺ channels in hypothalamic neurons of lean, but not obese rats. *Nat Neurosci* 2000; 3, 757.

21. Kim MS et al. Anti-obesity effects of alpha-lipoic acid mediated by suppression of hypothalamic AMP-activated protein kinase. *Nat Med* 2004; 10, 727.

22. Lee WJ et al. Alpha-lipoic acid prevents endothelial dysfunction in obese rats via activation of AMP-activated protein kinase. *Arterioscler Thromb Vasc Biol* 2005; 25, 2488.

23. Dalgaard LT and Pedersen O. Uncoupling proteins: functional characteristics and role in the pathogenesis of obesity and type II diabetes. *Diabetologia* 2001; 44, 946.

24. Schwartz MW, Woods SC, Porte D, Jr., Seeley RJ, and Baskin DG. Central nervous system control of food intake. *Nature* 2000; 404, 661.

25. Kahn BB, Alquier T, Carling D, and Hardie DG. AMP-activated protein kinase: ancient energy gauge provides clues to modern understanding of metabolism. *Cell Metab* 2005; 1, 15.

26. Kefas BA et al. AICA-riboside induces apoptosis of pancreatic beta cells through stimulation of AMP-activated protein kinase. *Diabetologia* 2003; 46, 250.

27. Diraison F et al. Over-expression of sterol-regulatory-element-binding protein-1c (SREBP1c) in rat pancreatic islets induces lipogenesis and decreases glucose-stimulated insulin release: modulation by 5-aminoimidazole-4-carboxamide ribonucleoside (AICAR). *Biochem J* 2004; 378, 769.

28. Andersson U et al. AMP-activated protein kinase plays a role in the control of food intake. *J Biol Chem* 2004; 279, 12005.

29. Minokoshi Y et al. AMP-kinase regulates food intake by responding to hormonal and nutrient signals in the hypothalamus. *Nature* 2004; 428, 569.

30. Lee WJ et al. Alpha-lipoic acid increases insulin sensitivity by activating AMPK in skeletal muscle. *Biochem Biophys Res Commun* 2005; 332, 885.

31. Yaworsky K, Somwar R, Ramlal T, Tritschler HJ, and Klip A. Engagement of the insulin-sensitive pathway in the stimulation of glucose transport by alpha-lipoic acid in 3T3-L1 adipocytes. *Diabetologia* 2000; 43, 294.

32. Cho KJ et al. Alpha-lipoic acid inhibits adipocyte differentiation by regulating pro-adipogenic transcription factors via mitogen-activated protein kinase pathways. *J Biol Chem* 2003; 278, 34823.

33. Torti FM, Torti SV, Larrick JW, and Ringold GM. Modulation of adipocyte differentiation by tumor necrosis factor and transforming growth factor beta. *J Cell Biol* 1989; 108, 1105.

34. Zhang B et al. Insulin- and mitogen-activated protein kinase-mediated phosphorylation and activation of peroxisome proliferator-activated receptor gamma. *J Biol Chem* 1996; 271, 31771.

35. Xia X and Serrero G. Inhibition of adipose differentiation by phosphatidylinositol 3-kinase inhibitors. *J Cell Physiol* 1999; 178, 9.

36. Font de Mora J, Porras A, Ahn N, and Santos E. Mitogen-activated protein kinase activation is not necessary for, but antagonizes, 3T3-L1 adipocytic differentiation. *Mol Cell Biol* 1997; 17, 6068.

37. Camp HS, Tafuri SR, and Leff T. c-Jun N-terminal kinase phosphorylates peroxisome proliferator-activated receptor-gamma1 and negatively regulates its transcriptional activity. *Endocrinology* 1999; 140, 392.

38. Engelman JA, Lisanti MP, and Scherer PE. Specific inhibitors of p38 mitogen-activated protein kinase block 3T3-L1 adipogenesis. *J Biol Chem* 1998; 273, 32111.

39. Schoonjans K et al. PPAR alpha and PPAR gamma activators direct a distinct tissue-specific transcriptional response via a PPRE in the lipoprotein lipase gene. *Embo J* 1996; 15, 5336.

40. Tontonoz P, Hu E, Graves RA, Budavari AI, and Spiegelman BM. mPPAR gamma 2: tissue-specific regulator of an adipocyte enhancer. *Genes Dev* 1994; 8, 1224.

41. Yang VW, Christy RJ, Cook JS, Kelly TJ, and Lane MD. Mechanism of regulation of the 422(aP2) gene by cAMP during preadipocyte differentiation. *Proc Natl Acad Sci USA* 1989; 86, 3629.

42. Reusch JE, Colton LA, and Klemm DJ. CREB activation induces adipogenesis in 3T3-L1 cells. *Mol Cell Biol* 2000; 20, 1008.

43. Ruan H, Hacohen N, Golub TR, Van Parijs L, and Lodish HF. Tumor necrosis factor-alpha suppresses adipocyte-specific genes and activates expression of preadipocyte genes in 3T3-L1 adipocytes: nuclear factor-kappaB activation by TNF-alpha is obligatory. *Diabetes* 2002; 51, 1319.

44. Camp HS and Tafuri SR. Regulation of peroxisome proliferator-activated receptor gamma activity by mitogen-activated protein kinase. *J Biol Chem* 1997; 272, 10811.

45. Hsi LC, Wilson L, Nixon J, and Eling TE. 15-lipoxygenase-1 metabolites down-regulate peroxisome proliferator-activated receptor gamma via the MAPK signaling pathway. *J Biol Chem* 2001; 276, 34545.

46. Reginato MJ, Krakow SL, Bailey ST, and Lazar MA. Prostaglandins promote and block adipogenesis through opposing effects on peroxisome proliferator-activated receptor gamma. *J Biol Chem* 1998; 273, 1855.

47. Distel RJ, Ro HS, Rosen BS, Groves DL, and Spiegelman BM. Nucleoprotein complexes that regulate gene expression in adipocyte differentiation: direct participation of c-fos. *Cell* 1987; 49, 835.

48. Homma H et al. Estrogen suppresses transcription of lipoprotein lipase gene. Existence of a unique estrogen response element on the lipoprotein lipase promoter. *J Biol Chem* 2000; 275, 11404.

49. Renard P and Raes M. The proinflammatory transcription factor NF-kappa-B: a potential target for novel therapeutical strategies. *Cell Biol Toxicol* 1999; 15, 341.

50. Schmid E, Hotz-Wagenblatt A, Hacj V, and Droge W. Phosphorylation of the insulin receptor kinase by phosphocreatine in combination with hydrogen peroxide: the structural basis of redox priming. *FASEB J* 1999; 13, 1491.

51. Mahadev K, Zilbering A, Zhu L, and Goldstein BJ. Insulin-stimulated hydrogen peroxide reversibly inhibits protein-tyrosine phosphatase 1b *in vivo* and enhances the early insulin action cascade. *J Biol Chem* 2001; 276, 21938.

52. Sen CK, Roy S, Han D, and Packer L. Regulation of cellular thiols in human lymphocytes by alpha-lipoic acid: a flow cytometric analysis. *Free Radic Biol Med* 1997; 22, 1241.

53. Czech MP, Lawrence JC, Jr., and Lynn WS. Evidence for the involvement of sulfhydryl oxidation in the regulation of fat cell hexose transport by insulin. *Proc Natl Acad Sci USA* 1974; 71, 4173.

54. Wilden PA and Pessin JE. Differential sensitivity of the insulin-receptor kinase to thiol and oxidizing agents in the absence and presence of insulin. *Biochem J* 1987; 245, 325.

55. Clark S and Konstantopoulos N. Sulphydryl agents modulate insulin- and epidermal growth factor (EGF)-receptor kinase via reaction with intracellular receptor domains: differential effects on basal versus activated receptors. *Biochem J* 1993; 292, 217.

56. Chen JJ, Kosower NS, Petryshyn R, and London IM. The effects of N-ethylmaleimide on the phosphorylation and aggregation of insulin receptors in the isolated plasma membranes of 3T3-F442A adipocytes. *J Biol Chem* 1986; 261, 902.

57. Jhun BH, Hah JS, and Jung CY. Phenylarsine oxide causes an insulin-dependent, GLUT4-specific degradation in rat adipocytes. *J Biol Chem* 1991; 266, 22260.

58. Yang J, Clark AE, Harrison R, Kozka IJ, and Holman GD. Trafficking of glucose transporters in 3T3-L1 cells. Inhibition of trafficking by phenylarsine oxide implicates a slow dissociation of transporters from trafficking proteins. *Biochem J* 1992; 281, 809.

19 Effects of Conjugated Linoleic Acid and Lipoic Acid on Insulin Action in Insulin-Resistant Obese Zucker Rats

Erik J. Henriksen

CONTENTS

ABSTRACT

Insulin-resistant conditions such as pre-diabetes and type 2 diabetes are characterized by defects in the ability of insulin to activate glucose transport in skeletal muscle. One animal model that has proven useful in elucidating the multifactorial etiology of skeletal muscle insulin resistance is the obese Zucker (fa/fa) rat, characterized by complete leptin resistance, massive central obesity, hyperinsulinemia, dyslipidemia, and oxidative stress (the imbalance between exposure of tissue to an oxidant stress and cellular antioxidant defenses). Studies published by our research group addressed the utility of two nutriceutical compounds, conjugated linoleic acid (CLA) and alpha-lipoic acid (ALA), both of which possess antioxidant properties, in improving the metabolic

conditions of obese Zucker rats. These studies indicate that chronic administration of the R-enantiomer of ALA (R-ALA) to obese Zucker rats improved
whole-body insulin sensitivity and enhanced the insulin-stimulated skeletal
muscle glucose transport system, at least in part by up-regulation of IRS-1-
dependent insulin signaling. CLA treatment of obese Zucker rats improved
glucose tolerance and insulin-stimulated glucose transport activity, attributable
exclusively to the trans-10, cis-12 isomer and associated with reductions in
oxidative stress and muscle lipid levels. Significant interactions exist between
CLA and R-ALA for enhancement of insulin action on skeletal muscle glucose
transport in the obese Zucker rat, also ascribed to reductions in muscle oxidative stress and lipid storage. Collectively, these investigations support the
fundamental concept that oxidative stress is an important component of the
etiology of insulin resistance that can be beneficially modulated by antioxidant
interventions, including CLA and ALA.

INTRODUCTION

Insulin resistance of skeletal muscle glucose disposal is a primary defect associated with the development of the pre-diabetic state and with the conversion from
a pre-diabetic state to overt type 2 diabetes.[1,2] Pre-diabetes is defined as a condition in which fasting blood glucose levels are elevated above normal (100 to
125 mg/dl) but do not exceed the threshold for the diagnosis of diabetes (>126
mg/dl).[3] Pre-diabetes is frequently characterized by several other cardiovascular
risk factors including visceral obesity, hyperinsulinemia, and dyslipidemia,
encompassing a condition of elevated cardiovascular disease risk known as metabolic syndrome (see Chapter 1). It is estimated over 40 million individuals in
the United States have pre-diabetes.[3] In the face of further metabolic destabilization (primarily via ß cell dysfunction), this population could increase substantially the number of overt type 2 diabetics (presently estimated to be >20 million)
in the U.S. While the etiology of the skeletal muscle insulin resistance present
in pre-diabetic and type 2 diabetic individuals is clearly multifactorial, accumulating evidence indicates that one factor that can cause insulin resistance and
exacerbate existing insulin resistance is oxidative stress, the imbalance between
exposure of tissues to oxidant stresses and cellular antioxidant defenses.[4]

A major challenge to the scientific community and to healthcare professionals
is the design and implementation of effective interventions to reduce insulin
resistance in these pre-diabetic and type 2 diabetic populations. Increased physical
activity and loss of visceral fat have long been the preferred interventions for
ameliorating insulin resistance,[5] but these lifestyle changes are difficult to imple-
ment and maintain in human pre-diabetic and type 2 diabetic subjects. Therefore,
pharmaceutical and nutriceutical treatments are seen as viable alternative therapies for reducing insulin resistance in these individuals.

Two nutriceutical compounds that received considerable attention recently
as interventions in insulin-resistant states are conjugated linoleic acid (CLA)

and alpha-lipoic acid (ALA).[6,7] Both substances have antioxidant properties, and their utility in the context of oxidative stress-associated insulin resistance of skeletal muscle glucose disposal will be the focus of this chapter. The normal regulation of the glucose transport system in skeletal muscle will be described first, followed by a brief section summarizing the fundamental defects of the glucose transport system, including the contribution of oxidative stress, that lead to the insulin-resistant state and ultimately to the development of overt type 2 diabetes. Finally, a discussion of potential physiological and pharmacological interventions in the prevention and treatment of insulin resistance is included, culminating with more detailed coverage of how CLA and ALA, individually and in combination, can modify oxidative stress and reduce insulin resistance of skeletal muscle glucose transport in the pre-diabetic state. This discussion will focus on the obese Zucker (fa/fa) rat, an animal model that displays complete leptin resistance, massive visceral obesity, glucose intolerance, insulin resistance, hyperinsulinemia, dyslipidemia, and oxidative stress, and clearly reflects many of the pathophysiological conditions present in humans with pre-diabetes and metabolic syndrome.

REGULATION OF SKELETAL MUSCLE GLUCOSE TRANSPORT SYSTEM

The acute regulation of the insulin-dependent glucose transport system in mammalian skeletal muscle occurs via the sequential activation of several intracellular proteins.[8,9] This cascade is initiated by insulin binding to the α subunit of the insulin receptor (IR), thereby activating the intrinsic tyrosine kinase of the transmembrane IR ß subunits. The activated IR tyrosine kinase can phosphorylate a number of important intracellular substrates, including isoforms of insulin receptor substrate (IRS1-4). The most important isoforms in skeletal muscle are primarily IRS-1 and IRS-2. Tyrosine-phosphorylated IRS-1 or IRS-2 can then interact with the p85 regulatory subunit of phosphatidylinositol-3-kinase (PI3-kinase), increasing the catalytic activity of the p110 subunit of PI3-kinase. Phosphoinositide moieties produced by PI3-kinase subsequently activate 3-phosphoinositide-dependent kinases (PDKs) that then phosphorylate and activate Akt, a serine/threonine kinase, and atypical protein kinase C isoforms. The activation of these steps eventually results in the translocation of the glucose transporter protein isoform GLUT-4 to the sarcolemmal membrane where glucose transport takes place via a facilitative diffusion process.

Glucose transport in skeletal muscle can also be activated by contractile activity.[10] The intracellular mechanisms responsible for this insulin-independent stimulation of glucose transport include the activation of 5-adenosine-monophosphate-activated protein kinase (AMP kinase), an enzyme stimulated by a decrease in cellular energy charge,[11-14] and activation of calcium and calmodulin-dependent protein kinase.[14,15] Ultimately, these contraction-associated signaling factors stimulate GLUT-4 translocation[16,17] and enhanced glucose transport activity.

ETIOLOGY OF INSULIN RESISTANCE IN OBESE ZUCKER RATS

The obese Zucker (fa/fa) rat develops numerous pathophysiological conditions because of a mutation in the leptin receptor gene.[18–20] This defect results in chronic hyperphagia and eventually in obesity, especially extensive visceral obesity. Skeletal muscle insulin resistance in obese Zucker rats is reflected in diminished insulin-stimulated GLUT-4 protein translocation[21,22] and glucose transport activity.[22–24] The compromised glucose transport system in muscles of obese Zucker rats likely results from specific defects in the insulin signaling cascade including reduced IRS-1 protein expression[25–27] and diminished insulin-stimulated IRS-1 tyrosine phosphorylation and PI3-kinase activation.[25,27] Moreover, the obese Zucker rat is markedly dyslipidemic and displays reduced activity of beneficial protein kinase C isoforms[28] and overactivation of deleterious protein kinase C isoforms[29,30] that may be important in the etiology of insulin resistance in this pre-diabetic animal model.

This abnormal functionality of the insulin signaling cascade in skeletal muscle is also characteristic of humans with pre-diabetes and overt type 2 diabetes. Defects in human insulin-resistant skeletal muscle include impaired insulin stimulation of tyrosine phosphorylation of IR and IRS-1 and of the activities of PI3-kinase and Akt[31–34] and reduced GLUT-4 protein translocation.[35,36]

The obese Zucker rat also displays characteristics of oxidative stress that may be associated with its well established insulin-resistant state. For example, intramuscular triglycerides and protein carbonyls — markers of oxidative damage[37] — are elevated in the skeletal muscles, cardiac muscles, and livers of these animals[38–40] and are associated with diminutions of insulin action on the IR/IRS-1-dependent signaling pathway.[27] We have also demonstrated that *in vitro* oxidant stress (~90 µM H_2O_2) can directly impair insulin action on IR/IRS-1-dependent signaling, glucose transport, and glycogen synthesis in otherwise insulin-sensitive skeletal muscles from lean Zucker rats (Kim, J.S., Saengsirisuwan, V., and Henriksen, E.J., manuscript submitted), providing additional support for the concept that oxidative stress can play a role in the etiology of the insulin-resistant state in skeletal muscle.

NUTRICEUTICAL INTERVENTIONS: CONJUGATED LINOLEIC ACID AND α-LIPOIC ACID

Effects of CLA in Obese Zucker Rats

CLA is a dienoic derivative of linoleic acid that exists primarily as one of two isomers — the cis-10, trans-12 isomer and the trans-10, cis-12 isomer, with numerous other minor isoforms possible. CLA possesses significant antioxidant properties[41] and has been used as an intervention against cancer and heart disease.[42] Chronic CLA administration diminished visceral fat in obese Zucker rats[39] and adult humans.[43] In the Zucker diabetic fatty (ZDF) rat, a model of overt type

2 diabetes, the trans-10, cis-12 isomer of CLA enhanced glucose tolerance, reduced fasting plasma glucose, insulin, and free fatty acids, and increased insulin-stimulated skeletal muscle glucose transport activity and glycogen synthase activity.[44,45] This supports its potential as an intervention for treatment of type 2 diabetes, possibly as an activator of the peroxisome proliferator-activated receptor-γ in a variety of tissues.[44]

Our research group conducted investigations of the metabolic actions of CLA in pre-diabetic obese Zucker rats. In this model of pre-diabetes, as in diabetic ZDF rats,[44,45] CLA treatment led to enhanced glucose tolerance, increased whole-body insulin sensitivity, decreased fasting plasma glucose, insulin, and free fatty acid levels, and improved insulin-mediated skeletal muscle glucose transport activity.[39,40] Importantly, CLA treatment caused a significant decrease in visceral fat in this grossly obese animal model.[39]

These metabolic actions of CLA can be attributed solely to the trans-10, cis-12 isomer, as the cis-10, trans-12 isomer is metabolically neutral in this animal model. Moreover, these trans-10, cis-12-CLA-induced metabolic improvements in the obese Zucker rat are highly correlated with reductions in skeletal muscle protein carbonyl levels and intramuscular triglyceride levels.[39] This latter finding is of importance in light of the well-recognized role of elevated muscle lipid in the etiology of insulin resistance.[46] Collectively, these results highlight the potential of the trans-10, cis-12 isomer of CLA as a nutriceutical intervention in specific conditions of glucose intolerance and insulin resistance, and indicate that the beneficial metabolic actions of CLA may be related to its ability to reduce intramuscular lipid levels and local indices of oxidative stress.

EFFECTS OF ALA IN OBESE ZUCKER RATS

The metabolic actions of ALA in pre-diabetic obese Zucker rats have been extensively investigated by our research group in the last decade, and most of these findings are summarized in recent review articles and chapters.[4,6,47] The highlights of these studies will be briefly covered in this section.

The vast majority of the beneficial metabolic actions of ALA in obese Zucker rats can be attributed to the R-enantiomer, with the S-enantiomer being either metabolically neutral or even eliciting some metabolic worsening such as decreased muscle GLUT4 glucose transporter protein expression.[48] R-ALA acutely enhances glucose transport and intracellular glucose metabolism in both insulin-sensitive[49,50] and insulin-resistant[48,50,51] skeletal muscle preparations. Chronic administration (2 to 6 weeks) of R-ALA to obese Zucker rats resulted in improvements of whole-body glucose tolerance and insulin sensitivity, reductions in fasting plasma insulin and lipids, and enhanced insulin-stimulated glucose transport activity in isolated skeletal muscle.[27,38]

The increased insulin action on skeletal muscle glucose transport associated with chronic R-ALA treatment of obese Zucker rats is correlated with a proportional reduction of oxidative stress, as reflected by muscle protein carbonyl levels.[38] This antioxidant intervention is associated with reductions in cardiac and liver protein carbonyl levels.[38] Chronic treatment of obese Zucker rats with R-ALA also induced dose-dependent reductions in plasma free fatty acids that correlated with improvements in whole-body glucose tolerance and insulin sensitivity and with improved insulin action on the skeletal muscle glucose transport system.[38,52] It now appears that R-ALA treatment can improve functionality of the IRS-1-dependent insulin signaling pathway in skeletal muscles of obese Zucker rats. Chronic administration of R-ALA to this insulin-resistant, pre-diabetic animal model induces an enhancement of IRS-1 protein expression and increased insulin-stimulated IRS-1 associated with the p85 subunit of phosphatidylinositol-3-kinase, a surrogate measure of PI3-kinase activation.[27]

These results on the beneficial effects of chronic treatment with ALA in obese Zucker rats were confirmed recently in another rodent model of obesity-associated insulin resistance and glucose dysregulation, the Otsuka Long-Evans Tokushima Fatty (OLETF) rat.[53] OLETF rats treated with ALA displayed reduced body weights, decreased blood glucose levels, and lower intramuscular triglyceride concentrations.[53] These investigators made the novel finding that the reduction in muscle lipids and the increase in muscle glucose uptake in short-term ALA-treated OLETF rats may be associated with the activation of AMP kinase.[54]

Several clinical investigations generally supported the animal model-based findings on the effects of ALA on glucoregulation. The acute infusion of ALA in type 2 diabetic human subjects resulted in ~30% improvement in glucose disposal rate during a euglycemic, hyperinsulinemic clamp.[55,56] Moreover, the chronic intravenous or oral administration of ALA to type 2 diabetic subjects also elicited significant improvements in whole-body insulin sensitivity.[57–59] The results from this limited number of clinical investigations, combined with the findings from animal models, provide further support for additional clinical studies on the effectiveness of ALA in the treatment of insulin resistance in pre-diabetic and type 2 diabetic humans.

INTERACTIONS OF CLA AND ALA IN OBESE ZUCKER RATS

To our knowledge, there is only one published investigation related to potential interactive effects of CLA and ALA to modulate insulin action in insulin-resistant skeletal muscle. The study, performed by our research group, was published in 2003.[40] Obese Zucker rats were treated chronically with either a mixture of CLA isomers, with R-ALA, or with a combination of R-ALA and CLA, at low, minimally effective doses, or at high, maximally effective doses.

While the high doses of CLA and R-ALA individually induced significant metabolic improvements in these pre-diabetic animals, including increased insulin-stimulated glucose transport activity and reduced muscle oxidative stress and lipid levels in skeletal muscle, the combination of these high doses of CLA and R-ALA did not provide for any further metabolic enhancements above those brought about by the individual agents. In contrast, the combination of low doses of CLA and R-ALA (which individually produced minimal metabolic actions) resulted in substantial improvements in glucose tolerance and whole-body insulin sensitivity and caused significant increases in insulin-stimulated glucose transport activity in skeletal muscle that were closely associated with reductions in muscle oxidative stress and lipid levels. This novel synergistic interaction between the fatty acid CLA and the metabolic antioxidant R-ALA supports the use of combined CLA and R-ALA in the modulation of whole-body and skeletal muscle insulin resistance.

PERSPECTIVES AND FUTURE DIRECTIONS

The information discussed in this chapter supports the beneficial role of the R-ALA and CLA nutriceutical compounds in the modulation of insulin action in insulin-resistant, pre-diabetic obese Zucker rats. With regard to these effects of R-ALA, the general consensus is that this antioxidant and reactive sulfhydryl-containing agent elicits improvements in glucose tolerance, insulin sensitivity, IR/IRS-1-dependent insulin signaling, and muscle glucose transport capacity. While some variability in the absolute effect of ALA administration on these variables in both animal models and in human type 2 diabetics does exist in the literature, this can generally be attributed to differences among studies in the ALA dose given, the route of ALA administration, the release rate of the ALA, and the initial degree of insulin resistance present in the experimental subjects.

On the other hand, several investigations utilizing both animal models and human subjects presented contradictory results arising from chronic CLA administration. In some investigations, CLA administration actually worsened insulin sensitivity in obese mice[60,61] and in abdominally obese men.[62,63] At least part of this controversy can be accounted for by differential responses to the CLA in the inherently different animal models employed. For example, the chronic administration of trans-10, cis-12-CLA to obese Zucker rats[39] or ZDF rats[45] caused a relatively small decrease in fat mass and an increase in whole-body and skeletal muscle insulin sensitivity, whereas this same CLA administration to obese C57BL/6 mice elicited a relatively large fat mass loss but induced insulin resistance.[60,61] The fundamentally different metabolic response to CLA in the mouse model can likely be ascribed to the development of a state of lipodystrophy marked by an extreme fat mass deficit and impaired insulin action. In contrast, the CLA-treated obese Zucker and ZDF rats retained a morbidly obese phenotype. The specific animal model employed and human population under investigation must be taken into consideration when assessing the potential for CLA to bring about beneficial modulation of metabolic parameters.

Finally, based on the results of our study in obese Zucker rats,[40] an important area for future clinical investigation would be the assessment of the effects of combined low doses of CLA and R-ALA on metabolic parameters in obese humans with insulin resistance, glucose intolerance, and dyslipidemia. There are novel interactions of these two nutriceutical compounds that may be exploited in the treatment of human insulin-resistant states, especially those states associated with visceral adiposity.

REFERENCES

1. Warram, J.H. et al. Slow glucose removal rate and hyperinsulinemia precede the development of type II diabetes in the offspring of diabetic parents. *Ann. Intern. Med.* 113, 909, 1990.
2. Ferrannini, E. Insulin resistance versus insulin deficiency in non-insulin-dependent diabetes mellitus: problems and prospects. *Endocr. Rev.* 19, 477, 1998.
3. American Diabetes Association. Diagnosis and classification of diabetes mellitus. *Diabetes Care* 27, S5, 2004.
4. Henriksen, E.J. Oxidative stress and antioxidant treatment: effects on muscle glucose transport in animal models of type 1 and type 2 diabetes, in *Antioxidants in Diabetes Management*, Packer, L. et al., Eds., Marcel Dekker, New York, 2000.
5. Henriksen, E.J. Invited review: effects of acute exercise and exercise training of insulin resistance. *J. Appl. Physiol.* 93, 788, 2002.
6. Henriksen, E.J. Therapeutic effects of lipoic acid on hyperglycemia and insulin resistance, in *Handbook of Antioxidants*, Cadenas, E. and Packer, L., Eds., Marcel Dekker, New York, 2001.
7. Belury, M.A. Dietary conjugated linoleic acid in health: physiological effects and mechanisms of action. *Ann. Rev. Nutr.* 22, 505, 2002.
8. Shepherd, P.R. and Kahn, B.B. Glucose transporters and insulin action: implications for insulin resistance and diabetes mellitus. *New Engl. J. Med.* 341, 248, 1999.
9. Zierath, J.R., Krook, A., and Wallberg-Henriksson, H. Insulin action and insulin resistance in human skeletal muscle. *Diabetologia* 43, 821, 2000.
10. Jessen, N. and Goodyear, L.J. Contraction signaling to glucose transport in skeletal muscle. *J. Appl. Physiol.* 99, 330, 2005.
11. Kurth-Kraczek, E.J. et al. 5 AMP-activated protein kinase activation causes GLUT4 translocation in skeletal muscle. *Diabetes* 48, 1667, 1999.
12. Mu, J. et al. A role for AMP-activated protein kinase in contraction- and hypoxia-regulated glucose transport in skeletal muscle. *Molecular Cell* 7, 1085, 2001.
13. Winder, W.W. Energy-sensing and signaling by AMP-activated protein kinase in skeletal muscle. *J. Appl. Physiol.* 91, 1017, 2001.
14. Wright, D.C. et al. Ca^{2+} and AMPK both mediate stimulation of glucose transport by muscle contractions. *Diabetes* 53, 330, 2004.
15. Holloszy J.O. and Hansen, P.A. Regulation of glucose transport into skeletal muscle. *Rev. Physiol. Biochem. Pharmacol.* 128, 99, 1996.
16. Goodyear, L.J., Hirshman, M.F., and Horton, E.S. Exercise-induced translocation of skeletal muscle glucose transporters. *Am. J. Physiol. Endocrinol. Metab.* 261, E795, 1991.
17. Gao, J. et al. Additive effect of contractions and insulin on GLUT-4 translocation into the sarcolemma. *J. Appl. Physiol.* 77, 1587, 1994.
18. Phillips, M.S. et al. Leptin receptor missense mutation in the fatty Zucker rat. *Nat. Genetics* 13, 18, 1996.
19. Iida, M. et al. Substitution at codon 269 (glutamine proline) of the leptin receptor (OB-R) cDNA is the only mutation found in the Zucker fatty (fa/fa) rat. *Biochem. Biophys. Res. Commun.* 224, 597, 1996.
20. Takaya, K. et al. Molecular cloning of rat leptin receptor isoform complementary DNAs: identification of a missense mutation in Zucker fatty (fa/fa) rats. *Biochem. Biophys. Res. Commun.* 225, 75, 1996.

21. King, P.A. et al. Exercise, unlike insulin, promotes glucose transporter transloca-
    tion in obese Zucker rat muscle. *Am. J. Physiol. Regulatory Integrative Comp.
    Physiol.* 265, R447, 1993.
22. Etgen, G.J. et al. Glucose transport and cell surface GLUT-4 protein in skeletal
    muscle of the obese Zucker rat. *Am. J. Physiol. Endocrinol. Metab.* 271, E294,
    1996.
23. Crettaz, M. et al. Insulin resistance in soleus muscle from obese Zucker rats:
    involvement of several defective sites. *Biochem. J.* 186, 525, 1980.
24. Henriksen, E.J. and Jacob, S. Effects of captopril on glucose transport activity in
    skeletal muscle of obese Zucker rats. *Metabolism* 44, 267, 1995.
25. Anai, M. et al. Altered expression levels and impaired steps in pathways to
    phosphatidylinositol-3-kinase activation via insulin receptor substrates 1 and 2 in
    Zucker fatty rats. *Diabetes* 47, 13, 1998.
26. Hevener, A.L., Reichart, D., and Olefsky, J., Exercise and thiazolidinedione ther-
    apy normalize insulin action in the obese Zucker fatty rat. *Diabetes* 49, 2154, 2000.
27. Saengsirisuwan, V. et al. Interactions of exercise training and R-(+)-alpha-lipoic
    acid on insulin signaling in skeletal muscle of obese Zucker rats. *Am. J. Physiol.
    Endocrinol. Metab.* 287, E529, 2004.
28. Cooper, D.R., Watson, J.E., and Dao, M.L. Decreased expression of protein kinase-
    C alpha, beta, and epsilon in soleus muscle of Zucker obese (fa/fa) rats. *Endocri-
    nology* 133, 2241, 1993.
29. Avignon, A. et al. Chronic activation of protein kinase C in soleus muscles and
    other tissues of insulin-resistant type II diabetic Goto-Kakizaki (GK), obese/aged,
    and obese/Zucker rats: a mechanism for inhibiting glycogen synthesis. *Diabetes*
    45, 1396, 1996.
30. Qu, X., Seale, J.P., and Donnelly, R. Tissue and isoform-selective activation of
    protein kinase C in insulin-resistant obese Zucker rats: effects of feeding. *J.
    Endocrinol.* 162, 207, 1999.
31. Goodyear, L.J. et al. Insulin receptor phosphorylation, insulin receptor substrate-
    1 phosphorylation, and phosphatidylinositol 3-kinase activity are decreased in
    intact skeletal muscle strips from obese subjects. *J. Clin. Invest.* 95, 2195, 1995.
32. Björnholm, M. et al. Insulin receptor substrate-1 phosphorylation and phosphati-
    dylinositol-3-kinase activity in skeletal muscle from NIDDM subjects after *in vivo*
    insulin stimulation. *Diabetes* 46, 524, 1997.
33. Krook, A. et al. Insulin-stimulated Akt kinase activity is reduced in skeletal muscle
    from NIDDM subjects. *Diabetes* 47, 1281, 1998.
34. Krook, A. et al. Characterization of signal transduction and glucose transport in
    skeletal muscle from type 2 diabetic patients. *Diabetes* 49, 284, 2000.
35. Zierath, J.R. et al. Insulin action on glucose transport and plasma membrane
    GLUT4 content in skeletal muscle from patients with NIDDM. *Diabetologia* 39,
    1180, 1996.
36. Ryder, J.W. et al. Use of a novel impermeable biotinylated photolabeling reagent
    to assess insulin- and hypoxia-stimulated cell surface GLUT4 content in skeletal
    muscle from type 2 diabetic patients. *Diabetes* 49, 647, 2000.
37. Reznick, A.Z. and Packer, L. Oxidative damage to proteins: spectrophotometric
    method for carbonyl assay. *Methods Enzymol.* 233, 357, 1994.
38. Saengsirisuwan, V. et al. Interactions of exercise training and alpha-lipoic acid on
    glucose transport in obese Zucker rat. *J. Appl. Physiol.* 91, 145, 2001.

39. Henriksen, E.J. et al. Isomer-specific actions of conjugated linoleic acid on muscle glucose transport in the obese Zucker rat. *Am. J. Physiol. Endocrinol. Metab.* 285, E98, 2003.

40. Teachey, M.K. et al. Interactions of conjugated linoleic acid and alpha-lipoic acid on insulin action in the obese Zucker rat. *Metabolism* 52, 1167, 2003.

41. Leung, Y.H. and Liu, R.H. Trans-10, cis-12-conjugated linoleic acid isomer exhibits stronger oxyradical scavenging capacity than cis-9, trans-11-conjugated linoleic acid isomer. *J. Agricult. Food Chem.* 48, 5469, 2000.

42. Moya-Camarena, S.Y. and Belury, M.A. Species differences in the metabolism and regulation of gene expression by conjugated linoleic acid. *Nutr. Rev.* 57, 336, 1999.

43. Blankson, H. et al. Conjugated linoleic acid reduces body fat mass in overweight and obese humans. *J. Nutr.* 130, 2943, 2000.

44. Houseknecht, K.L. et al. Dietary conjugated linoleic acid normalizes impaired glucose tolerance in the Zucker diabetic fatty fa/fa rat. *Biochem. Biophys. Res. Commun.* 244, 678, 1998.

45. Ryder, J.W. et al. Isomer-specific antidiabetic properties of conjugated linoleic acid. Improved glucose tolerance, skeletal muscle insulin action, and UCP-2 gene expression. *Diabetes* 50, 1149, 2001.

46. Shulman, G.I. Cellular mechanisms of insulin resistance. *J. Clin. Invest.* 106, 171, 2000.

47. Henriksen, E.J. Exercise training and the antioxidant alpha-lipoic acid in the treatment of insulin resistance and type 2 diabetes. *Free Rad. Biol. Med.* 40, 3, 2006.

48. Streeper, R.S. et al. Differential effects of lipoic acid stereoisomers on glucose metabolism in insulin-resistant skeletal muscle. *Am. J. Physiol. Endocrinol. Metab.* 273, E185, 1997.

49. Haugaard, N. and Haugaard, E.S. Stimulation of glucose utilization by thioctic acid in rat diaphragm incubated *in vitro*. *Biochim. Biophys. Acta* 222, 583, 1970.

50. Henriksen, E.J. et al. Stimulation by α-lipoic acid of glucose transport activity in skeletal muscle of lean and obese Zucker rats. *Life Sci.* 61, 805, 1997.

51. Jacob, S. et al. The antioxidant α-lipoic acid enhances insulin stimulated glucose metabolism in insulin-resistant rat skeletal muscle. *Diabetes* 45, 1024, 1996.

52. Peth, J.A. et al. Effects of a unique conjugate of alpha-lipoic acid and gamma-linolenic acid on insulin action in the obese Zucker rat. *Am. J. Physiol. Regulatory Integrative Comp. Physiol.* 278, R453, 2000.

53. Song, K.H. et al. Alpha-lipoic acid prevents diabetes mellitus in diabetes-prone obese rats. *Biochem. Biophys. Res. Commun.* 326, 197, 2005.

54. Lee, W.J. et al. Alpha-lipoic acid increases insulin sensitivity by activating AMPK in skeletal muscle. *Biochem. Biophys. Res. Commun.* 332, 885, 2005.

55. Rett, K. et al. Effect of acute infusion of thioctic acid on oxidative and non-oxidative metabolism in obese subjects with NIDDM (abstract). *Diabetologia* 38, A41, 1995.

56. Jacob, S. et al. Enhancement of glucose disposal in patients with type 2 diabetes by alpha-lipoic acid. *Drug Res.* 45: 872, 1995.

57. Jacob, S. et al. Improvement of insulin-stimulated glucose disposal in type 2 diabetes after repeated parenteral administration of thioctic acid. *Exp. Clin. Endocrinol. Diab.* 104, 284, 1996.

58. Jacob, S. et al. Oral administration of rac-α-lipoic acid modulates insulin sensitivity in patients with type 2 diabetes mellitus: a placebo controlled pilot trial. *Free Rad. Biol. Med.* 27, 309, 1999.

59. Konrad, T. et al. Alpha-lipoic acid treatment decreases serum lactate and pyruvate concentrations and improves glucose effectiveness in lean and obese patients with type 2 diabetes. *Diabetes Care* 22, 280, 1999.

60. Tsuboyama-Kasaoka, N. et al. Conjugated linoleic acid supplementation reduces adipose tissue by apoptosis and develops lipodystrophy in mice. *Diabetes* 49, 1534, 2000.

61. Roche, H.M. et al. Isomer-dependent metabolic effects of conjugated linoleic acid: insights from molecular markers sterol regulatory element-binding protein-1c and LXRα. *Diabetes* 51, 2037, 2002.

62. Riserus, U. et al. Treatment with dietary trans-10, cis-12 conjugated linoleic acid causes isomer-specific insulin resistance in obese men with the metabolic syndrome. *Diabetes Care* 25, 1516, 2002.

63. Riserus, U. et al. Supplementation with conjugated linoleic acid causes isomer-dependent oxidative stress and elevated C-reactive protein: a potential link to fatty acid-induced insulin resistance. *Circulation* 106, 1925, 2002.

20 Trivalent Chromium Supplementation Inhibits Oxidative Stress, Protein Glycosylation, and Vascular Inflammation in High Glucose-Exposed Human Erythrocytes and Monocytes

Sushil K. Jain

CONTENTS

ABSTRACT

Epidemiological studies have shown lower levels of chromium among men with diabetes and cardiovascular disease (CVD) compared with healthy control subjects. The mechanism by which chromium may decrease the incidence of CVD and insulin resistance is not known. This study demonstrates that chromium inhibits glycosylation of proteins, oxidative stress, and pro-inflammatory cytokine secretion, all of which are risk factors in the development of CVD. Erythrocytes or monocytes isolated from fresh human blood were treated with high concentrations of glucose (mimicking diabetes) in the presence or absence of chromium chloride in a medium at 37°C for 24 hr. We observed that chromium supplementation prevented increases in protein glycosylation and oxidative stress caused by high levels of glucose in erythrocytes. In monocytes, chromium supplementation inhibited the high glucose-induced secretion of interleukin (IL)-6 and tumor necrosis factor (TNF)-α. This study demonstrates that chromium supplementation protects against oxidative stress and vascular inflammation caused by exposure to high glucose in blood cells and provides evidence for a novel mechanism by which chromium supplementation may decrease the incidence of CVD in diabetic patients.

INTRODUCTION

Vascular inflammation and CVD are the leading causes of morbidity and mortality in the diabetic population and remain major public health issues. Hyperglycemia is one of the major risk factors in the development of vascular complications.[1] Intensive blood glucose control dramatically reduces the devastating complications that result from poorly controlled diabetes. Diabetes is treated with diet and insulin administration. However, for many patients, achieving tight glucose control is difficult with current regimens. The risk of CVD in diabetics is three- to five-fold greater than that of the general population.

PRO-INFLAMMATORY CYTOKINES AND VASCULAR INFLAMMATION IN DIABETES

TNF-α and IL-6 are pro-inflammatory cytokines produced by macrophages and other cell types in response to various stimuli.[2,3] The levels of these cytokines are elevated in the blood of many diabetic patients.[4–10] Increases in circulating levels of TNF-α and IL-6 decrease insulin sensitivity,[2,10–16] which can necessitate higher doses of insulin to maintain glycemic control in type 1 diabetic patients. Increased levels of insulin administration carry a risk of hypoglycemia. In addition, elevated circulating levels of TNF-α and IL-6 can induce expression of adhesion molecules and thus monocyte–endothelial cell adhesion, now recognized as an early and rate-limiting step in vascular inflammation and the development of vascular disease.[17–23]

Different human endothelial cells may have distinct ICAM forms and expression mechanisms.[24] Expression of VCAM-1 independent of systemic levels of TNF-α has been shown in pulmonary endothelial cells.[17] Several genes associated with adhesion molecules (ICAM-1, VCAM-1) and inflammatory cytokines are regulated by NFκB.[25,26] The agents that suppress NFκB activation can diminish cell adhesion and vascular inflammation.

OXIDATIVE STRESS AND VASCULAR INFLAMMATION

Oxidative stress can also influence the expression of multiple genes in vascular cells, including signaling molecules such as PKC, NFκB, and ERK.[27–32] Over-expression of these genes may lead to endothelial dysfunction and ultimately to microvascular and macrovascular disease. High glucose (HG) can up-regulate expression of transcription factors, such as NFκB, activating protein-1 and the TNF-α genes in monocytes.[31] TNF-α accumulation in conditioned media increased 10-fold and mRNA levels were increased 11.5-fold by HG.[31] This indicates that both NFκB and AP-1 mediated enhanced TNF-α transcription by HG.[31]

This expression of the TNF-α gene is mediated by the protein kinases p38 and JNK-1, which are respectively dependent on and independent of oxidative stress pathways.[30,31] Several studies advocate the importance of the p38 pathway in diabetes.[30] cAMP-dependent protein kinases (PKAs) can activate phosphory-lation of substrate proteins and cross-talk with MAPK pathways and proteins involved in signal transduction pathways, leading to altered gene expression and modulation of physiological processes.[32–37] cAMP is known to modulate cytokine production in a number of cell types.[32–37] This suggests that oxidative stress plays a key role in the regulatory pathway that progresses from elevated glucose to monocyte and endothelial cell activation in the enhanced vascular inflammation of diabetes.[21,38]

TRIVALENT CHROMIUM AND DIABETES

Trivalent chromium, the reduced form of the element, is required for insulin action.[39–44] A chromium-containing oligopeptide present in insulin-sensitive cells binds to the insulin receptor, markedly increasing the activity of the insulin-stimulated tyrosine kinase.[39,40] Overt chromium deficiency has been demonstrated in patients receiving total parenteral nutrition without chromium supplementa-tion.[45] It is characterized by hyperglycemia, glycosuria, and weight loss that cannot be controlled with insulin.[45,46] As a consequence, total parenteral nutrition solutions are regularly supplemented with chromium.[47] Intraperitoneal injections of potassium chromate also reversed atherosclerotic plaques in rabbits.[48,49]

The main route of exposure to chromium in the general population is dietary intake. Most chromium in the diet is trivalent, and any hexavalent chromium in food or water is reduced to the trivalent form in the acidic environment of the stomach.[50,51] Foods with high chromium concentrations include whole grain prod-ucts, green beans, broccoli, and bran cereals.[52] The chromium contents of meats, poultry, and fish vary widely because chromium may be introduced during trans-port, processing, and fortification.[52] Foods rich in refined sugars are low in chromium and actually promote chromium loss.[53]

Based on the chromium contents of well balanced diets,[52] adequate intake values in adults have been established at 35 μg/day for men and 25 μg/day for women.[51] Although there are no national survey data covering chromium intake,[51] a study of self-selected diets of U.S. adults indicated that the chromium intakes of a substantial proportion of subjects may be well below the adequate intake.[54] Similar results have been shown in the United Kingdom, Finland, Canada, and New Zealand.[55] Thus, subclinical chromium deficiency may be a contributor to glucose intolerance, insulin resistance, and cardiovascular disease, particularly in aging populations and populations that have increased chromium requirements because of high sugar diets.[43]

Concentrations of chromium in the blood, lenses, and toenails are lower in diabetic patients compared with concentrations in the normal population.[46,56,57] Several studies have suggested that chromium supplementation may be beneficial in individuals with type 2 and type 1 diabetes, gestational diabetes, or steroid-induced diabetes, as evidenced by decreased blood glucose values or decreased insulin requirements.[42,43,58–67] Although the majority of research on diabetes has focused on type 2, a few small studies have tested the efficacy of Cr^{3+} on type 1 diabetes and found it effective.[64] One study supplemented 162 patients (48 with type 1 diabetes, the remainder with type 2) with 200 μg/day Cr^{3+} picolinate. Seventy-one percent of the type 1 patients responded positively, allowing 30% decreases in insulin doses. Blood sugar fluctuations also responded positively, decreasing as early as 10 days after treatment. Supplementation of chromium as the picolinate (600 μg/day) in a 28-year-old woman with an 18-year history of type 1 diabetes reduced HbA_{1C} from 11.3 to 7.9% in 3 months.[65] When patients receiving total parenteral therapy were supplemented with chromium, their diabetes symptoms reversed and they required smaller doses of exogenous insulin.[66]

In atherosclerotic rabbits, an injection of chromium chloride resulted in marked reductions in the plaques covering the aortic intimal surfaces, aortic weights, and cholesterol contents.[49] Chromium can reduce elevated cholesterol and triglycerides in a dose-dependent relationship.[68] Results from two case-control studies suggest an inverse association between chromium levels in toenails and risks of myocardial infarction in the general population.[69] Similarly, a recent report of the Health Professionals Follow-up Study found lower levels of toenail chromium among men with diabetes and CVD compared with healthy control subjects.[56]

However, randomized trials of chromium supplementation in diabetes have not been definitive. Many studies have not been blinded, used inappropriate glucose metabolism assessment parameters, or included heterogeneous and poorly characterized patient populations. More rigorous blinded and well controlled studies are needed to fully assess the efficacy and mechanism of the action of Cr^{3+} supplementation as an adjuvant therapy for diabetic patients.

Figure 20.1 illustrates the effects of high glucose (HG) and trivalent chromium on TNF-α secretion in peripheral blood mononuclear cells (PBMCs) isolated from normal human blood. HG resulted in a significant increase in TNF-α secretion in PBMCs. Supplementation with chromium caused a

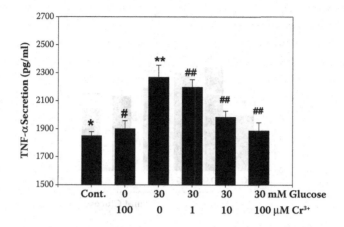

FIGURE 20.1 Effect of high glucose and Cr^{3+} on TNF-α secretion in PBMCs isolated from normal human volunteers. Values represent mean ± SE (n = 4). Differences in values (* versus**, ** versus ##) are significant (p <0.01). (*Source:* Jain SK and Lim G. *Horm Metabol Res* 38, 60, 2006.)

concentration-dependent inhibition of TNF-α secretion. Mannitol used as an osmolarity control did not cause any changes in TNF-α secretion compared with basal levels (data not shown). Similarly, Figure 20.2 illustrates the effect of trivalent chromium on inhibition of IL-6 secretion in PBMCs exposed to HG. HG treatment with or without chromium had no effect on cell growth (data not given). This suggests that inhibition of TNF-α and IL-6 secretion is not due to cellular toxicity caused by chromium supplementation.

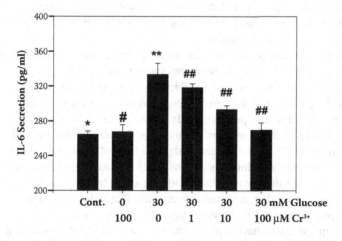

FIGURE 20.2 Effect of high glucose and Cr^{3+} on IL-6 secretion in PBMCs isolated from normal human volunteers. Values represent mean ± SE (n = 4). Differences in values (* versus**, ** versus ##) are significant (p <0.01). (*Source:* Jain SK and Lim G. *Horm Metabol Res* 38, 60, 2006.)

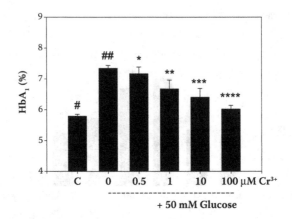

FIGURE 20.3 Effects of different chromium concentrations on hemoglobin glycation in high glucose-treated erythrocytes. Values represent mean ± SE (n = 4). Differences in values (# versus ##, ## versus **, ## versus ***, and ## versus ****) are significant (p <0.05). (*Source:* Jain SK, Patel, P, and Rogier K, *Antioxidant Redox Signaling* 8, 238, 2006.)

Figure 20.3 illustrates the effects of different concentrations of chromium on high glucose-induced hemoglobin glycosylation. Increasing concentrations of glucose caused increases in hemoglobin glycosylation that were inhibited by supplementation with chromium. This shows that the inhibitory effect of chromium can be seen at concentrations as low as 0.5 μM. Figure 20.4 shows that chromium supplementation prevented increases in lipid peroxidation caused by high glucose concentrations. The effect of chromium on inhibition of lipid peroxidation was seen even at 0.5 μM. RBCs exposed to normal glucose concentrations did not show any changes in glycosylated hemoglobin or lipid peroxidation values with or without chromium.

High glucose concentrations can result in increased oxidative stress from excessive oxygen radical production from auto-oxidation of glucose, glycosylated proteins or stimulation of cytochrome P450-like activity by the excessive NADPH produced by glucose metabolism.[70–75] Novel data reveal the inhibition of glycosylation by chromium. A previous study showed that oxidative stress alone can increase glycosylation of proteins.[76] This suggests that the inhibition of glycosylation may be mediated by an antioxidative effect of chromium. However, the precise mechanism by which chromium inhibits oxidative stress is not known. For instance, chromium may reduce oxidative stress by activation of glutathione reductase or some other antioxidative enzyme that detoxifies oxygen radicals, thereby reducing the oxidative stress caused by high glucose.

Other investigators have reported that Cr^{3+} supplementation lowers the blood levels of oxidative stress markers in animal models as well as in diabetic patients.[77–82] The effect of chromium on HG-induced oxidative stress, up-regulation of adhesion molecules, transcription factors such as NFkB and activating

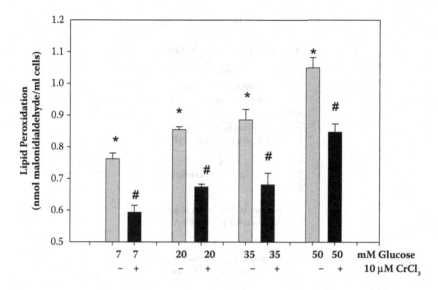

FIGURE 20.4 Effect of chromium on lipid peroxidation in erythrocytes treated with different concentrations of glucose. Values represent mean ± SE (n = 4). Differences in values (* versus #) are significant (p <0.05). (*Source:* Jain SK, Patel, P, and Rogier K, *Antioxidant Redox Signaling* 8, 238, 2006.)

protein-1, and the TNF-α gene in PBMCs is not known. Investigations are needed to elucidate the molecular mechanisms by which chromium can effect pro-inflammatory cytokine secretion and vascular inflammation. Whether Cr^{3+} supplementation prevents the effect of diabetes on the over-expression of regulatory genes in vascular cells including signaling molecules such as PKC, NFκB, and ERK, thereby preventing vascular inflammation in diabetes, is not known.

CONCLUSION

Figure 20.5 shows that hyperglycemia and ketosis can increase oxygen radical generation and oxidative stress in diabetes.[83] Oxidative stress can cause over-expression of regulatory genes in vascular cells, increases in the adhesivity of monocytes to endothelium, and vascular disease in diabetes. Chromium supplementation has the potential to decrease cellular oxidative stress, lower blood levels of pro-inflammatory cytokines (TNF-α and IL-6) and glycosylation of proteins, and thereby the adhesivity of monocytes to endothelium. Both glycosylation of proteins and endothelial-monocyte adhesivity have been proposed to be involved in the pathogenesis of cellular dysfunction leading to the vascular complications of diabetes.[21]

The evidence that chromium can inhibit these markers of oxidative stress and vascular inflammation must be explored at the clinical level to see whether widely used supplements such as chromium picolinate and chromium niacinate can lower

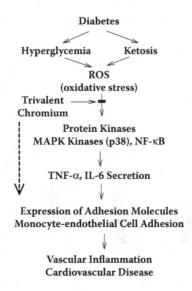

FIGURE 20.5 Proposed model for role of trivalent chromium supplementation in prevention of oxidative stress and vascular disease in diabetes.

levels of pro-inflammatory cytokines and thereby improve insulin sensitivity and glycemic control among the diabetic patient population. If so, chromium supplementation may be used as an adjuvant therapy for reduction of vascular inflammation and CVD in diabetes.

ACKNOWLEDGMENTS

The author is supported by a grant from NIDDK and the Office of Dietary Supplements of the National Institutes of Health (RO1 DK064797), and thanks Georgia Morgan for excellent editing of this manuscript.

REFERENCES

1. Klein R. Hyperglycemia and microvascular and macrovascular disease in diabetes. *Diab Care* 18, 258, 1995.
2. Rader DJ. Inflammatory markers of coronary risk. *New Engl J Med* 343, 1179, 2000.
3. Ming WF, Bersani L, and Mantovani A. Tumor necrosis factor-α is chemotactic for monocytes and polymorphonuclear leukocytes. *J Immunol* 138, 1469, 1987.
4. Jain SK, Kannan K, Lim G, McVie R, and Bocchini JA. Hyperketonemia increases TNF-α secretion in cultured U937 monocytes and type-1 diabetic patients. *Diabetes* 51, 2287, 2002.

5. Lechleitner M, Koch T, Herold M, Dzien A, and Hoppichler F. Tumour necrosis factor-alpha plasma level in patients with type I diabetes mellitus and its association with glycaemic control and cardiovascular risk factors. *J Intern Med* 248, 67, 2000.

6. Jain SK et al. Elevated blood interleukin-6 levels in hyperketonemic type 1 diabetic patients and secretion by acetoacetate-treated cultured U937 monocytes. *Diab Care* 26, 2139, 2003.

7. Hussain MF et al. Elevated serum levels of macrophage-derived cytokines precede and accompany the onset of IDDM. *Diabetologia* 39, 60, 1996.

8. Cavallo MG et al. Cytokines in sera from insulin-dependent diabetic patients at diagnosis. *Clin Exp Immunol* 86, 256, 1991.

9. Ohno Y, Aoki N, and Nishimura A. *In vitro* production of interleukin-1, interleukin-6, and tumor necrosis factor-α in insulin-dependent diabetes mellitus. *J Clin Endocrinol Metab* 77, 1072, 1993.

10. Morohoshi M, Fujisawa K, Uchimura I, and Numano F. The effect of glucose and advanced glycosylation end products on IL-6 production by human monocytes. *Ann NY Acad Sci* 748, 562, 1995.

11. Andreozzi, F et al. Plasma interleukin-6 levels are independently associated with insulin secretion in a cohort of Italian Caucasian nondiabetic subjects. 55, 2021, 2006.

12. Kern PA, Ranganathan S, Li C, Wood L, and Ranganathan G. Adipose tissue tumor necrosis factor and interleukin-6 expression in human obesity and insulin resistance. *Am J Physiol* 280, E745, 2001.

13. Halse R, Pearson SL, McCormack JG, Yeaman SJ, and Taylor R. Effects of tumor necrosis factor-α on insulin action in cultured human muscle cells. *Diabetes* 50, 1102, 2001.

14. Bruun JM, Verdich C, Toubro S, Astrup A, and Richelsen B. Association between measures of insulin sensitivity and circulating levels of interleukin-8, interleukin-6 and tumor necrosis factor-alpha: effect of weight loss in obese men. *Eur J Endocrinol.* 148, 535, 2003.

15. Vozarova B et al. The interleukin-6 (−174) G/C promoter polymorphism is associated with type 2 diabetes mellitus in Native Americans and Caucasians. *Hum Genet* 112, 409, 2003.

16. Rotter V, Nagaev I, and Smith U. Interleukin-6 induces insulin resistance in 3T3-L1 adipocytes and is like IL-8 and tumor necrosis factor-α overexpressed in human fat cells from insulin-resistant subjects. *J Biol Chem* 46, 45777, 2003.

17. Gamble JR, Harlan JM, Klebanoff SF, and Vadas MA. Stimulation of the adherence of neutrophils to umbilical vein and endothelium by human recombinant tumor necrosis factor. *Proc Natl Acad Sci USA* 82, 8667, 1985.

18. Cavender D, Saegusa Y, and Ziff M. Stimulation of endothelial cell binding of lymphocytes by tumor necrosis factor. *J Immunol* 139, 1855, 1987.

19. Huber SA, Sakkinen P, Conze D, Hardin N, and Tracy R. Interleukin-6 exacerbates early atherosclerosis in mice. *Arterioscler Thromb Vasc Biol* 19, 2364, 1999.

20. Zhang WJ and Frei B. Intracellular metal ion chelators inhibit TNF-alpha-induced SP-1 activation and adhesion molecule expression in human aortic endothelial cells. *Free Radic Biol Med* 34, 674, 2003.

21. Cardelli M, Andreozzi F, Laretta E et al. Plasma interleukin-6 levels are increased in subjects with impaired glucose tolerance but not in those with impaired fasting glucose in a cohort of Italian Caucasians. *Diab Metab Res Rev* 23, 141, 2007.

22. Granger DN, Vowinked T, and Petnehazy T. Modulation of the inflammatory reponse in cardiovascular disease. *Hypertension* 43, 924, 2004.

23. Kim JA, Berliner JA, Natarajan RD, and Nadler JL. Evidence that glucose increases monocyte binding to human aortic endothelial cells. *Diabetes* 43, 1103, 1994.

24. Carley W et al. Distinct ICAM-1 forms and expression pathways in synovial microvascular endothelial cells. *Cell Mol Biol* 45, 79, 1999.

25. Natarajan R, Reddy MA, Malik KU, Fatima S, and Khan BV. Signaling mechanisms of nuclear factor-kappa-b-mediated activation of inflammatory genes by 13-hydroperoxyoctadecadienoic acid in cultured vascular smooth muscle cells. *Arterioscler Thromb Vasc Biol* 21, 1408, 2001.

26. Bharti AC and Aggrawal BB. Nuclear factor-kappa B and cancer: its role in the prevention and therapy. *Biochem Pharmacol* 64, 883, 2002.

27. Li N and Karin M. Is NF-kB the sensor of oxidative stress? *FASEB J* 13, 1137, 1999.

28. Suzuki YJ, Forman HJ, and Sevanian A. Oxidants as stimulators of signal transduction. *Free Rad Biol Med* 22, 269, 1997.

29. Brownlee M. The pathogenesis of diabetic complications: a unifying mechanism. *Diabetes* 54, 1615, 2005.

30. Igarashi M et al. Glucose or diabetes activates p38 mitogen-activated protein kinase via different pathways. *J Clin Invest* 103, 185, 1999.

31. Guha M, Bai W, Nadler JL, and Natarajan R. Molecular mechanisms of tumor necrosis factor α gene expression in monocytic cells via hyperglycemia-induced oxidant stress-dependent and independent pathways. *J Biol Chem* 275, 17728, 2000.

32. Reddy MA et al. The oxidized lipid and lipoxygenase product 12(S)-hydroxyeicosatetraenoic acid induces hypertrophy and fibronectin transcription in vascular smooth muscle cells via p38 MAPK and cAMP response element-binding protein activation: mediation of angiotensin II effects. *J Biol Chem* 277, 9920, 2002.

33. Daniel PB, Walker WH, and Habener JF. Cyclic AMP signaling and gene regulation. *Ann Rev Nutr* 18, 353, 1988.

34. Chen CH, Zhang DH, LaPorte JM, and Ray A. Cyclic AMP activates p38 mitogen-activated protein kinase in Th2 cells, phosphorylation of GATA-3 and stimulation of Th2 cytokine gene expression. *J Immunol* 165, 5597, 2000.

35. Serkkola E and Hurme M. Activation of NF-kappa B by cAMP in human myeloid cells. *FEBS Lett* 234, 327, 1993.

36. Ollivier V, Parry GC, Cobb RR, de Prost D, and Mackman N. Elevated cyclic AMP inhibits NF-kappa-B-mediated transcription in human monocytic cells and endothelial cells. *J Biol Chem* 271, 20828, 1996.

37. Treebak JT et al. AMP mediated AS-160 phosphorylation in skeletal muscle is dependent on AMPK catalytic and regulatory subunits. *Diabetes* 55, 2051, 2006.

38. Singh U, Devraj S, and Jialal, I. Vitamin E, oxidative stress and inflammation. *Ann Rev Nutr* 25, 151, 2005.

39. Vincent JB. Recent advances in the nutritional biochemistry of trivalent chromium. *Proc Nutr Soc* 63, 41, 2004.

40. Vincent JB. Mechanisms of chromium action: low-molecular weight chromium-binding substance. *J Am Coll Nutr* 18, 6, 1999.

41. Offenbacher EG and Pi-Sunyer FX. Beneficial effect of chromium-rich yeast on glucose tolerance and blood lipids in elderly subjects. *Diabetes* 29, 919, 1980.

42. Cefalu WT and Hu FB. Role of chromium in human health and in diabetes. *Diab Care* 27, 2741, 2004

43. Anderson RA. Chromium, glucose intolerance and diabetes. *J Am Coll Nutr* 17, 548, 1998.

44. Chromium and Diabetes Workshop, Office of Dietary Supplements, National Institutes of Health, Bethesda, 1999. http://ods.nih.gov/news/conferences/chromium_diabetes. html

45. Jeejeebhoy KN. Chromium and parenteral nutrition. *J Trace Elem Exp Med* 12, 85, 1999.

46. Morris BW et al. Chromium homeostasis in patients with type II (NIDDM) diabetes. *J Trace Elem Med Biol* 13, 57, 1999.

47. Department of Foods and Nutrition, American Medical Association. Guidelines for essential trace element preparations for parenteral use: statement by an expert panel. *JAMA* 241, 2051, 1979.

48. Abraham AS et al. The effect of chromium on established atherosclerotic plaques in rabbits. *Am J Clin Nutr* 33, 2294, 1980.

49. Abraham AS, Brooks BA, and Eylath U. Chromium- and cholesterol-induced atherosclerosis in rabbits. *Ann Nutr Metab* 35, 203, 1991.

50. Stoecker BJ et al., Eds. *Modern Nutrition in Health and Disease*, Williams & Wilkins, Baltimore, 1999, p. 277.

51. Panel on Micronutrients, Food and Nutrition Board, National Academy of Sciences. Dietary reference intakes for vitamin A, vitamin K, arsenic, boron, chromium, copper, iodine, iron, manganese, molybdenum, nickel, silicon, vanadium, and zinc. National Academy Press, Washington, 2001.

52. Anderson RA, Bryden NA, and Polansky MM. Dietary chromium intake, freely chosen diets, institutional diet, and individual foods. *Biol Trace Elem Res* 32, 117, 1992.

53. Kozlovsky AS et al. Effects of diets high in simple sugars on urinary chromium losses. *Metabolism* 35, 515, 1986.

54. Anderson RA and Kozlovsky AS. Chromium intake, absorption and excretion of subjects consuming self-selected diets. *Am J Clin Nutr* 41, 1177, 1985.

55. Anderson RA. Chromium, in *Trace Elements in Human and Animal Nutrition*, Mertz W, Ed, Academic Press, Orlando, 1987, p. 225.

56. Rajpathak S et al. Lower toenail chromium in men with diabetes and cardiovascular disease compared with healthy men. *Diab Care* 27, 2211, 2004.

57. Pineau A, Guillard O, and Risse JF. A study of chromium in human cataractous lenses and whole blood of diabetics, senile, and normal population. *Biol Trace Elem Res* 32, 1338, 1992.

58. Anderson RA et al. Elevated intakes of supplemental chromium improve glucose and insulin variables in individuals with type 2 diabetes. *Diabetes* 46, 1786, 1997.

59. Mossop RT. Effects of chromium III on fasting blood glucose, cholesterol and cholesterol HDL levels in diabetics. *Cent Afr J Med* 29, 80, 1983.

60. Cheng N et al. Follow-up survey of people in China with type 2 diabetes mellitus consuming supplemental chromium. *J Trace Elem Exp Med* 12, 55, 1999.

61. Jovanovic-Peterson L, Gutierrez M., and Peterson, C. Chromium supplementation for gestational diabetes women improves glucose tolerance and decreases hyperinsulinemia. *Diab Care* 45 (Suppl. 2), 237A, 1996.

62. Ravina, A, Slezak, L, Mirsky, N, Bryden, NA, and Anderson, RA. Reversal of corticosteroid-induced diabetes mellitus with supplemental chromium. *Diab Med* 16, 164, 1999.

63. Abraham AS, Brooks BA, and Eylat U. The effect of chromium supplementation on serum glucose and lipids in patients with non-insulin dependent diabetes mellitus. *Med Clin Exp* 41, 768, 1992.

64. Ravina A et al. Clinical use of the trace element chromium (III) in the treatment of diabetes mellitus. *J Trace Elem Exp Med* 8, 183, 1995.

65. Fox GN and Sabovic Z. Chromium picolinate supplementation for diabetes mellitus. *J Fam Pract* 46, 83, 1998.

66. Anderson RA. Chromium and parenteral nutrition. *Nutrition* 11 (Suppl 1), 83, 1995.

67. Lee NA and Reasner CA. Beneficial effect of chromium supplementation on serum triglyceride levels in NIDDM. *Diab Care* 17, 1449, 1994.

68. Lamson DM and Plaza SM. The safety and efficacy of high-dose chromium. *Alternative Med Rev* 7, 218, 2002.

69. Guallar EJF et al. Toenail chromium and risk of myocardial infarction. *Am J Epidemiol* 162, 157, 2005.

70. Jain SK. Hyperglycemia can cause membrane lipid peroxidation and osmotic fragility in human red blood cells. *J Biol Chem* 264, 21340, 1989.

71. Rajeswari P, Natarajan R, Nadler JL, and Kumar, D. Glucose induces lipid peroxidation and inactivation of membrane associated iron transport enzymes in human erythrocytes *in vivo* and *in vitro*. *J. Cell Physiol* 149, 100, 1991.

72. Natarajan R, Lanting L, Gonzales N, and Nadler J. Formation of an F2-isoprostane in vascular smooth muscle cells by elevated glucose and growth factors. *Am J Physiol* 271, 159, 1996.

73. Giugliano D, Paolisso G, and Ceriello A. Oxidative stress and diabetic vascular complications. *Diab Care* 19, 257, 1996.

74. Jain SK, McVie R, Duett J, and Herbst JJ. Erythrocyte membrane lipid peroxidation and glycosylated hemoglobin in diabetes. *Diabetes* 38, 1539, 1989.

75. Hunter SJ et al. Demonstration of glycated insulin in human diabetic plasma and decreased biological activity assessed by euglycemic-hyperinsulinemic clamp technique in humans. *Diabetes* 52, 492, 2003.

76. Jain SK and Palmer M. The effect of oxygen radical metabolites and vitamin E on glycosylation of proteins. *Free Rad Biol Med* 22, 593, 1997.

77. Jain SK and Kannan K. Chromium chloride inhibits oxidative stress and TNF-α secretion caused by exposure to high glucose in cultured monocytes. *Biochem Biophys Res Commun* 289, 687, 2001.

78. Bahijiri SM, Mira SA, Mufti AM, and Ajabnoor MA. The effects of inorganic chromium and brewer's yeast supplementation on glucose tolerance, serum lipids and drug dosage in individuals with type 2 diabetes. *Saudi Med J* 21, 831, 2000.

79. Vinson JA, Mandarano MA, Shuta DL, Bagchi M, and Bagchi D. Beneficial effects of a novel IH636 grape seed proanthocyanidin extract and a niacin-bound chromium in a hamster atherosclerosis model. *Mol Cell Biochem* 240, 99, 2002.

80. Preuss HG, Montamarry S, Echard B, Scheckenbach R, and Bagchi D. Long-term effects of chromium, grape seed extract, and zinc on various metabolic parameters of rats. *Mol Cell Biochem* 223, 95, 2001.

81. Cheng HH, Lai MH, Hou WC, and Huang CL. Antioxidant effects of chromium supplementation with type 2 diabetes mellitus and euglycemic subjects. *J Agric Food Chem* 52, 1385, 2004.
82. Anderson RA et al. Potential antioxidant effects of zinc and chromium supplementation in people with type 2 diabetes mellitus. *J Am Coll Nutr* 20, 212, 2001.
83. Jain SK, McVie R, and Bocchini JA. Ketosis, oxidative stress and type 1 diabetes. *Pathophysiology* 13, 163, 2006.
84. Jain SK and Lim G. Chromium chloride inhibits TNF-α and IL-6 secretion in isolated human blood monoclear cells exposed to high glucose. *Horm Metab Res* 38, 60, 2006.
85. Jain SK, Patel, P, and Rogier K. Trivalent chromium inhibits protein glycation and lipid peroxidation in high glucose-treated erythrocytes. *Antioxidant Redox Signaling* 8, 238, 2006.

Index

A

Abdominal obesity thresholds, 10
  gender comparisons, 10
Adenosine triphosphate, uncoupling protein 2,
    pancreatic beta cell function
  beta cell mass, survival, 218–219
  insulin secretion, 216–218
Adenosine triphosphate III
  initial report, metabolic syndrome
    definition, 5
  metabolic syndrome, 6
  revised criteria/guidelines, 7–9
Adhesion molecules, clinical marker of
    inflammation, 152
Adipokines, 116–118
Adiponectin
  clinical marker of inflammation, 147–148
  oligomeric composition, 167–176
    adipocytes, 168–169
    central nervous system effects,
      172–173
    multimers, biological activities of,
      169–171
    thermogenesis, 172
    weight loss results, 171–172
Adipose tissue, nutritional inflammation
    modulation, 231–232
Alpha lipoic acid, 273–288. *See also*
    Lipoic acid
  adipocyte differentiation regulation, 280
    MAPK signaling pathways, 281–283
    transcription factor modulation,
      281–282
  diabetes mellitus prevention, 261–272
  hypothalamic AMPK, 277–279
  insulin resistance, 289–300
    etiology of insulin resistance, 292
    nutriceutical interventions, 292–295
    skeletal muscle glucose transport
      system, regulation, 291
  structure, 274
American Association of Clinical
    Endocrinologists
  metabolic syndrome guidelines, 5–7

Anti-inflammatory effect, insulin, pro-
    inflammatory effect,
  macronutrients, balance,
  metabolic syndrome, 15–32
Antioxidants
  oxidative stress, 71–92
    childbirth, 75–78
      perinatal asphyxia, 76–77
      resuscitation, use of pure oxygen,
        77–78
    human milk, 82–84
    in overweight, obesity, 38–39
    perinatal period, 72–73
    pregnancy, 74–75
      fetal development, reactive oxygen
        species, 74
      preeclampsia, 75
    premature infants, 78–82
      bronchopulmonary dysplasia, 79
      neonatal necrotizing enterocolitis,
        80
      periventricular leukomalacia, 80–81
      retinopathy of prematurity, 81–82
    reactive oxygen species, tissue-
      produced, 72
  perinatal period, oxidative stress, 71–92
Atherosclerosis, inflammation, 139–166
  adhesion molecules, 152
  adiponectin, 147–148
  C-reactive protein, 144–146, 157–159
  cardiovascular risk, 141–143
  clinical markers, 143–152
  hemostatic factors, 159–160
  hemostatic parameters, 150–151
  interleukin-1, 152
  interleukin-6, 152
  interleukin-10, 152
  leptin, 146–147
  modification, 156–160
  resistin, 151–152
  tumor necrosis factor-alpha, 148–150, 159
  visfatin, 152
Atherosclerotic plaque destabilization, lipid-
    induced macrophage death,
    251–260
  atherosclerosis, 254–255

315

Printed in the United States
by Baker & Taylor Publisher Services